Progress in Nonequilibrium Green's Functions V

(PNGF V)

Journal of Physics: Conference Series Volume 427

Jyvaskyla, Finland
27-31 August 2012

ISBN: 978-1-62748-304-9
ISSN: 1742-6588

Printed from e-media with permission by:

Curran Associates, Inc.
57 Morehouse Lane
Red Hook, NY 12571

proceedings.com

Some format issues inherent in the e-media version may also appear in this print version.

Printed by Curran Associates, Inc. (2013)

Institute of Physics
Dirac House, Temple Back
Bristol BS1 6BE UK

Phone: 44 1 17 929 7481
Fax: 44 1 17 920 0979

techtracking@iop.org

Additional copies of this publication are available from:

Curran Associates, Inc.
57 Morehouse Lane
Red Hook, NY 12571 USA
Phone: 845-758-0400
Fax: 845-758-2634
Email: curran@proceedings.com
Web: www.proceedings.com

TABLE OF CONTENTS

Preface

The fifth interdisciplinary conference "Progress in Nonequilibrium Green's Functions" (PNGF5) was held at the University of Jyväskylä, Finland, on August 27th — 31st, 2012. The conference continued the successful tradition of its predecessors (Rostock 1999, Dresden 2002, Kiel 2005 and Glasgow 2009) to bring together different communities for an interdisciplinary exchange of recent results and theoretical concepts.

The conference focused on recent developments, current challenges and future perspectives in nonequilibrium Green's functions theory in various fields of physics but included also other many-body methods. Roughly 20 invited talks were given by some of the top scientists in the field, accompanied by 10 contributed talks.[1] Also a poster session was set up to enhance scientific discussions, building up new collaborations and enriching views and ideas. As at the previous meetings, the atmosphere was interactive and stimulating, benefitting both experienced scientists and young researchers and students.

The present volume contains 14 articles based on works presented at this conference. The articles partly have review character so they should be of use for an interdisciplinary community working or interested in nonequilibrium Green's functions. All papers were refereed according to high scientific standards.

The conference would not have been possible without financial support from the *Federation of Finnish Learned Societies*, *Finnish Academy of Science and Letters* and *Nanoscience Center* of the university of Jyväskylä which are greatly acknowledged. The local organizing committee is also grateful to the administration personnel, Marjut Hilska and Riitta-Liisa Kuittinen, for assistance. We further thank professor Antti-Pekka Jauho for invaluable help and advice during the organization of the conference.

The editors of the present conference proceedings acknowledge the authors for their excellent papers and all the referees for participating in a thorough peer-reviewing of the manuscripts. Finally, it is our pleasure to announce that the sixth conference "Progress in Nonequilibrium Green's Functions" (PNGF6) will be held in August 2015 at the University of Lund, Sweden.

Riku Tuovinen and Robert van Leeuwen, University of Jyväskylä
Michael Bonitz, University of Kiel
February 2013

[1] Slides of several presentations can be found online at `www.jyu.fi/physics/pngf5`.

Journal of Physics: Conference Series **427** (2013) 011001 doi:10.1088/1742-6596/427/1/011001

David C. Langreth — Obituary

Antti-Pekka Jauho
Lyngby, Denmark

David C. Langreth passed away on May 27th, 2011. He was an outstanding theoretical condensed matter physicist with a large number of important contributions in many areas, such as surface physics, density-functional theory, many-body physics, and transport theory. For the community working with nonequilibrium Green's functions his name is inseparably linked to "Langreth identities", a set of rules which elegantly and concisely allows one to extract real-time Green's functions from their contour-ordered counterparts. Using these rules one can very easily derive the (generalized) Baym–Kadanoff equations, or, equivalently, the Keldysh equation, which encode the standard paradigm for nonequilibrium transport phenomena. His famous lecture notes on nonequilibrium Green's functions [NATO ASI 17, (1976)] have been the cornerstone for generations of students learning nonequilibrium techniques, until text-books relying on these very same notes became available.

David learned the nonequilibrium techniques during his stay at Illinois with Leo Kadanoff, one of the originators of the formalism. His very first published paper [D. C. Langreth and L. P. Kadanoff, Phys. Rev. **133**, 1070 (1964)] on polaron mobility contains a precursor of the Langreth rules, though in equilibrium disguise. In subsequent works he developed these concepts further, and one can see the contours of the modern theory already in his paper from 1967 [D. C. Langreth, Phys. Rev. **159**, 717 (1967)], which addresses finite temperature effects on the polaron problem. His monumental 39-page long paper on spin resonance in magnetic alloys together with John Wilkins [D. C. Langreth and J. W. Wilkins, Phys. Rev. B **6**, 3189 (1972)] is even today perhaps the most complete application of the Kadanoff–Baym equations to a problem involving both continuous (space, time) and discontinuous (spin) variables: all the key concepts such as gradient expansion, nonequilibrium analytic continuation, and spin-sums are carefully explained and elaborated. In later years he used the nonequilibrium techniques from time to time, for example in studies of the out-of-equilibrium Kondo effect [R. Aguado and D. C. Langreth, Phys. Rev. Lett. **85**, 1946 (2000)]. All of his papers are examples of scholarly excellence.

The nonequilibrium Green's function community will miss sorely David Langreth, one of the deepest thinkers in our discipline, and whose work will continue to influence our research for many years to come.

David C. Langreth — In memoriam

Ulf von Barth and Bengt Lundqvist
Lund and Göteborg, Sweden

Ulf and Bengt had the pleasure of getting to know David Langreth very early on in their careers, i.e., in the early seventies. Ulf has had very few role models in his life and David has been one of them. And so he was for Bengt. What attracted our admiration was David's ability to combine his excellence and his deep knowledge of fundamental physics with a modest and humble personality. His integrity was unquestionable and David never overstated his case.

In the seventies David had several occasions to visit Lund, Chalmers and Nordita in Copenhagen. He was an appreciated guest scientist. He was also an evaluator of Swedish Solid State Physics. From 1975 to 1976 he was a guest Professor at Nordita. At that time, David had a strong interest in electrons at surfaces - an interest shared by the experimental and theoretical groups at Chalmers. An important work by David from this time is his sum rule for surfaces [Phys. Rev. B 11, 2155 (1975)]. This work led on to some of the most influential parts of David's research. In a collaboration with John Perdew, David engaged in a detailed study of the effects of electronic correlations in surfaces of varying steepness. In a series of papers from 1975 to 1985 [examples are: Phys. Rev. B **15**, 2884 (1977), Phys. Rev. B **21** 5469 (1980), and Phys. Rev. B **26**, 2810 (1982)] David and his collaborators laid down the many-body foundation of the subsequent generalized gradient approximations within density-functional theory (DFT). In our mind, this fact is often not sufficiently appreciated by the practitioners of DFT calculations. In the later stages of this period, Ulf worked at IBM, Yorktown Heights (NY) and had the privilege to be of help to David in numerically testing the Langreth–Mehl gradient functional [Phys. Rev. B **28**, 1809 (1983), Physica Scripta **32**, 391 (1985)] in atoms and solids.

The groups at Chalmers and in Lund strongly benefited from David's deep knowledge of DFT. In those days, however, the Lund group was mainly interested in the theoretical interpretation of X-ray and Auger spectra. Also in this area David had made major advances. In a famous paper from 1970 [Phys. Rev. B 1, 471] David gave a formally exact solution to the Green function of a long lived core hole in a fully interacting electron gas. Later important works in this area resulted from a collaboration with Shung, for instance Phys. Rev. B **28**, 4976 (1983) and Phys. Rev. B **33**, 2247 (1986).

Journal of Physics: Conference Series **427** (2013) 011001 doi:10.1088/1742-6596/427/1/011001

The first picture from 1981 shows David (on the left) as faculty opponent of the thesis of Ulf's first graduate student in Lund, Günter Grossmann. The overhead picture shows the energy dependent singularity exponents calculated by Grossmann within the 'independent Fermion model' of Mahan, Nozieres, and de Dominicis.

David's connections to Chalmers and particularly his friendship with Bengt go back some forty years.

The second picture shows David in 1988 at the helm of Bengt's sailing boat in the archipelago of Bohuslän.

In later years David was a visiting professor at Nordita in 1988 and at Chalmers in 1993. In 1994, David was made an honorary doctor to Chalmers University. The long time collaboration with Bengt and his group at Chalmers ("the Rutgers–Chalmers collaboration") has led to yet another very successful outcome of David's research, namely the use of the methods of DFT to treat the notoriously difficult problem of van-der-Waals interactions. The use of the non-local functionals resulting from this collaboration has been expanding very rapidly and broadly over the past ten years. Fortunately, David lived long enough to see the exponential growth of the citations to this part of his work. Examples of early works: Phys. Rev. **76**, 102 (1996), Phys. Rev. Lett. **77**, 2029 (1996), and Phys. Rev. B **59**, 4708 (1999) and of later works: Phys. Rev. Lett. **92**, 246401 (2004) and Phys. Rev. B **82**, 081101 (2010).

David's research interests have been very broad but still confined to the general framework of many-body physics. And David has been successful also in areas of less interest to the authors of the present notes. As highlighted by the notes by Professor Jauho, David made major contributions within the area of non-equilibrium physics and he also spent a lot of his talent on processes like charge transfer at surfaces. [Phys. Rev. B **43**, 2541 (1991)]. Yet another area with great contributions by David is Kondo physics with connections to problems at surfaces and quantum conduction. [Examples: Phys. Rev. B **43**, 2541 (1991), Phys. Rev. B **61**, R13341 (2000), and Phys. Rev. B **71**, 165321 (2005)].

Journal of Physics: Conference Series **427** (2013) 011001 doi:10.1088/1742-6596/427/1/011001

Conference participants

Arrachea, Liliana (*Buenos Aires, Argentina*)

Balzer, Karsten (*Hamburg, Germany*)

Bonitz, Michael (*Kiel, Germany*)

Brandbyge, Mads (*Lyngby, Denmark*)

Cunningham, Brian (*Belfast, UK*)

Eckstein, Martin (*Hamburg, Germany*)

Garny, Mathias (*Hamburg, Germany*)

Gomes da Rocha, Claudia (*Jyväskylä, Finland*)

Guo, Hong (*Montréal, Canada*)

Han, Jong (*Buffalo, USA*)

Harju, Ari (*Espoo, Finland*)

Hermanns, Sebastian (*Kiel, Germany*)

Hochstuhl, David (*Kiel, Germany*)

Hyrkäs, Markku (*Jyväskylä, Finland*)

Jahnke, Frank (*Bremen, Germany*)

Jauho, Antti-Pekka (*Lyngby, Denmark*)

Kainulainen, Kimmo (*Jyväskylä, Finland*)

Kinnunen, Jami (*Espoo, Finland*)

Marini, Andrea (*Rome, Italy*)

Ness, Hervé (*York, UK*)

Nozaki, Daijiro (*Dresden, Germany*)

Pastawski, Horacio (*Córdoba, Argentina*)

Pavlyukh, Yaroslav (*Halle, Germany*)

Rios, Arnau (*Surrey, UK*)

Stefanucci, Gianluca (*Rome, Italy*)

Säkkinen, Niko (*Jyväskylä, Finland*)

Tuovinen, Riku (*Jyväskylä, Finland*)

Uimonen, Anna-Maija (*Jyväskylä, Finland*)

van Leeuwen, Robert (*Jyväskylä, Finland*)

Verdozzi, Claudio (*Lund, Sweden*)

Wu, Ming-Wei (*Hefei, China*)

Journal of Physics: Conference Series **427** (2013) 011001 doi:10.1088/1742-6596/427/1/011001

List of talks and posters

(I) — Invited talk; (C) — Contributed talk; (P) — Poster

(I) Yaroslav Pavlyukh,
University of Halle, Germany,
ON THE INITIAL STAGE OF QUASIPARTICLE DECAY

(I) Ari Harju,
Aalto University, Finland,
ELECTRONIC TRANSPORT IN GRAPHENE STRUCTURES

(I) Kimmo Kainulainen,
University of Jyväskylä, Finland,
FLAVOURED QUANTUM TRANSPORT EQUATIONS AND THEIR APPICATIONS IN COSMOLOGY

(I) Karsten Balzer,
University of Hamburg, Germany,
FEW-ELECTRON DYNAMICS FROM THE KADANOFF–BAYM EQUATIONS. SINGLE AND DOUBLE EXCITATIONS IN QUANTUM WELLS AND HUBBARD CHAINS

(C) David Hochstuhl,
University of Kiel, Germany,
TIME-DEPENDENT RESTRICTED ACTIVE SPACE CONFIGURATION INTERACTION CALCULATIONS FOR THE PHOTOIONIZATION OF MULTIELECTRON ATOMS

(I) Mads Brandbyge,
Technical University of Denmark,
SEMI-CLASSICAL LANGEVIN AND NEGF APPROACH TO ATOMIC DYNAMICS IN NON-EQUILIBRIUM NANOSYSTEMS

(I) Liliana Arrachea,
University of Buenos Aires, Argentina,
NONEQUILIBRIUM GREEN'S FUNCTIONS IN THE STUDY OF HEAT TRANSPORT AND COOLING IN MESOSCOPIC DEVICES AND NANOMECHANICAL SYSTEMS

(C) Martin Eckstein,
University of Hamburg, Germany,
THE HUBBARD MODEL IN STRONG ELECTRIC FIELDS

(C) Anna-Maija Uimonen,
University of Jyväskylä, Finland,
CORRELATION EFFECTS ON THE TIME-DEPENDENT QUANTUM TRANSPORT: COMPARATIVE STUDY OF MANY-BODY PERTURBATION THEORY AND TIME-DEPENDENT DENSITY FUNCTIONAL THEORY

(I) Ming-Wei Wu,
University of Science and Technology of China,
SPIN DYNAMICS IN SEMICONDUCTORS AND GRAPHENE

(I) **Claudio Verdozzi,**
University of Lund, Sweden,
KADANOFF–BAYM DYNAMICS FOR LATTICE MODEL HAMILTONIANS: COMPARATIVE ASSESSMENT OF SOME CONSERVING APPROXIMATIONS TO EXACT AND TDDFT RESULTS

(C) **Claudia Gomes da Rocha,**
University of Jyväskylä, Finland,
CONTROLLING AC TRANSPORT IN GRAPHENE-BASED FABRY–PEROT DEVICES

(C) **Jami Kinnunen,**
Aalto University, Finland,
EXTENDED BRUECKNER THEORY FOR ULTRACOLD ATOMIC GASES

(C) **Arnau Rios,**
University of Surrey, UK,
TOWARDS A NON-EQUILIBRIUM GREEN'S FUNCTIONS APPROACH FOR NUCLEAR REACTIONS

(I) **Hong Guo,**
McGill University, Canada,
DEVICE MODELING FROM ATOMISTIC FIRST PRINCIPLES: THEORY OF THE NONEQUILIBRIUM VERTEX CORRECTION

(I) **Frank Jahnke,**
University of Bremen, Germany,
CONFIGURATION-PICTURE DESCRIPTION OF CARRIER SCATTERING IN SEMICONDUCTOR QUANTUM DOTS

(I) **Gianluca Stefanucci,**
University of Rome "Tor Vergata", Italy,
WICK THEOREM FOR NONEQUILIBRIUM SYSTEMS IN GENERAL INITIAL STATES

(I) **Andrea Marini,**
National Research Council, Italy,
PHONON-INDUCED DYNAMICS OF ELECTRONS AND EXCITONS IN SOLIDS DRIVEN OUT-OF-EQUILIBRIUM BY STRONG LASER PULSES: AN AB-INITIO APPROACH

(I) **Hervé Ness,**
University of York, UK,
MODELLING NON-EQUILIBRIUM AND MANY-BODY EFFECTS IN QUANTUM TRANSPORT THROUGH SINGLE-MOLECULE NANODEVICES

(I) **Horacio Pastawski,**
LaNAIS, Argentina,
DECOHERENT MANY-BODY INTERACTIONS IN QUANTUM TRANSPORT: FROM CONDUCTING POLYMERS AND GIANT MAGNETORESISTANCE TO QUANTUM DYNAMICAL PHASE TRANSITIONS IN DOUBLE DOTS AND HETEROGENEOUS CATALYSIS

(I) **Jong Han,**
University at Buffalo, USA,
QUANTUM SIMULATION OF STRONGLY CORRELATED QUANTUM DOTS OUT OF EQUILIBRIUM

(C) **Niko Säkkinen,**
University of Jyväskylä, Finland,
DOUBLE EXCITATIONS IN FINITE SYSTEMS WITH THE KADANOFF–BAYM FORMALISM

(I) **Mathias Garny,**
DESY, Germany,
RENORMALIZATION OF RELATIVISTIC KADANOFF–BAYM EQUATIONS

(C) **Daijiro Nozaki,**
Technical university of Dresden, Germany,
PREDICTION OF QUANTUM INTERFERENCE IN MOLECULAR JUNCTIONS USING A PARABOLIC DIAGRAM: UNDERSTANDING THE ORIGIN OF FANO AND ANTI- RESONANCES

(C) **Riku Tuovinen,**
University of Jyväskylä, Finland,
TRANSIENT DYNAMICS WITHOUT TIME PROPAGATION

(P) Michael Bonitz,
University of Kiel, Germany,
NONEQUILIBRIUM QUANTUM PAIR DISTRIBUTION FUNCTION FROM NONEQUILIBRIUM GREENS FUNCTIONS

(P) Claudia Gomes da Rocha,
University of Jyväskylä, Finland,
MANIFESTATION OF QUANTUM WAGON-WHEEL EFFECT IN CARBON BASED NANOSTRUCTURES

(P) Sebastian Hermanns,
University of Kiel, Germany,
NON-EQUILIBRIUM GREEN'S FUNCTION DESCRIPTION OF HUBBARD NANOCLUSTERS WITHIN THE GENERALIZED KADANOFF–BAYM ANSATZ

(P) Brian Cunningham,
Queen's university of Belfast, UK,
NON-CONSERVATIVE CURRENT-INDUCED FORCES: THE CONCEPT OF AN ATOMIC WATERWHEEL

(P) David Hochstuhl,
University of Kiel, Germany,
TIME-DEPENDENT RESTRICTED ACTIVE SPACE CONFIGURATION INTERACTION CALCULATIONS FOR THE PHOTOIONIZATION OF MULTIELECTRON ATOMS

(P) Arnau Rios,
University of Surrey, UK,
EQUILIBRIUM GREEN'S FUNCTION CALCULATION OF THE NUCLEON MEAN-FREE PATH

Journal of Physics: Conference Series **427** (2013) 011001 doi:10.1088/1742-6596/427/1/011001

Conference photographs

Group photo. Upper row from left to right: Hervé Ness, Liliana Arrachea, Gianluca Stefanucci, Claudio Verdozzi, Jong Han, David Hochstuhl, Michael Bonitz, Andrea Marini, Martin Eckstein, Markku Hyrkäs, Robert van Leeuwen, Yaroslav Pavlyukh, Sebastian Hermanns, Frank Jahnke, Ari Harju, Ming-Wei Wu, Jami Kinnunen, Antti-Pekka Jauho, Anna-Maija Uimonen, Niko Säkkinen and Riku Tuovinen.
Lower row from left to right: Daijiro Nozaki, Mads Brandbyge, Horacio Pastawski, Arnau Rios, Brian Cunningham, Karsten Balzer, Hong Guo and Claudia Gomes da Rocha.
Photo: Ville Kotimäki

Journal of Physics: Conference Series **427** (2013) 011001 doi:10.1088/1742-6596/427/1/011001

Waiting for the ship to arrive in the Jyväskylä harbour: Brian Cunningham, Horacio Pastawski, David Hochstuhl, Sebastian Hermanns, Karsten Balzer, Martin Eckstein, inter alios.

Hopping in the ship Suometar where the conference dinner was served: Claudia Gomes da Rocha and Horacio Pastawski, inter alios.

Under the Äijälänjoki bridge: Frank Jahnke, Jong Han, Antti-Pekka Jauho and Martin Eckstein.

Sunset and rainbow on the lake Päijänne: Riku Tuovinen, Michael Bonitz, Robert van Leeuwen and Yaroslav Pavlyukh.

Peer review statement

All papers published in this volume of *Journal of Physics: Conference Series* have been peer reviewed through processes administered by the proceedings Editors. Reviews were conducted by expert referees to the professional and scientific standards expected of a proceedings journal published by IOP Publishing.

Progress in Nonequilibrium Green's Functions V (PNGF V) IOP Publishing
Journal of Physics: Conference Series **427** (2013) 012001 doi:10.1088/1742-6596/427/1/012001

Equilibrium and nonequilibrium many-body perturbation theory: a unified framework based on the Martin-Schwinger hierarchy

Robert van Leeuwen[1] **and Gianluca Stefanucci**[2]

[1]Department of Physics, Nanoscience Center, FIN 40014, University of Jyväskylä, Jyväskylä, Finland

[2]Dipartimento di Fisica, Università di Roma Tor Vergata, Via della Ricerca Scientifica 1, 00133 Rome, Italy

Abstract. We present a unified framework for equilibrium and nonequilibrium many-body perturbation theory. The most general nonequilibrium many-body theory valid for general initial states is based on a time-contour originally introduced by Konstantinov and Perel'. The various other well-known formalisms of Keldysh, Matsubara and the zero-temperature formalism are then derived as special cases that arise under different assumptions. We further present a single simple proof of Wick's theorem that is at the same time valid in all these flavors of many-body theory. It arises simply as a solution of the equations of the Martin-Schwinger hierarchy for the noninteracting many-particle Green's function with appropriate boundary conditions. We further discuss a generalized Wick theorem for general initial states on the Keldysh contour and derive how the formalisms based on the Keldysh and Konstantinov-Perel'-contours are related for the case of general initial states.

1. Introduction

In many physical situations we are interested in knowing the expectation value of some observable quantity of a system in or out of equilibrium. For quantum systems of many identical and interacting particles a very convenient mathematical object to extract this information is the Green's function. Let $\hat{\rho}$ be the density matrix which describes the system at time, say, t_0 and $\hat{H}(t)$ be the Hamiltonian of the system for times $t > t_0$. The n-particle Green's function G_n is defined according to

$$G_n(1 \dots n; 1' \dots n') = \frac{1}{i^n} \mathrm{Tr}\left[\hat{\rho}\, T\left\{\hat{\psi}_H(1) \dots \hat{\psi}_H(n)\hat{\psi}^{\dagger}_H(n') \dots \hat{\psi}_H(1')\right\}\right]. \tag{1}$$

In this formula $1 = (x_1, t_1)$, $2 = (x_2, t_2)$, etc. are collective indices for the position-spin coordinates $x = \mathbf{r}, \sigma$ and time t, the symbol Tr denotes a trace over the Fock space, T is the time-ordering operator and $\hat{\psi}_H(j) = \hat{U}(t_0, t_j)\hat{\psi}(j)\hat{U}(t_j, t_0)$ are field operators in the Heisenberg picture with respect to the Hamiltonian $\hat{H}$ (hence $\hat{U}$ is the evolution operator). The quantum average of a n-body operator can be calculated from the equal-time Green's function G_n.

The direct evaluation of G_n from Eq. (1) is, in general, an impossible task. The first difficulty is brought by the Hamiltonian $\hat{H} = \hat{H}_0 + \hat{H}_{\mathrm{int}}$ which is typically the sum of a one-body operator $\hat{H}_0$ and a m-body operator $\hat{H}_{\mathrm{int}}$ with $m \geq 2$. For $\hat{H}_{\mathrm{int}} \neq 0$ the field operator $\hat{\psi}_H$ in the

Heisenberg picture is a complicated object and must be approximated in some clever way. The second difficulty consists in taking the trace over the Fock space with a density matrix $\hat{\rho}$. The density matrix is a self-adjoint, positive semi-definite operator with unit trace and, therefore, it can be written as $\hat{\rho} = e^{-\hat{X}}/\,\mathrm{Tr}[\hat{X}]$ where $\hat{X}$ is a self-adjoint operator. For instance for systems in equilibrium $\hat{X} = \beta\hat{H}^{\mathrm{M}}$ with β the inverse temperature and $\hat{H}^{\mathrm{M}} = \hat{H} - \mu\hat{N}$ the grand-canonical Hamiltonian. To make contact with this equilibrium situation we define $\hat{H}^{\mathrm{M}} = \hat{X}/\beta$ so that

$$\hat{\rho} = \frac{e^{-\beta\hat{H}^{\mathrm{M}}}}{Z} \tag{2}$$

with $Z = \mathrm{Tr}[e^{-\beta\hat{H}^{\mathrm{M}}}]$. In equilibrium Z is the partition function. In order to specify the initial preparation of the system we can assign either $\hat{\rho}$ or $\hat{H}^{\mathrm{M}}$ since there is a one-to-one correspondence between the two. If we now separate $\hat{H}^{\mathrm{M}} = \hat{H}_0^{\mathrm{M}} + \hat{H}_{\mathrm{int}}^{\mathrm{M}}$ into the sum of a one-body operator $\hat{H}_0^{\mathrm{M}}$ and a m-body operator $\hat{H}_{\mathrm{int}}^{\mathrm{M}}$ with $m \geq 2$ then the trace in Eq. (1) can easily be worked out for $\hat{H}_{\mathrm{int}}^{\mathrm{M}} = 0$ whereas we have to use suitable approximation schemes for $\hat{H}_{\mathrm{int}}^{\mathrm{M}} \neq 0$.

Different Many-Body Perturbation Theories (MBPT) have been put forward to overcome these difficulties. The most popular MBPT's are probably the zero-temperature (real-time) Green's Function Formalism (GFF) and the finite-temperature (imaginary-time) Matsubara GFF [1]. These two formalisms are limited to equilibrium situations. Systems driven out of equilibrium by an external field are usually studied within the (adiabatic real-time) Keldysh GFF [2, 3]. The Keldysh GFF, however, neglects the effect of initial correlations which are relevant in the short-time dynamics of general quantum systems, such as in transient dynamics in quantum transport or in the study of atoms and molecules in external laser fields. There exist two alternative GFF's to include initial correlations. The first is based on the idea of Konstantinov and Perel' [4] and consists in attaching the imaginary-time Matsubara track to the original Keldysh contour, see Refs. [3, 5, 6]. The second GFF does instead account for initial correlations through extra Feynman diagrams, the evaluation of which requires the knowledge of the reduced n-particle density matrices

$$\Gamma_n(x_1 \ldots x_n; x_1' \ldots x_n') = \mathrm{Tr}[\hat{\rho}\,\hat{\psi}^\dagger(x_1') \ldots \hat{\psi}^\dagger(x_n')\hat{\psi}(x_n) \ldots \hat{\psi}(x_1)] \tag{3}$$

where the symbol Tr signifies a trace over the Fock space, see Refs. [7, 8, 9, 10, 11, 12]. These last two formalisms are both exact and hence equivalent.

In all the aforementioned GFF's the dressed (interacting) G_n is expanded in powers of the interaction Hamiltonian ($\hat{H}_{\mathrm{int}}$ and/or $\hat{H}_{\mathrm{int}}^{\mathrm{M}}$), leading to an expansion of G_n in terms of the bare (noninteracting) Green's functions $G_{0,n}$. The appealing feature of any GFF is the possibility of reducing the $G_{0,n}$ to an (anti)symmetrized product of $G_0 \equiv G_{0,1}$ by means of Wick's theorem [13]. Even though the mathematical structure of all GFF's is identical, these formalisms are usually treated as independent probably due to the fact that the existing proofs of Wick's theorem are very much formalism-dependent. In this paper we show that Wick's theorem is the solution of a boundary problem for the Martin-Schwinger Hierarchy (MSH) [14] and that different GFF's correspond to different domains and parameters for the MSH [15]. In this way we can easily explain the common mathematical structure of every GFF and see how, e.g., the Keldysh GFF reduces to the zero-temperature GFF in equilibrium or the Konstantinov-Perel' GFF reduces to the Keldysh GFF under the adiabatic assumption. Our reformulation also allows us to prove a generalized Wick's theorem for interacting density matrices $\hat{\rho}$. This naturally leads to the diagrammatic expansion with extra Feynman diagrams previously mentioned. The generalized Wick expansion has a form identical to that of a Laplace expansion for permanents/determinants (for bosons/fermions). Consequently, the calculation of the various prefactors is both explicit and greatly simplified. In this contribution we only state

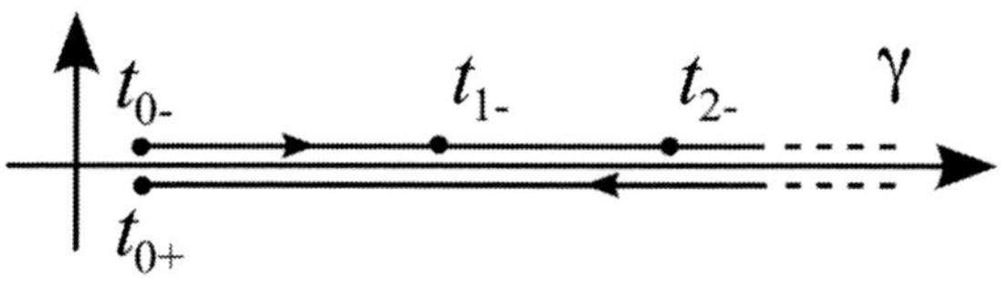

Figure 1. Contour γ of Eq. (4). The contour consists of a forward branch going from t_0 to ∞ (on this branch the points are denoted by $z = t_-$) and a backward branch going from ∞ to t_0 (on this branch the points are denoted by $z = t_+$).

the generalized Wick's theorem and refer the reader to Refs. [12, 15] for the proof. We will, however, discuss the equivalence between the GFF based on the generalized Wick's theorem and the Konstantinov-Perel' GFF.

2. General formula for the Green's function

The n-particle Green's function in Eq. (1) can also be written as

$$G_n(1 \ldots n; 1' \ldots n') = \frac{1}{i^n} \operatorname{Tr}\left[\hat{\rho}\, \mathcal{T}\left\{e^{-i\int_\gamma dz \hat{H}(z)} \hat{\psi}(1) \ldots \hat{\psi}(n)\hat{\psi}^\dagger(n') \ldots \hat{\psi}(1')\right\}\right]. \tag{4}$$

Let us explain this formula and discuss the equivalence with Eq. (1). In Eq. (4) the integral is over the contour γ of Fig. 1 which goes from t_0 to ∞ and back to t_0 whereas $\mathcal{T}$ is the contour ordering operator which rearranges operators with later contour arguments to the left. We denote by $z = t_\pm$ the points on γ lying on the lower/upper branch at a distance t from the origin and define the field operators with arguments on the contour as

$$\hat{\psi}(x, z) = \hat{\psi}(x). \tag{5}$$

More generally every operator $\hat{O}(t)$ with a real-time argument can be converted into an operator $\hat{O}(z)$ with a contour-time argument according to the rule $\hat{O}(t_+) = \hat{O}(t_-) = \hat{O}(t)$. In particular $\hat{H}(t_-) = \hat{H}(t_+) = \hat{H}(t)$. The reason to keep the contour argument in Eq. (4) even for operators that do not have an explicit time dependence (like the field operators) stems from the need of specifying their position along the contour, thus rendering unambiguous the action of $\mathcal{T}$. Once the operators are ordered we can omit the time arguments if there is no time dependence. For instance if $t_1 < t_2$ then

$$\begin{aligned}\mathcal{T}\left\{e^{-i\int_\gamma dz \hat{H}(z)} \hat{\psi}(x_1, t_{1-})\hat{\psi}^\dagger(x_2, t_{2-})\right\} &= \pm\hat{U}(t_0, \infty)\hat{U}(\infty, t_2)\hat{\psi}^\dagger(x_2)\hat{U}(t_2, t_1)\hat{\psi}(x_1)\hat{U}(t_1, t_0) \\ &= \mathcal{T}\left\{\hat{\psi}_H(x_1, t_1)\hat{\psi}_H^\dagger(x_2, t_2)\right\},\end{aligned} \tag{6}$$

where the $\pm$ sign in the first equality is for bosons/fermions. One can verify that Eq. (6) is valid also for $t_1 > t_2$. This example can easily be generalized to many field operators. We conclude that Eq. (4) is equivalent to Eq. (1) for contour arguments on the upper branch of γ. The G_n in Eq. (4) is, however, more general since the contour arguments can lie either on the upper or lower branch of γ. Quantities like photoemission currents, hyper-polarizabilities and more generally high-order response properties require the knowledge of this more general Green's function.

The density matrix in Eq. (4) can be incorporated into the contour ordering operator if we extend γ as illustrated in Fig. 2 and define the Hamiltonian with imaginary-time arguments as $\hat{H}(t_0 - i\tau) = \hat{H}^{\mathrm{M}}$. Since

$$e^{-\beta\hat{H}^{\mathrm{M}}} = e^{-i\int_{t_0}^{t_0 - i\beta} \hat{H}(z)} = \mathcal{T}\left\{e^{-i\int_{t_0}^{t_0-i\beta} \hat{H}(z)}\right\} \tag{7}$$

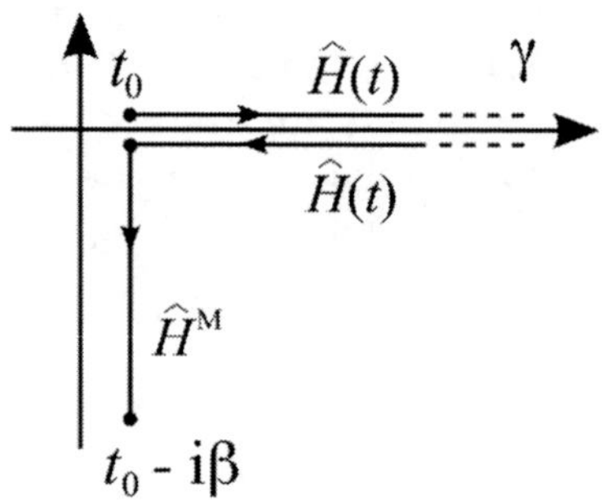

Figure 2. Contour γ and Hamiltonian along the contour γ to get the exact (Konstantinov-Perel') Green's function from Eq. (8).

we have

$$G_n(1\ldots n;1'\ldots n') = \frac{1}{i^n}\frac{\mathrm{Tr}\left[\mathcal{T}\left\{e^{-i\int_\gamma dz\hat{H}(z)}\hat{\psi}(1)\ldots\hat{\psi}(n)\hat{\psi}^\dagger(n')\ldots\hat{\psi}(1')\right\}\right]}{\mathrm{Tr}\left[\mathcal{T}\left\{e^{-i\int_\gamma dz\hat{H}(z)}\right\}\right]} \tag{8}$$

where in the denominator we took into account that $\mathcal{T}\{e^{-i\int_{t_0-}^{t_0+}dz\hat{H}(z)}\} = \hat{U}(t_0,\infty)\hat{U}(\infty,t_0) = 1$. Equation (8) is, by construction, equivalent to Eq. (4). It gives the *exact* Green's function provided that the integral is done along the contour γ of Fig. 2 and provided that the Hamiltonian changes along the contour as illustrated in the same figure. We now show that the Green's function of every GFF can be written as in Eq. (8), the only difference being the shape of γ and the Hamiltonian along γ.

We mentioned in the introduction that the calculation of the trace simplifies if the density matrix is of the form $\hat{\rho}_0 = e^{-\beta\hat{H}_0^{\rm M}}/Z_0$, with $\hat{H}_0^{\rm M}$ a one-body operator and $Z_0 = \mathrm{Tr}[e^{-\beta\hat{H}_0^{\rm M}}]$. It is possible to turn a trace with $\hat{\rho}$ into a trace with $\hat{\rho}_0$ if the adiabatic assumption is fulfilled. According to the adiabatic assumption one can generate the density matrix $\hat{\rho}$ with Hamiltonian $\hat{H}^{\rm M} = \hat{H}_0^{\rm M} + \hat{H}_{\rm int}^{\rm M}$ starting from the density matrix $\hat{\rho}_0$ with Hamiltonian $\hat{H}_0^{\rm M}$ and then switching on $\hat{H}_{\rm int}^{\rm M}$ adiabatically, i.e.,

$$\hat{\rho} = \frac{e^{-\beta\hat{H}^{\rm M}}}{Z} = \hat{U}_\eta(t_0,-\infty)\frac{e^{-\beta\hat{H}_0^{\rm M}}}{Z_0}\hat{U}_\eta(-\infty,t_0) = \hat{U}_\eta(t_0,-\infty)\,\hat{\rho}_0\,\hat{U}_\eta(-\infty,t_0), \tag{9}$$

where $\hat{U}_\eta$ is the real-time evolution operator with Hamiltonian

$$\hat{H}_\eta(t) = \hat{H}_0^{\rm M} + e^{-\eta|t-t_0|}\hat{H}_{\rm int}^{\rm M}, \tag{10}$$

and η is an infinitesimally small positive constant. This Hamiltonian is equal to $\hat{H}_0^{\rm M}$ when $t \to -\infty$ and is equal to the full interacting $\hat{H}^{\rm M}$ when $t = t_0$. In general the validity of the adiabatic assumption should be checked case by case. Under the adiabatic assumption we can rewrite Eq. (4) as (omitting the arguments of G_n)

$$G_n = \frac{1}{i^n}\mathrm{Tr}\left[\hat{\rho}_0\,\hat{U}_\eta(-\infty,t_0)\mathcal{T}\left\{e^{-i\int_\gamma dz\hat{H}(z)}\hat{\psi}(1)\ldots\hat{\psi}(n)\hat{\psi}^\dagger(n')\ldots\hat{\psi}(1')\right\}\hat{U}_\eta(t_0,-\infty)\right]. \tag{11}$$

We now see that if we construct the contour γ of Fig. 3 and let the Hamiltonian change along the contour as

$$\hat{H}(t_\pm) = \begin{cases} \hat{H}_\eta(t) = \hat{H}_0^{\rm M} + e^{-\eta|t-t_0|}\hat{H}_{\rm int}^{\rm M} & \text{for } t < t_0 \\ \\ \hat{H}(t) = \hat{H}_0(t) + \hat{H}_{\rm int}(t) & \text{for } t > t_0 \end{cases}$$

$$\hat{H}(z\in\gamma^{\rm M}) = \hat{H}_0^{\rm M} = \hat{H}_0 - \mu\hat{N},$$

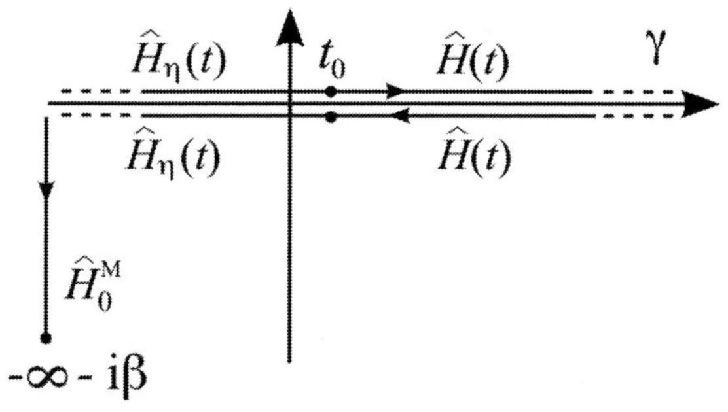

Figure 3. Contour γ and Hamiltonian along the contour γ to get the adiabatic (Keldysh) Green's function from Eq. (8).

then Eq. (11) takes the same form as Eq. (8). We will refer to this way of calculating G_n as the adiabatic formula. This is exactly the formula used by Keldysh in his original paper [2]. The adiabatic formula is correct only provided that the adiabatic assumption is fulfilled.

We can derive yet another expression of G_n for systems in equilibrium at zero temperature. In equilibrium $\hat{H}^{\rm M} = \hat{H} - \mu\hat{N}$ and therefore $\hat{H}_0^{\rm M} = \hat{H}_0 - \mu\hat{N}$ and $\hat{H}_{\rm int}^{\rm M} = \hat{H}_{\rm int}$. Assuming that $\hat{H}_0$ and $\hat{H}_{\rm int}$ commute with $\hat{N}$ the evolution operator $\hat{U}_\eta$ in Eq. (9) can be calculated with Hamiltonian

$$\hat{H}_\eta(t) = \hat{H}_0 + e^{-\eta|t-t_0|}\hat{H}_{\rm int} \tag{12}$$

since the addition of $-\mu\hat{N}$ corresponds to multiplying $\hat{U}_\eta$ by a phase factor. In Eq. (9) this phase factor cancels out since $\hat{U}_\eta(-\infty, t_0) = [\hat{U}_\eta(t_0, -\infty)]^\dagger$. Furthermore, for any finite contour-times in G_n we can approximate the evolution operator $\hat{U}$ in the field operators $\hat{\psi}_H$ with the evolution operator U_η since we can always choose $\eta \ll 1/|t - t_0|$ and hence $\hat{H}_\eta \sim \hat{H}$. Thus Eq. (11) becomes

$$G_n = \frac{1}{i^n}\,{\rm Tr}\left[\hat{\rho}_0\,\mathcal{T}\left\{e^{-i\int_\gamma dz\hat{H}(z)}\hat{\psi}(1)\ldots\hat{\psi}(n)\hat{\psi}^\dagger(n')\ldots\hat{\psi}(1')\right\}\right] \tag{13}$$

where γ is a contour that goes from $-\infty$ to ∞ and back to $-\infty$ and $\hat{H}(t_\pm) = \hat{H}_\eta(t)$ is the Hamiltonian of Eq. (12). Next we observe that the interacting $\hat{\rho}$ can also be generated starting from $\hat{\rho}_0$ and then propagating *backward* in time from ∞ to t_0 using the same evolution operator $\hat{U}_\eta$ since $\hat{H}_\eta(t_0 - t) = \hat{H}_\eta(t_0 + t)$. In other words

$$\hat{\rho} = \hat{U}_\eta(t_0, \infty)\,\hat{\rho}_0\,\hat{U}_\eta(\infty, t_0). \tag{14}$$

Comparing this equation with Eq. (9) we conclude that

$$\hat{\rho}_0 = \hat{U}_\eta(-\infty, \infty)\,\hat{\rho}_0\,\hat{U}_\eta(\infty, -\infty). \tag{15}$$

If the ground state $|\Phi_0\rangle$ of $\hat{H}_0 - \mu\hat{N}$ is nondegenerate then the zero-temperature $\hat{\rho}_0 = |\Phi_0\rangle\langle\Phi_0|$ is a pure state and Eq. (15) implies that

$$\langle\Phi_0|\hat{U}_\eta(\infty, -\infty) = e^{i\alpha_0}\langle\Phi_0|. \tag{16}$$

We will refer to the adiabatic assumption in combination with equilibrium at zero temperature and with the condition of no ground-state degeneracy as the *zero-temperature assumption*. The zero-temperature assumption can be used to manipulate Eq. (13) a bit more. We have

$$\hat{\rho}_0 = |\Phi_0\rangle\langle\Phi_0| = \frac{|\Phi_0\rangle\langle\Phi_0|\hat{U}_\eta(\infty, -\infty)}{\langle\Phi_0|\hat{U}_\eta(\infty, -\infty)|\Phi_0\rangle} = \lim_{\beta\to\infty}\frac{e^{-\beta\hat{H}_0^{\rm M}}\hat{U}_\eta(\infty, -\infty)}{{\rm Tr}\left[e^{-\beta\hat{H}_0^{\rm M}}\hat{U}_\eta(\infty, -\infty)\right]}. \tag{17}$$

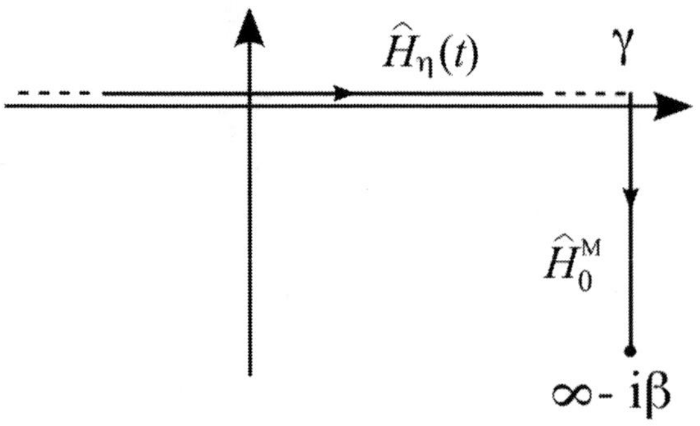

Figure 4. Contour γ and Hamiltonian along the contour γ to get the zero-temperature Green's function from Eq. (8).

Inserting this result into Eq. (13) we find that the zero-temperature Green's function can again be written as in Eq. (8) with the contour γ that starts at $-\infty$, goes all the way to ∞ and then down to $\infty - i\beta$, see Fig. 4, and with the Hamiltonian $\hat{H}(z)$ that varies along the contour as illustrated in the same figure. It is worth noticing that the contour γ has the special property of having only a forward branch and that for the zero-temperature assumption to make sense the Hamiltonian of the system must be time independent. There is indeed no reason to expect that by switching on and off the interaction the system goes back to the same state in the presence of external driving fields.

To summarize the exact (Konstantinov-Perel'), adiabatic (Keldysh) and zero-temperature Green's functions have the same mathematical structure, given by Eq. (8). What changes is the contour and the Hamiltonian along the contour.

3. Wick's theorem and Many-Body Perturbation Theory

To be concrete we specialize the discussion to interaction Hamiltonians $\hat{H}_{\rm int}$ and $\hat{H}^{\rm M}_{\rm int}$ which are two-body operators. Higher order n-body operators lead to more voluminous equations but do not rise conceptual complications. Thus we write

$$\hat{H}_{\rm int}(z) = \frac{1}{2}\int dx_1 dx_2\; v(x_1,x_2;z)\hat{\psi}^\dagger(x_1)\hat{\psi}^\dagger(x_2)\hat{\psi}(x_2)\hat{\psi}(x_1). \tag{18}$$

In the exact (Konstantinov-Perel') formula the interaction $v(x_1,x_2;z) = v(x_1,x_2)$ is the interparticle interaction for z on the horizontal branches whereas $v(x_1,x_2;z)$ depends on the initial preparation for z on the vertical track. For instance in equilibrium $v(x_1,x_2;t_0-i\tau) = v(x_1,x_2)$. On the other hand in the adiabatic (Keldysh) and zero-temperature formula $v(x_1,x_2;z) = e^{-\eta|t-t_0|}v(x_1,x_2)$ for z on the horizontal branches whereas $v(x_1,x_2;z) = 0$ for z on the vertical track. Let us consider Eq. (8) and write the exponential of $\hat{H}$ as the product of the exponentials of $\hat{H}_0$ and $\hat{H}_{\rm int}$:

$$G_n(1\ldots n;1'\ldots n') = \frac{1}{i^n}\frac{\mathrm{Tr}\left[\mathcal{T}\left\{e^{-i\int_\gamma dz\hat{H}_0(z)}e^{-i\int_\gamma dz\hat{H}_{\rm int}(z)}\hat{\psi}(1)\ldots\hat{\psi}(n)\hat{\psi}^\dagger(n')\ldots\hat{\psi}(1')\right\}\right]}{\mathrm{Tr}\left[\mathcal{T}\left\{e^{-i\int_\gamma dz\hat{H}_0(z)}e^{-i\int_\gamma dz\hat{H}_{\rm int}(z)}\right\}\right]}. \tag{19}$$

The expansion in powers of $\hat{H}_{\rm int}$ leads to an expansion of G_n in terms of noninteracting Green's functions $G_{0,n}$. The $G_{0,n}$ are obtained from Eq. (19) by setting $\hat{H}_{\rm int}(z) = 0$ for all $z \in \gamma$. For instance for $n = 1$ we get

$$G(a;b) = \frac{\sum\limits_{k=0}^{\infty}\frac{1}{k!}\left(\frac{i}{2}\right)^k\int v(1;1')\ldots v(k;k')G_{0,2k+1}(a,1,1',\ldots;b,1^+,1'^+,\ldots)}{\sum\limits_{k=0}^{\infty}\frac{1}{k!}\left(\frac{i}{2}\right)^k\int v(1;1')\ldots v(k;k')G_{0,2k}(1,1',\ldots;1^+,1'^+,\ldots)}, \tag{20}$$

for $n = 2$ we get

$$G_2(a,b;c,d) = \frac{\sum_{k=0}^{\infty} \frac{1}{k!}\left(\frac{i}{2}\right)^k \int v(1;1')\dots v(k;k')G_{0,2k+2}(a,b,1,1'\dots;c,d,1^+,1'^+,\dots)}{\sum_{k=0}^{\infty} \frac{1}{k!}\left(\frac{i}{2}\right)^k \int v(1;1')\dots v(k;k')G_{0,2k}(1,1',\dots;1^+,1'^+,\dots)}, \tag{21}$$

etc. In these equations $a = (x_a, t_a)$, $b = (x_b, t_b)$ are collective indices like $1,\ 2, \dots$, the interaction $v(j;j') \equiv \delta(z_j, z_j')v(x_j, x_j'; z_j)$ and the integrals are over $1, 1', \dots, k, k'$.

The appealing feature of any GFF is the possibility of reducing the noninteracting $G_{0,n}$ to a (anti)symmetrized product for (fermions) bosons of one-particle Green's functions G_0. This reduction is called Wick's theorem. The existing proofs of Wick's theorem are rather laborious and differ depending on whether one is working with the zero-temperature or Matsubara or Keldysh Green's functions. Below we give a simple and general proof of Wick's theorem which applies to all cases.

We consider a one-body Hamiltonian of the form

$$\hat{H}_0(z) = \int dx\, \hat{\psi}^\dagger(x)h(x,z)\hat{\psi}(x). \tag{22}$$

The more general case of a nondiagonal $h(x, x', z)$ can be treated in a similar manner. The Green's functions $G_{0,n}$ satisfy the noninteracting MSH

$$\left[i\frac{d}{dz_k} - h(k)\right] G_{0,n}(1\dots n;1'\dots n') = \sum_{j=1}^{n} (\pm)^{k+j}\, \delta(k;j')\, G_{0,n-1}(1\dots \overset{\sqcap}{k} \dots n;1'\dots \overset{\sqcap}{j'} \dots n') \tag{23}$$

$$G_{0,n}(1\dots n;1'\dots n')\left[-i\frac{\overleftarrow{d}}{dz_k'} - h(k')\right] = \sum_{j=1}^{n} (\pm)^{k+j}\, \delta(j;k')\, G_{0,n-1}(1\dots \overset{\sqcap}{j} \dots n;1'\dots \overset{\sqcap}{k'} \dots n') \tag{24}$$

where the hook over the arguments in $G_{0,n-1}$ means that those variables are missing. The MSH is a set of coupled differential equations to be solved on the contour γ of the Green's function of interest (exact, adiabatic, or zero-temperature). In all cases from the definition Eq. (8) it follows that the $G_{0,n}$ satisfy the Kubo-Martin-Schwinger (KMS) relations, i.e., the $G_{0,n}$ are (anti)periodic along the contour γ with respect to all their contour arguments. Therefore we can calculate the $G_{0,n}$ by solving the MSH with KMS relations. We now show that the solution is given by the Wick theorem

$$G_{0,n}(1,\dots,n;1',\dots,n') = \begin{vmatrix} G_0(1;1') & \dots & G_0(1;n') \\ \vdots & & \vdots \\ G_0(n;1') & \dots & G_0(n;n') \end{vmatrix}_\pm \tag{25}$$

where the symbol $|\dots|_\pm$ signifies the permanent/determinant for the case of bosons/fermions and G_0 is the solution of Eqs. (23) and (24) with $n = 1$, i.e.,

$$\left[i\frac{d}{dz_1} - h(1)\right] G_0(1;1') = \delta(1;1'), \qquad G_0(1;1')\left[-i\frac{\overleftarrow{d}}{dz_1'} - h(1')\right] = \delta(1;1') \tag{26}$$

with KMS boundary conditions. Expanding the permanent/determinant along row, say, k we get

$$G_{0,n}(1,\dots,n;1',\dots,n') = \sum_{j=1}^{n}(\pm)^{k+j}G_0(k,j')G_{0,n-1}(1\dots \overset{\sqcap}{k} \dots n;1'\dots \overset{\sqcap}{j'} \dots n') \tag{27}$$

which is clearly a solution of Eq. (23). Similarly, we can readily verify that Eq. (25) is also solution of Eq. (24) by expanding the permanent/determinant along column k. It remains to check that the $G_{0,n}$ in Eq. (25) fulfills the KMS relations. The contour argument z_k appears in all the G_0 of the k-th row of Eq. (25) and nowhere else. Therefore when we move z_k from the starting to the ending point of γ all entries of row k pick up a $(\pm)$ sign. Since the permanent/determinant of a matrix in which we multiply a row by $(\pm)$ is $(\pm)$ the permanent/determinant of the original matrix we conclude that $G_{0,n}$ is (anti)periodic with respect to the first n contour arguments. With a similar reasoning one can prove that $G_{0,n}$ is (anti)periodic with respect to the last n contour arguments. This concludes the proof.

The Wick theorem has been proven without any assumption on the shape of the contour and without any assumption on the form of the single-particle Hamiltonian $h(x, z)$ along the contour. Inserting Eq. (25) into, e.g., Eq. (20) we get the MBPT formula for the one-particle Green's function

$$G(a;b) = \frac{\sum_{k=0}^{\infty} \frac{1}{k!}\left(\frac{i}{2}\right)^k \int v(1;1')..v(k;k') \begin{vmatrix} G_0(a;b) & G_0(a;1^+) & \dots & G_0(a;k'^+) \\ G_0(1;b) & G_0(1;1^+) & \dots & G_0(1;k'^+) \\ \vdots & \vdots & \ddots & \vdots \\ G_0(k';b) & G_0(k';1^+) & \dots & G_0(k';k'^+) \end{vmatrix}_\pm}{\sum_{k=0}^{\infty} \frac{1}{k!}\left(\frac{i}{2}\right)^k \int v(1;1')..v(k;k') \begin{vmatrix} G_0(1;1^+) & G_0(1;1'^+) & \dots & G_0(1;k'^+) \\ G_0(1';1^+) & G_0(1';1'^+) & \dots & G_0(1';k'^+) \\ \vdots & \vdots & \ddots & \vdots \\ G_0(k';1^+) & G_0(k';1'^+) & \dots & G_0(k';k'^+) \end{vmatrix}_\pm} \tag{28}$$

which is an exact expansion of the interacting G in terms of the noninteracting G_0. The MBPT for higher order Green's functions can be derived similarly. In the next Section we discuss how the variuos GFF's follow from Eq. (28).

4. Matsubara, Keldysh and zero-temperature formalisms

In the Konstantinov-Perel' formalism the Green's function G is given by Eq. (28) where the z-integrals run on the contour of Fig. 2. It is worth stressing that if the times t_a and t_b in $G(a;b)$ are smaller than a maximum time T then it is sufficient to perform the z integrals over a shrunken contour like the one illustrated in Fig. 5. This is a direct consequence of the fact that if the contour is longer than T then the terms with integrals after T cancel off [15].

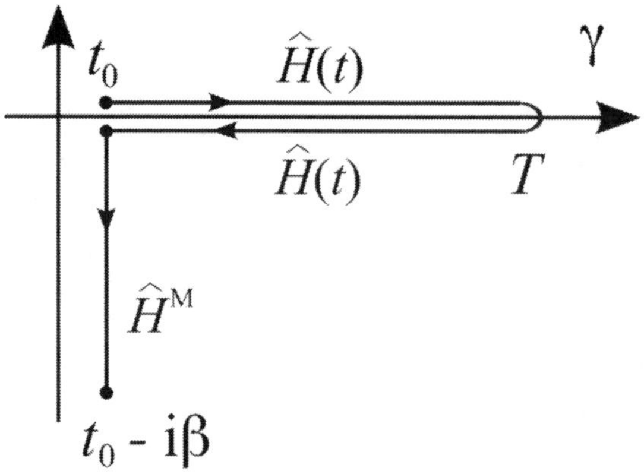

Figure 5. The shrunken contour γ that can be used in Eq. (28) to obtain the Green's function with real-time smaller than T.

The Matsubara GFF is used to calculate Green's function with imaginary times and it is typically applied to systems in equilibrium at finite temperature. For this reason the Matsubara

GFF is also called the "finite-temperature formalism". To calculate G with imaginary-time arguments we can choose $T = t_0$ in Fig. 5 and hence shrink the horizontal branches to a point leaving only the vertical track. Therefore the Matsubara GFF consists of expanding the Green's function as in Eq. (28) with the z-integrals restricted to the vertical track. It is important to realize that no assumptions, like the adiabatic or the zero-temperature assumption, are made in this formalism. The Matsubara GFF is exact but limited to initial (or equilibrium) averages. Equivalently we can say that the Matsubara G is the same as the Konstantinov-Perel' G on the vertical track.

The formalism originally used by Keldysh was based on the adiabatic assumption. The Keldysh Green's functions are again given by Eq. (28) but the z-integrals are done over the contour of Fig. 3 and the Hamiltonian changes along the contour as illustrated in the same figure. The important simplification of the Keldysh GFF is that the interaction v is zero on the vertical track. Consequently in Eq. (28) we can restrict the z-integrals to the horizontal branches. Like the Konstantinov-Perel' formalism, the Keldysh GFF can be used to deal with nonequilibrium situations in which the external perturbing fields are switched on after time t_0. In the special case of no external fields we can calculate interacting equilibrium Green's functions at any finite temperature with real-time arguments.

The zero-temperature formalism relies on the zero-temperature assumption. As we already discussed this assumption makes sense only in the absence of external fields. The corresponding zero-temperature Green's function is given by Eq. (28) in which the z-integrals are done over the contour of Fig. 4 and the Hamiltonian changes along the contour as illustrated in the same figure. Like in the Keldysh GFF the interaction v vanishes along the vertical track and hence the z-integrals can be restricted to a contour that goes from $-\infty$ to ∞. The contour ordering operator is then the same as the standard time-ordering operator. For this reason the zero-temperature Green's function is also called time-ordered Green's function. The zero-temperature GFF allows us to calculate the interacting G in equilibrium at zero temperature with real-time arguments. It cannot, however, be used to study systems out of equilibrium and/or at finite temperature. In some cases, however, the zero-temperature formalism is used also at finite temperatures (finite β) as the finite temperature corrections are small. This approximated formalism is sometimes referred to as the *real-time finite temperature formalism* [1]. We emphasize that in the real-time finite-temperature formalism (like in the Keldysh formalism) the temperature enters in Eq. (28) only through G_0 which satisfies the KMS relations. In the Konstantinov-Perel' formalism, on the other hand, the temperature enters through G_0 and through the contour integrals since the interaction is nonvanishing along the vertical track.

5. Generalized Wick's theorem

The MBPT of the exact GFF requires the knowledge of the operator $\hat{H}^{\mathrm{M}}$ on the vertical track. In many physical situations, however, it is easier to specify the initial state (or initial density matrix) instead of $\hat{H}^{\mathrm{M}}$. In these cases the preliminary step to apply MBPT consists in obtaining $\hat{H}^{\mathrm{M}}$ from $\hat{\rho}$, something that can be rather awkward. For instance if $\hat{\rho} = |\Psi\rangle\langle\Psi|$ is a pure state then $\hat{H}^{\mathrm{M}}$ is an operator with $|\Psi\rangle$ as the ground state. The existence of a generalized Wick's theorem that uses directly $\hat{\rho}$ would be of very valuable. In this Section we will show how to construct such a generalized framework.

We consider again Eq. (4) but this time we do not incorporate $\hat{\rho}$ in the contour ordering. In Eq. (4) the contour γ is that of Fig. 1 and the Hamiltonian is the physical Hamiltonian which, for simplicity, we take as the sum of $\hat{H}_0$ in Eq. (22) and $\hat{H}_{\mathrm{int}}$ in Eq. (18). We write the exponential in Eq. (4) as the product of two exponentials, one containing $\hat{H}_0$ and the other containing $\hat{H}_{\mathrm{int}}$, like we did in Eq. (19). The subsequent expansion of G in powers of $\hat{H}_{\mathrm{int}}$ leads to the expansion

$$G(a;b) = \sum_{k=0}^{\infty} \frac{1}{k!} \left(\frac{i}{2}\right)^k \int v(1;1') \dots v(k;k') g_{2k+1}(a,1,1',\dots;b,1^+,1'^+,\dots) \tag{29}$$

and similarly for higher order Green's function. In Eq. (29) the Green's functions g_n are noninteracting Green's functions averaged with an arbitrary density matrix $\hat{\rho}$

$$g_n(1\dots n;1'\dots n') = \frac{1}{i^n} \mathrm{Tr}\left[\hat{\rho}\, \mathcal{T}\left\{e^{-i\int_\gamma dz \hat{H}_0(z)} \hat{\psi}(1)\dots\hat{\psi}(n)\hat{\psi}^\dagger(n')\dots\hat{\psi}(1')\right\}\right]. \tag{30}$$

We will now prove a generalized Wick theorem to write these g_n in terms of the one-particle Green's function $g \equiv g_1$ and the n-particle reduced density matrices Γ_n defined in Eq. (3).

The Green's functions g_n satisfy the noninteracting MSH on the contour of Fig. 1. The problem in solving the MSH to obtain the g_n's is that we cannot use the KMS relations as boundary conditions. Indeed it is easy to verify that the g_n are not (anti)periodic along the contour. A convenient choice of boundary conditions follows directly from the definition of g_n and reads

$$(\pm i)^n \lim_{z_k, z'_j \to t_{0-}} g_n(1\dots n;1',\dots n') = \Gamma_n(x_1\dots x_n;x'_1\dots x'_n) \tag{31}$$

where the limit is taken with the order $z_1 < \dots < z_n < z'_n < \dots < z'_1$ of the contour arguments. The permanent/determinant

$$g_n(1\dots n;1'\dots n') = \begin{vmatrix} g(1;1') & & g(1;n') \\ \vdots & \vdots & \vdots \\ g(n;1') & & g(n;n') \end{vmatrix}_\pm \equiv |g|_n(1\dots n;1'\dots n') \tag{32}$$

is a solution of the MSH but, in general, with the wrong boundary conditions. In Eq. (32) the symbol $|\dots|_\pm$ signifies the permanent/determinant of the matrix inside the vertical bars. The particular solution must be supplied with the solution of the homogeneous equations

$$\left[i\frac{d}{dz_k} - h(k)\right] \tilde{g}_n(1\dots n;1'\dots n') = 0 \tag{33}$$

$$\tilde{g}_n(1\dots n;1'\dots n') \left[-i\frac{\overleftarrow{d}}{dz'_k} - h(k')\right] = 0 \tag{34}$$

to satisfy the correct boundary conditions. We observe that $\tilde{g}_n$ is not discontinuous when its contour arguments cross each other since in the right hand side of Eqs. (33) and (34) there is no δ-function. Consequently the equal-time limit of $\tilde{g}_n$ is independent of the order of the contour arguments.

Let us start by showing how to solve the MSH for g_2. We write $g_2 = |g|_2 + \tilde{g}_2$ where g satisfies the first equation of the MSH with boundary conditions

$$(\pm i) \lim_{z_1, z'_1 \to t_{0-}} g(1;1') = \Gamma_1(x_1;x'_1) \equiv \Gamma(x_1;x'_1). \tag{35}$$

The boundary conditions for $\tilde{g}_2$ follow directly from Eq. (31) and read

$$(\pm i)^2 \lim_{z_k, z'_j \to t_{0-}} \tilde{g}_2 = (\pm i)^2 \lim_{z_k, z'_j \to t_{0-}} (g_2 - |g|_2) = \Gamma_2 - |\Gamma|_2 \equiv C_2. \tag{36}$$

Next we consider the spectral function on the contour

$$A(1;1') = i\left[g^{>}(1;1') - g^{<}(1;1')\right]. \tag{37}$$

This function takes the same value for $z_1 = t_{1\pm}$ and $z_1' = t_{1\pm}'$ and satisfies the equations

$$\left[i\frac{d}{dz_1} - h(1)\right] A(1;1') = A(1;1')\left[-i\frac{\overleftarrow{d}}{dz_1'} - h(1')\right] = 0. \tag{38}$$

Furthermore, due to the (anti)commutation rules of the field operators

$$A(x_1, z; x_1', z) = \delta(x_1 - x_1'). \tag{39}$$

Therefore

$$\begin{aligned}\tilde{g}_2(1,2;1',2') &= \int d\bar{x}_1 d\bar{x}_2 d\bar{x}_1' d\bar{x}_2' A(1;\bar{x}_1, t_{0-}) A(2;\bar{x}_2, t_{0-}) C_2(\bar{x}_1, \bar{x}_2; \bar{x}_1', \bar{x}_2') \\ &\times \; A(\bar{x}_1', t_{0-}; 1') A(\bar{x}_2', t_{0-}; 2')\end{aligned} \tag{40}$$

is clearly the solution of the homogeneous MSH with the correct boundary conditions, see Eq. (36). We can manipulate Eq. (40) by introducing a linear combination of δ-functions on the contour

$$\delta_-(z) \equiv \delta(z, t_{0-}) - \delta(z, t_{0+}). \tag{41}$$

The spectral function appearing in Eq. (40) can be written as

$$A(1; x_1', t_{0-}) = i\int_\gamma d\bar{z}\; g(1; x_1', \bar{z})\delta_-(\bar{z}) \tag{42}$$

and

$$A(x_1, t_{0-}; 1') = -i\int_\gamma d\bar{z}\; \delta_-(\bar{z}) g(x_1, \bar{z}; 1'). \tag{43}$$

Inserting these expressions into Eq. (40) we find

$$\tilde{g}_2(1,2;1',2') = \int d\bar{1} d\bar{2} d\bar{1}' d\bar{2}' g(1;\bar{1}) g(2;\bar{2}) C_2(\bar{1},\bar{2};\bar{1}',\bar{2}') g(\bar{1}';1') g(\bar{2}';2') \tag{44}$$

where

$$C_2(\bar{1},\bar{2};\bar{1}',\bar{2}') = \delta_-(\bar{z}_1)\delta_-(\bar{z}_2) C_2(\bar{x}_1, \bar{x}_2; \bar{x}_1', \bar{x}_2') \delta_-(\bar{z}_1')\delta_-(\bar{z}_2') \tag{45}$$

is the two-particle initial-correlation function. In conclusion $\tilde{g}_2$ can be written in terms of g and Γ_n with $n \le 2$. Since $g_2 = |g|_2 + \tilde{g}_2$ this result provides a decomposition of g_2 in terms of g and reduced n-particle denity matrices.

The generalization of Wick's theorem to g_n reads

$$g_n = |g|_n + \sum_{l=1}^{n-2} \sum_{PQ} (\pm)^{|P+Q|}\, |g|_l(P;Q')\, \tilde{g}_{n-l}(\breve{P};\breve{Q}') \;+\; \tilde{g}_n. \tag{46}$$

In this formula P and Q are a subset of l ordered indices between 1 and n whereas $\breve{P}$ and $\breve{Q}$ is the ordered complementary subset. For instance if $n = 3$ and $l = 1$ then we can have $P = 1$ and hence $\breve{P} = (2,3)$, or $P = 2$ and hence $\breve{P} = (1,3)$, or $P = 3$ and hence $\breve{P} = (1,2)$. The sign of the various terms is given by $|P + Q| = \sum_{i=1}^{l}(p_i + q_i)$ where p_i and q_i are the indices in the

l-tuple P and Q. The solution of the homogeneous MSH with the correct boundary conditions is

$$\tilde{g}_n(1\ldots n;1'\ldots n') = \int g(1;\bar{1})\ldots g(n;\bar{n})C_n(1\ldots n;1'\ldots n')g(\bar{n}';n')\ldots g(\bar{1}';1') \quad (47)$$

where the integral is over all barred variables and the n-particle initial-correlation functions are given by

$$C_n(1\ldots n;1'\ldots n') = \delta_-(z_1)\ldots\delta_-(z_n)C_n(x_1\ldots x_n;x'_1\ldots x'_n)\delta_-(z'_1)\ldots\delta_-(z'_n) \quad (48)$$

with

$$C_n = \Gamma_n - \sum_{l=1}^{n-2}\sum_{PQ}(\pm)^{|P+Q|}|\Gamma|_l(X_P;X'_Q)C_{n-l}(X_{\breve{P}};X'_{\breve{Q}}) - |\Gamma|_n. \quad (49)$$

This is a recursive formula for the C_n. The collective coordinate $X_P = (x_{p_1}\ldots x_{p_l})$ is a subset of the coordinates $(x_1\ldots x_n)$ and similarly the collective coordinate $X'_Q = (x'_{q_1}\ldots x'_{q_l})$ is a subset of the coordinates $(x'_1\ldots x'_n)$.

We defer the reader to Ref. [12] for the proof of the generalized Wick theorem. Here we observe that with the generalized Wick theorem we can express g_n in terms of g and Γ_n. Since Γ_n can easily be calculated from $\hat{\rho}$ the generalized Wick theorem is especially suited to do MBPT when we know $\hat{\rho}$ instead of $\hat{H}^{\rm M}$. We further observe that the generalized Wick theorem has the same mathematical structure of the Laplace expansion for the permanent/determinant of the sum of two matrices A and B [15]

$$|A+B|_m = |A|_m + \sum_{l=1}^{m-1}\sum_{PQ}(\pm)^{|P+Q|}|A|_l(P;Q)|B|_{m-l}(\breve{P};\breve{Q}) + |B|_m \quad (50)$$

where $|A|_l(P;Q)$ is the permanent/determinant of the $l\times l$ matrix obtained with the rows P and the columns Q of the matrix A. The same notation has been used for the matrix B. With the identification $A_{kj} = g(k;j')$ and $|B|_{m-l}(\breve{P};\breve{Q}) = \tilde{g}_{m-l}(\breve{P};\breve{Q}')$ for $l = 1\ldots m-2$ and the definition $\tilde{g}_1 \equiv 0$ Eqs. (46) and (50) become identical. We can thus symbolically write the generalized Wick theorem as

$$g_m = |g+\tilde{g}|_m \quad (51)$$

whose precise meaning is given by Eq. (46).

6. Relation with the Konstantinov-Perel' formalism

In Eq.(29) we have seen how we can expand the Green's function G into noninteracting Green's functions g_n satisfying the generalized Wick theorem (46). This can be used to define a diagrammatic expansion of the Green's function. Let us see what terms we get when we insert Eq.(46) into Eq.(29). The first term $|g|_n$ in Eq.(46) generates the usual series of connected diagrams for the Green's function in powers of the interaction (the disconnected diagrams are zero since they are integrated from t_0 to t_0 and have no external points). The remaining terms in Eq.(46) are linear in the functions $\tilde{g}_m$ with $m = 2,\ldots,n$. As a consequence of Eq.(47) these functions can be diagrammatically represented by blocks C_m with m ingoing and m outgoing g lines. Since the Wick expression (46) is linear in the $\tilde{g}_m$ each diagram contribution to the Green's function contains at most one correlation block. In Fig. 6 we display the diagrams for the Green's function up to first order in the interaction involving four diagrams with a C_2 block and one diagram with a C_3 block. The last three diagrams, however, vanish since for those diagrams there are internal time-integrations for Green's funtion lines that enter and leave a C_m block that can be reduced to a point, since the initial correlation block only exists at time t_{0-}. The fact that the C_m blocks are not repeated within a single diagram prevents us from deriving

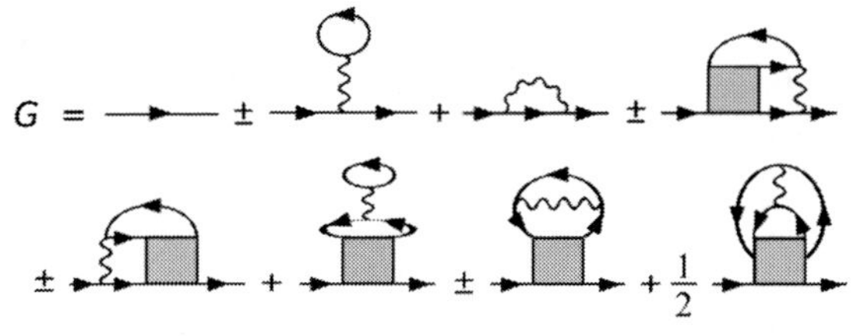

Figure 6. Diagrammatic expansion of the Green's function to first order in the interaction.

an irreducible self-energy in terms of them. We can, however, define a reducible self-energy σ_r and write the Green's function as

$$G(1;2) = g(1;2) + \int_{\bar{\gamma}} d\bar{1}d\bar{2}\, g(1;\bar{1})\sigma_r(\bar{1};\bar{2})g(\bar{2};2) \tag{52}$$

where we used the notation $\bar{\gamma}$ for the contour of Fig.1 to distinguish it from the extended contour γ that we will use later. The reducible self-energy can be split into three contributions. These are the sets of self-energy diagrams that start with an interaction line at their entrance vertices, and end with a correlation block at their exit vertices, denoted by σ_r^L, the diagrams that start with a correlation block and end with an interaction line, denoted by σ_r^R, and the remaining diagrams (which either contain no correlation block or contain a correlation block only attached to internal vertices) which we will denoted by $\tilde{\sigma}_r$. The labels L and R therefore refer to the location of an interaction line at the entrance vertices. We can thus write

$$\sigma_r = \tilde{\sigma}_r + \sigma_r^L + \sigma_r^R \tag{53}$$

where

$$\sigma_r^L(1;2) \;=\; \sigma_r^L(1;x_2)\delta_-(z_2) \tag{54}$$

$$\sigma_r^R(1;2) \;=\; \delta_-(z_1)\sigma_r^R(x_1;2). \tag{55}$$

We emphasize that σ_r is a reducible self-energy and therefore it should not be confused with the self-energy of Ref. [7]. We will now proceed to connect the reducible self-energy σ_r to the irreducible self-energy appearing in the equation of motion for the Green's function on the extended contour. As discussed in the introduction the initial ensemble is of the form $\hat{\rho} = e^{-\hat{X}}/\text{Tr}[e^{-\hat{X}}]$, where $\hat{X} = \sum_m \hat{X}_m$ is in general a sum of m-body operators of the form

$$\hat{X}_m = \frac{1}{m!}\int dx_1 \ldots dx'_m\, v_m(x_1,\ldots,x_m,x'_1,\ldots,x'_m)\hat{\psi}^\dagger(x_1)\ldots\hat{\psi}^\dagger(x_m)\hat{\psi}(x'_m)\ldots\hat{\psi}(x'_1) \tag{56}$$

The functions v_m must be chosen in such a way that the pre-scribed density matrices of Eq.(3) are obtained. This is, in general, a difficult task. If we, however, assume that we have succeeded in this task we can define the Matsubara Hamiltonian $\hat{H}^{\text{M}} = \hat{X}/\beta$ and use the contour of Fig. (2) and the expression for the Green's function (8) to expand the Green's function diagrammatically in powers of G_0 using the standard Wick theorem of Eq.(25) [3, 5, 6]. The diagrammatic rules for the Green's function in the case of m-body operators are given in Ref. [5]. If we collect the irreducible pieces of this expansion in an irreducible self-energy Σ then we obtain a Dyson equation on the extended contour

$$G(1;2) = G_0(1;2) + \int_\gamma d3d4\, G_0(1;3)\Sigma(3;4)G(4,2). \tag{57}$$

It will be more convenient to write this as equations of motion

$$(i\partial_{z_1} - h(1))G(1;2) \;=\; \delta(1;2) + \int_\gamma d3\, \Sigma(1;3)G(3;2) \tag{58}$$

$$G(1;2)(-i\overleftarrow{\partial_{z_2}} - h(2)) \;=\; \delta(1;2) + \int_\gamma d3\, G(1;3)\Sigma(3;2) \tag{59}$$

When split into various components these are simply the Kadanoff-Baym equations on the extended contour. Let us now write the contour γ as $\gamma = \bar{\gamma} \oplus \gamma^{\mathrm{M}}$, where γ^{M} denotes the vertical (or Matsubara) track of the contour and $\bar{\gamma}$ the remaining piece, which is identical to the contour of Fig. 1. The last term on the r.h.s of Eq.(58) can then be written as

$$\int_\gamma d3\, \Sigma(1;3)G(3;2) = \int_{\bar{\gamma}} d3\, \Sigma(1;3)G(3;2) - i \int dx_3 \int_0^\beta d\tau\, \Sigma^\rceil(1;x_3\tau)G^\lceil(x_3\tau;2) \tag{60}$$

where we introduced the parametrization $z = t_0 - i\tau$ on the vertical track γ^{M}. For a general function $A(z,z')$ on the contour (spatial coordinates suppressed) we further defined

$$\begin{aligned} A^\rceil(z = t_\pm, \tau) &= A(z = t_\pm, t_0 - i\tau) &(61)\\ A^\lceil(\tau, z = t_\pm) &= A(t_0 - i\tau, z = t_\pm). &(62) \end{aligned}$$

From the Dyson equation (57) and the Langreth rules on the contour γ we can further derive that [15, 16]

$$G^\lceil(1;2) = -i \int d\bar{x}\, G^{\mathrm{M}}(1;\bar{x}t_0)G^{\mathrm{A}}(\bar{x}t_0;2) + [G^{\mathrm{M}} \star \Sigma^\lceil \cdot G^{\mathrm{A}}](1;2) \tag{63}$$

where $\star$ denotes a convolution between t_0 and $t_0 - i\beta$ on the vertical track and $\cdot$ denotes a convolution between t_0 and ∞. We further defined the advanced and Matsubara Green's functions as

$$\begin{aligned} G^{\mathrm{A}}(1;2) &= -\theta(t_2 - t_1)[G^>(1;2) - G^<(1;2)] &(64)\\ G^{\mathrm{M}}(1;2) &= G(x_1t_0 - i\tau_1; x_2t_0 - i\tau_2) &(65) \end{aligned}$$

If we further use that

$$\int_{\bar{\gamma}} dz\, \delta_-(z)G(\bar{x}z; xt_\pm) = G^{\mathrm{A}}(\bar{x}t_0; xt) \tag{66}$$

we find by inserting Eq.(63) into the last term of Eq.(60) that

$$\begin{aligned} &-i \int dx_3 \int_0^\beta d\tau\, \Sigma^\rceil(1;x_3\tau)G^\lceil(x_3\tau;2) = \\ &\quad \int_{\bar{\gamma}} d3\, [\Sigma^\rceil \star G^{\mathrm{M}} \star \Sigma^\lceil](1;3)G(3;2) - i \int_{\bar{\gamma}} d3\, [\Sigma^\rceil \star G^{\mathrm{M}}](1;x_3t_0)\delta_-(z_3)G(3;2) \qquad (67) \end{aligned}$$

If we therefore define Σ_{L} by

$$\Sigma_{\mathrm{L}}(1;2) = -i[\Sigma^\rceil \star G^{\mathrm{M}}](1;x_2t_0)\delta_-(z_2), \tag{68}$$

we can rewrite the equation of motion (58) for the Green's function as

$$(i\partial_{z_1} - h(1))G(1;2) = \delta(1;2) + \int_{\bar{\gamma}} d3\, [\Sigma + \Sigma^\rceil \star G^{\mathrm{M}} \star \Sigma^\lceil + \Sigma_{\mathrm{L}}](1;3)G(3;2). \tag{69}$$

A similar procedure can be carried out for the adjoint equation (59). We find

$$G(1;2)(-i\overleftarrow{\partial}_{z_2} - h(2)) = \delta(1;2) + \int_{\bar{\gamma}} d3\, G(1;3)[\Sigma + \Sigma^\rceil \star G^{\mathrm{M}} \star \Sigma^\lceil + \Sigma_{\mathrm{R}}](3;2) \tag{70}$$

where we defined

$$\Sigma_{\mathrm{R}}(1;2) = -i\delta_-(z_1)[G^{\mathrm{M}} \star \Sigma^\lceil](x_1t_0;2). \tag{71}$$

Given the self-energy and the Green's function with arguments on the imaginary track γ^{M}, we can regard Eqs. (69) and (70) as equations of motion for the Green's function G on the contour $\bar{\gamma}$. These equations can be integrated using the noninteracting Green's function g of Eq.(35) since it satisfies

$$(\pm i) \lim_{z_1, z_1' \to t_{0-}} g(1;1') = (\pm i) \lim_{z_1, z_1' \to t_{0-}} G(1;1') = \Gamma(x_1; x_1'). \tag{72}$$

If we, therefore, define the total self-energy as

$$\Sigma_{\mathrm{tot}} = \Sigma + \Sigma^{\rceil} \star G^{\mathrm{M}} \star \Sigma^{\lceil} + \Sigma_{\mathrm{L}} + \Sigma_{\mathrm{R}} \tag{73}$$

we can write G in terms of two equivalent Dyson equations

$$G(1;2) = g(1;2) + \int_{\bar{\gamma}} d3d4\, g(1;3)\Sigma_{\mathrm{tot}}(3;4)G(4;2) \tag{74}$$

$$G(1;2) = g(1;2) + \int_{\bar{\gamma}} d3d4\, G(1;3)\Sigma_{\mathrm{tot}}(3;4)g(4;2) \tag{75}$$

To check that these equations are equivalent to the Eqs. (69) and (70) we need to be careful. The standard approach is to act with the operator of the form $i\partial_z - h$ and its adjoint on both Dyson equations and use the equation of motion for g

$$(i\partial_{z_1} - h(1))g(1;2) = \delta(1;2) \tag{76}$$

$$g(1;2)(-i\overleftarrow{\partial}_{z_2} - h(2)) = \delta(1;2) \tag{77}$$

We need to be careful, however, since we cannot change integration and differentiation in the presence of delta-functions under the integral sign. The relevant integrals over the delta-functions need to be done first before we use Eqs.(76) and (77). In Eq.(74) we have an integral of the form

$$\int_{\bar{\gamma}} d3\, g(1;3)\Sigma_{\mathrm{R}}(3;4) = -i \int dx_3 [g^{>}(1; x_3 t_0) - g^{<}(1; x_3 t_0)][G^{\mathrm{M}} \star \Sigma^{\lceil}](x_3 t_0; 4) \tag{78}$$

On the right hand side of this equation we recognize the contour spectral function of Eq.(37) that satisfies Eq. (38). We therefore see that

$$(i\partial_{z_1} - h(1)) \int_{\bar{\gamma}} d3\, g(1;3)\Sigma_{\mathrm{R}}(3;4) = 0 \tag{79}$$

Similarly we have

$$\left(\int_{\bar{\gamma}} d3\, \Sigma_{\mathrm{L}}(3;4) g(4;2) \right) (-i\overleftarrow{\partial}_{z_2} - h(2)) = 0 \tag{80}$$

Then by acting with $i\partial_{z_1} - h(1)$ on Eq.(74) we see that we recover Eq.(69). Similarly by acting with $-i\overleftarrow{\partial}_{z_2} - h(2)$ from the left on Eq.(75) we recover Eq.(70). It only remains to check that the Dyson Eqs.(74) and (75) satisfy the correct boundary conditions. Since in the limit $z_1, z_2 \to t_{0-}$ the contribution for the integrals on the r.h.s. of the equations vanish we see that the condition (72) is indeed satisfied.

Now we are ready to discuss the connection between the formulation based on the initial correlation blocks and the formalism based on integrations along the imaginary track. By comparing Eq.(74) to Eq.(52) we see that

$$\sigma_r = \Sigma_{\mathrm{tot}} + \Sigma_{\mathrm{tot}} g \Sigma_{\mathrm{tot}} + \Sigma_{\mathrm{tot}} g \Sigma_{\mathrm{tot}} g \Sigma_{\mathrm{tot}} + \ldots = \Sigma_{\mathrm{tot}} \frac{1}{1 - g\Sigma_{\mathrm{tot}}} \tag{81}$$

and hence

$$\Sigma_{\rm tot} = \sigma_r(1 - g\Sigma_{\rm tot}) = \sigma_r - \sigma_r g \sigma_r + \sigma_r g \sigma_r g \sigma_r - \ldots = \sigma_r \frac{1}{1 + g\sigma_r} \tag{82}$$

This yields the expansion of the irreducible self-energy $\Sigma_{\rm tot}$ in terms of the Green's functions g and the correlation blocks C_m. We would like to mention that $\Sigma_{\rm tot}$ could in principle be calculated from the appropriate extention of the Hedin equations to include initial correlations [8]. This would lead to an expansion of $\Sigma_{\rm tot}$ in terms of the dressed Green's function G and correlation blocks. However, the iterative solution of these equations depend on the starting point. In particular if we start with a self-energy which contains only a C_2-block then the iterative procedure cannot generate diagrams with C-blocks of higher order.

7. Conclusions

We presented a unified framework for equilibrium and nonequilibrium many-body perturbation theory. The most general formalism for nonequilibrium many-body theory for general initial states is based on the Keldysh contour to which we attach a vertical track describing a general initial state. This idea goes back to the works of Konstantinov and Perel' (who considered equilibrium initial states), Danielewicz and Wagner. On this contour we can straightforwardly prove a Wick theorem by solving the noninteracting Martin-Schwinger hierarchy for the noninteracting many-body Green's functions with KMS boundary conditions. This short proof of Wick's theorem does not need any of the usually introduced theoretical concepts such as normal ordering and contractions. The statement is simply that the noninteracting m-particle Green's function is a determinant or permanent of one-particle Green's functions. We showed how the various other well-known formalisms of Keldysh, Matsubara and the zero-temperature formalism can be derived as special cases that arise under different assumptions. We further discussed a generalized Wick theorem for general initial states on the Keldysh contour. It again arises as a solution of the noninteracting Martin-Schwinger hierarchy for the noninteracting many-body Green's functions but this time with initial conditions specified by initial m-body density matrices. The final result of Eq.(51) is an elegant alternative to the Wick theorem of Eq.(25) for KMS boundary conditions. We finally showed how the formalisms based on the Keldysh and Konstantinov-Perel'-contours are related for the case of general initial states.

References

[1] A. L. Fetter and J. D. Walecka, *Quantum Theory of Many-Particle Systems* (McGraw-Hill, New York, 1971).
[2] L. V. Keldysh, JETP **20**, 1018 (1965).
[3] P. Danielewicz, Ann. Phys. (N.Y.) **152**, 239 (1984).
[4] O. V. Konstantinov and V. I. Perel', Sov. Phys. JETP **12**, 142 (1961).
[5] M. Wagner, Phys. Rev. B **44**, 6104 (1991).
[6] V. G. Mozorov and G. Röpke, Ann. Phys. (N.Y.) **278** , 127 (1998)
[7] A. G. Hall, J. Phys. A: Math. Gen. **8**, 214 (1975).
[8] D. Semkat, D.Kremp and M.Bonitz, Phys. Rev. E **59**, 1557 (1999)
[9] D. Semkat, D.Kremp and M.Bonitz, J. Math. Phys. **41**, 7458 (2000)
[10] M. Bonitz, *Quantum Kinetic Theory* (Teubner, 1998).
[11] M. Garny and M. M. Müller, Phys. Rev. D**80**, 085011 (2009)
[12] R. van Leeuwen and G. Stefanucci, Phys. Rev. B **85**, 115119 (2012).
[13] G. C. Wick, Phys. Rev. **80**, 268 (1950).
[14] P. C. Martin and J. Schwinger, Phys. Rev. **115**, 1342 (1959).
[15] G. Stefanucci and R. van Leeuwen, *Nonequilibrium Many-Body Theory of Quantum Systems: A Modern Introduction* (Cambridge University Press, 2013).
[16] G. Stefanucci and C.-O. Almbladh, Phys. Rev. B **69**, 195318 (2004)

Progress in Nonequilibrium Green's Functions V (PNGF V) IOP Publishing
Journal of Physics: Conference Series **427** (2013) 012002 doi:10.1088/1742-6596/427/1/012002

Nonequilibrium Green function approach to the pair distribution function of quantum many-body systems out of equilibrium

M Bonitz[1], **S Hermanns**[1], **K Kobusch**[1], **and K Balzer**[2]

[1] Institut für Theoretische Physik und Astrophysik, Universität Kiel, D-24098 Kiel, Germany
[2] University of Hamburg, Max Planck Research Department for Structural Dynamics Building 99 (CFEL), Luruper Chaussee 149, D-22761 Hamburg, Germany

E-mail: `bonitz@theo-physik.uni-kiel.de`

Abstract. The pair distribution function (PDF) is a key quantity for the analysis of correlation effects of a quantum system both in equilibrium and far from equilibrium. We derive an expression for the PDF in terms of the single-particle Green functions—the solutions of the Keldysh/Kadanoff-Baym equations in the two-time plane—for a one- or two-component system. The result includes initial correlations and generalizes previous density matrix expressions from single-time quantum kinetic theory. Explicit expressions for the PDF are obtained in second Born approximation.

1. Introduction

In recent years, the interest in correlated quantum many-body systems has increased steadily due to the observation of correlation effects such as liquid and crystal formation, superfluidity of Bose systems or bound states. Examples of systems include nuclear matter [1, 2], solid state systems [3] and dense astrophysical, laboratory dusty plasmas or the quark gluon plasma, for a recent overview see [4, 5]. A quantity sensitive to spatial correlations is the pair distribution function (PDF) $h(r)$—the probability to find an arbitrary pair of particles at a distance r.

Of particular current interest is the short-time behavior of correlated systems following external perturbation such as excitation by intense radiation, e.g., [6, 7]. In quantum systems such as atoms or solids, this is often connected with rapid electron thermalization coupled to the dynamics of electronic correlations. Time and space-resolved measurement techniques detecting chemical reaction products, nuclear collision fragments or electrons and ions produced by laser ionization, e.g., [8], are becoming increasingly powerful. This brings the direct measurement of time-dependent pair correlations described by the nonequilibrium generalization of the PDF, $h(r, t)$, within reach and increases the need for theoretical and computational tools that are able to predict $h(r, t)$.

From the theory side, PDFs are routinely computed for nonideal classical systems using, e.g., Monte Carlo or molecular dynamics simulations, e.g., [4]. Extensions to quantum systems in equilibrium are available, e.g., in the frame of quantum Monte Carlo methods. However, there is still a high demand for accurate nonequilibrium simulations. Theoretical access to nonequilibrium PDFs is straightforwardly reached within density operator theory, e.g., [9, 10, 11]. An alternative approach to quantum many-body systems out of equilibrium is based on nonequilibrium Green functions (NEGF). Apart from exact time-dependent numerical calculations, e.g., [12], which are limited to a few particles, numerical results for the two-time NEGF are expected to be the most accurate ones for strongly correlated quantum systems in nonequilibrium. This concerns, in particular, full two-time calculations, i.e. direct solutions of the Keldysh/Kadanoff-Baym equations (KBE) which, after the pioneering work of Danielewicz [1], have now become routine [2, 13, 14, 15, 16, 17, 18]. They yield the single-particle Green functions and all one-particle observables, including the spectral function, as a function of time. A number of two-particle quantities can also be computed, taking advantage of the two-time structure of the equations, most importantly the interaction energy. However, in this approach, the two-particle Green function is eliminated by introduction of the selfenergy Σ (the integration contour $\mathcal{C}$ and the arguments of the functions will be explained in Sec. 2),

$$\pm\mathrm{i}\int_{\mathcal{C}} \mathrm{d}2\, V(1-2)G(12,1'2^+) = \int_{\mathcal{C}} \mathrm{d}2\,\Sigma(1,2)G(2,1')\,. \tag{1}$$

This concept of the selfenergy has proven extremely successful for reducing the many-particle problem to an effective single-particle one (quasiparticle picture) which forms the basis of Green function theory and the Feynman diagram technique. At the same time, the direct access to the pair distribution function—which contains important additional information beyond the one carried by the single-particle Green function—is lost. Due to this reason, so far, no NEGF results for the full pair distribution function have been obtained which go beyond the computation of double occupancies, accessible from Eq. (1) [19].

In this paper we solve this problem. Starting from the Bethe-Salpeter equation for the two-particle Green function, we demonstrate in Sec. 2 how, from a given time-dependent solution for the single-particle Green function in a chosen approximation for the selfenergy Σ, the PDF in the same approximation can be reconstructed. In this paper, we choose, as an example, the second order Born selfenergy approximation $\Sigma^{2\mathrm{B}}$ which includes the interaction up to second order ($\Sigma^{2\mathrm{B}} \propto V^2$). Our results include the standard PDF $h(r)$ in equilibrium, and its extension $h(\mathbf{r}_1,\mathbf{r}_2)$ to inhomogeneous systems and to multicomponent systems where the PDF becomes a matrix $h^{\mathrm{ab}}(\mathbf{r}_1,\mathbf{r}_2)$, a,b labeling the particle species. Further, we obtain results for an arbitrary nonequilibrium system where the PDF becomes time-dependent, $h^{\mathrm{ab}} \to h^{\mathrm{ab}}(t)$. Finally, the results are directly generalized to two-time PDF's, $h^{\mathrm{ab}}(\mathbf{r}_1,\mathbf{r}_2) \to h^{\mathrm{ab}}(\mathbf{r}_1,\mathbf{r}_2,t)$. Our results selfconsistently include the influence of a correlated initial state, and they describe the decay of initial correlations which is particularly important at the initial stage of the relaxation. This problem was studied in detail for the single-time PDF within density operator theory [9, 10, 11, 20], and these results are contained in our theory as a special case.

In Sec. 6, we discuss, as an illustration, the possible application of the PDF reconstruction algorithm to a two-component system of electrons and holes in a bilayer structure. This system has attracted substantial interest in recent years, because it exhibits strong electron-electron and hole-hole correlations leading to spatial ordering and, at the same time, electron-hole bound states (excitons). There have been detailed investigations of the phase diagram [21, 22, 23, 24], of exciton and hole crystallization [22, 25, 26], of collective excitations [27, 28] and of exciton Bose condensation and superfluidity, e.g., [25, 29]. We conclude with a summary and discussion in Sec. 7.

2. Bethe-Salpeter equation

In nonequilibrium Green functions theory, the PDF of particle species "a" and "b" is obtained from the two-particle correlation function [5] taken at four equal times $t_1 = t_2 = t'_1 = t'_2 = t$,

$$\begin{aligned} h^{\mathrm{ab}}(\mathbf{r}_1,\mathbf{r}_2,\mathbf{r}'_1,\mathbf{r}'_2,t) &= \mathrm{i}^2 g^{\mathrm{ab},<}(1,2,1',2') \\ &= \langle \Psi^\dagger_{\mathrm{a}}(1')\Psi^\dagger_{\mathrm{b}}(2')\Psi_{\mathrm{b}}(2)\Psi_{\mathrm{a}}(1)\rangle\,, \end{aligned} \tag{2}$$

Journal of Physics: Conference Series **427** (2013) 012002 doi:10.1088/1742-6596/427/1/012002

where $\Psi_{\mathrm{a}}^{\dagger}$ and Ψ_{a} are fermionic or bosonic creation and annihiliation operators, and we introduced the short-hand notation $1 = \mathbf{r}_1, s_{z1}, t_1$. To determine the two-particle correlation function $g^{\mathrm{ab},<}$ in nonequilibrium, we start from the more general two-particle Green function g^{ab} on the Schwinger-Keldysh contour $\mathcal{C}$ which consists of a Hartree-Fock and a correlation part

$$g^{\mathrm{ab}}(12,1'2') = g^{\mathrm{ab}}_{\mathrm{HF}}(12,1'2') + g^{\mathrm{ab}}_{\mathrm{corr}}(12,1'2')\,, \tag{3}$$

$$g^{\mathrm{ab}}_{\mathrm{HF}}(12,1'2') = g^{\mathrm{a}}(1,1')g^{\mathrm{b}}(2,2') \pm \delta_{\mathrm{ab}} g^{\mathrm{a}}(1,2')g^{\mathrm{b}}(2,1')\,, \tag{4}$$

where we allow for four different time arguments, t_1, t_2, t_1', t_2', all located on the contour $\mathcal{C}$. The correlation part, $g^{\mathrm{ab}}_{\mathrm{corr}}$, obeys the Bethe-Salpeter equation, e.g., [34],

$$g^{\mathrm{ab}}_{\mathrm{corr}}(12,1'2') = \mathrm{i}\int_{\mathcal{C}} \mathrm{d}\bar{1}\mathrm{d}\bar{2}\mathrm{d}\tilde{1}\mathrm{d}\tilde{2}\, g^{\mathrm{a}}(1,\bar{1})g^{\mathrm{b}}(2,\bar{2})K^{\mathrm{ab}}(\bar{1}\bar{2},\tilde{1}\tilde{2})\, g^{\mathrm{ab}}(\tilde{1}\tilde{2},1'2')\,, \tag{5}$$

which presents a formal closure of the second equation of the Martin-Schwinger hierarchy for the two-particle Green function. Compared to Ref. [34], we also restored the exchange term where plus (minus) refers to bosons (fermions). In the language of Feynman diagrams, the effective interaction kernel K^{ab} contains all irreducible two-particle diagrams. This approach on the two-particle level can directly be used to formulate the corresponding Dyson equation for the one-particle Green function, involving a selfenergy Σ^{ab}. For a given K^{ab}, the corresponding Σ^{ab} is consistently determined by, cf. [5],

$$\pm\int_{\mathcal{C}} \mathrm{d}\bar{1}\mathrm{d}\bar{2}\, g^{\mathrm{a}}(1,\bar{1})\frac{\delta\Sigma^{\mathrm{ab}}(\bar{1}\bar{2})}{\delta V^{\mathrm{ab}}(2'2)}g^{\mathrm{b}}(\bar{2},1') = \int_{\mathcal{C}} \mathrm{d}\bar{1}\mathrm{d}\bar{2}\mathrm{d}\tilde{1}\mathrm{d}\tilde{2}\, g^{\mathrm{a}}(1,\bar{1})g^{\mathrm{b}}(2,\bar{2})K^{\mathrm{ab}}(\bar{1}\bar{2},\tilde{1}\tilde{2})g^{\mathrm{ab}}(\tilde{1}\tilde{2},1'2')\,, \tag{6}$$

where V^{ab} is the particle interaction.

Returning to Eq. (5), it is evident that the solution of the BSE for $g^{\mathrm{ab}}_{\mathrm{corr}}(t_1 = t_2 = t_1' = t_2')$ at four equal times yields the PDF, cf. Eq. (2). The direct solution of the BSE is though highly involved, since, even in the mentioned case of equal times, it requires the knowledge of the time-offdiagonal three-time elements $g^{\mathrm{ab}}(t_{\tilde{1}}, t_{\tilde{2}}, t_1' t_1')$. To reduce this complexity and make a reconstruction from single-particle quantities possible, in the following, we will simplify the structure of the general dynamic interaction K^{ab}.

1.) The first simplification is to introduce the screened ladder approximation, $K^{\mathrm{ab}}(\bar{1}\bar{2},\tilde{1}\tilde{2}) \to V^{\mathrm{ab}}(\bar{1}\bar{2})\delta(\bar{1}-\tilde{1})\delta(\bar{2}-\tilde{2})$. Suppressing for a moment the space integration and spin summation (they are implied by the repeated arguments $\bar{1}$ and $\bar{2}$ under the integral), we obtain from Eq. (5)

$$g^{\mathrm{ab}}_{\mathrm{corr}}(12,1'2') = \mathrm{i}\int_{\mathcal{C}} \mathrm{d}\bar{t}_1\mathrm{d}\bar{t}_2\, g^{\mathrm{a}}(1,\bar{1})g^{\mathrm{b}}(2,\bar{2})V^{\mathrm{ab}}(\bar{1}\bar{2})\, g^{\mathrm{ab}}(\bar{1}\bar{2},1'2')\,. \tag{7}$$

We mention that this approximation is crucial for the successful reconstruction of $g^{\mathrm{ab}}_{\mathrm{corr}}(12,1'2')$ from single-particle quantities, since, according to Eq. (6), the knowledge of $\Sigma^{\mathrm{ab}}(1,1')$ only implicitly determines two degrees of freedom of $K^{\mathrm{ab}}(\bar{1}\bar{2},\tilde{1}\tilde{2})$. This is sufficient only in the case $K^{\mathrm{ab}}(\bar{1}\bar{2},\tilde{1}\tilde{2}) \propto \delta(\bar{1}-\tilde{1})\delta(\bar{2}-\tilde{2})$. Nevertheless, this approximation already yields rich physics corresponding to the screened ladder approximation. Despite this simplification, Equation (7) still contains strong binary correlations (T-matrix) and dynamical screening effects leading to a complicated integro-differential equation.

2.) Our next simplification is to neglect dynamic effects in the potential (dynamical screening, related to the GW approximation) which leads to the replacement $V^{\mathrm{ab}}(\bar{1}\bar{2}) \to V^{\mathrm{ab}}(\bar{r}_{12})\delta(\bar{t}_1 - \bar{t}_2)$, where $\bar{r}_{12} = |\mathbf{r}_1 - \mathbf{r}_2|$,

$$g^{\mathrm{ab}}_{\mathrm{corr}}(12,1'2') = \mathrm{i}\int_{\mathcal{C}} \mathrm{d}\bar{t}\, g^{\mathrm{a}}(1,\bar{1})g^{\mathrm{b}}(2,\bar{2})V^{\mathrm{ab}}(\bar{r}_{12})\, g^{\mathrm{ab}}(\bar{1}\bar{2},1'2')\,, \tag{8}$$

and under the integral $\bar{t}_1 = \bar{t}_2 = \bar{t}$. This equation corresponds to a static T-matrix approximation for the two-particle Green function and the pair correlations, i.e. to a complete summation of the Born series. It can be noted that, for times $t_1 = t_2$ and $t'_1 = t'_2$, Eq. 8 now constitutes a closed equation of Dyson-type for the two-time correlation Green function $g^{\rm ab}_{\rm corr}(t_1, t_1, t'_1 t'_1)$.

3.) Since our goal is to reconstruct the pair distributions from a solution of the two-time KBE for the single-particle Green functions in second order Born approximation, we limit ourselves to the first iteration of the integral equation (8), i.e. we replace, under the integral, the two-particle Green function by the Hartree-Fock approximation:

$$g^{\rm ab}(12, 1'2') = {\rm i} \int_C {\rm d}\bar{t}\, g^{\rm a}(1, \bar{1}) g^{\rm b}(2, \bar{2}) V^{\rm ab}(\bar{r}_{12}) g^{\rm ab}_{\rm HF}(12, 1'2'), \quad \text{with} \quad \bar{t}_1 = \bar{t}_2 = \bar{t}. \tag{9}$$

We can bring this equation to a more compact form by introducing the following new functions

$$G^{\rm ab}_0(12, 1'2') = g^{\rm a}(1, 1') g^{\rm b}(2, 2'), \tag{10}$$

$$\Sigma^{\rm ab}_0(12, 1'2') = V^{\rm ab}(r_{12}) \left\{ G^{\rm ab}_0(12, 1'2') \pm \delta_{\rm ab} G^{\rm ab}_0(12, 2'1') \right\}, \tag{11}$$

where $G^{\rm ab}_0$ is the Hartree approximation for the two-particle Green function ($g^{\rm ab}_{\rm HF}$ without exchange terms) and $\Sigma^{\rm ab}_0$ the first-order two-particle selfenergy. With these definitions, Eq. (9) becomes

$$g^{ab}_{\rm corr}(12, 1'2') = {\rm i} \int_C {\rm d}\bar{t}\, G^{\rm ab}_0(12, \bar{1}\bar{2}) \Sigma^{\rm ab}_0(\bar{1}\bar{2}, 1'2'), \quad \text{with} \quad \bar{t}_1 = \bar{t}_2 = \bar{t}. \tag{12}$$

We underline that all expressions, so far, are written on the Schwinger-Keldysh contour, i.e. all single-particle Green functions are 3×3 matrices whereas the two-particle functions contain 3^4 Keldysh matrix elements. To obtain the nonequilibrium pair correlation function, we now have to extract, from Eq. (12), functions depending on the real physical time. In particular, according to Eq. (2), we need only the two-particle correlation function $g^{\rm ab,<}$ which we consider in the following.

3. Reduction of the time structure of the two-particle correlation function $g^{\rm ab,<}$

In order to compute the pair distribution function, Eq. (2), the two-particle correlation function $g^{\rm ab,<}$ at four equal times is needed. To this end, we first determine the equation of motion for the two-time two-particle correlation function $g^{\rm ab,<}(t, t')$ and then specialize to the one-time two-particle correlation function $g^{\rm ab,<}(t)$.

3.1. Two-time two-particle correlation function $g^{\rm ab,<}(t, t')$

The problem of the large number of Keldysh matrix elements of $g^{\rm ab}$ can be simplified drastically in the static Born approximation. As a first step in simplifying the time dependencies we specialize to functions with pairwise equal time arguments, $t_1 = t_2 = t$, and $t'_1 = t'_2 = t'$. Then it is clear from Eq. (12) that the functions $G^{\rm ab}_0$ and $\Sigma^{\rm ab}_0$ depend only on two times in the following way

$$G^{\rm ab}_0(\mathbf{r}_1\mathbf{r}_2; \mathbf{r}'_1\mathbf{r}'_2; tt') = g^{\rm a}(\mathbf{r}_1 t; \mathbf{r}'_1 t') g^{\rm b}(\mathbf{r}_2 t; \mathbf{r}'_2 t'), \tag{13}$$

$$\Sigma^{\rm ab}_0(\mathbf{r}_1\mathbf{r}_2; \mathbf{r}'_1\mathbf{r}'_2; tt') = V^{\rm ab}(r_{12}) \{ G^{\rm ab}_0(\mathbf{r}_1\mathbf{r}_2; \mathbf{r}'_1\mathbf{r}'_2; tt') \pm \delta_{\rm ab} G^{\rm ab}_0(\mathbf{r}_1\mathbf{r}_2; \mathbf{r}'_2\mathbf{r}'_1; tt') \}, \tag{14}$$

where we also restored the coordinate arguments (spin variables are not written explicitly). Then Eq. (12) turns into

$$g^{\rm ab}_{\rm corr}(\mathbf{r}_1\mathbf{r}_2; \mathbf{r}'_1\mathbf{r}'_2; tt') = {\rm i} \int_C {\rm d}\bar{t}\, G^{\rm ab}_0(\mathbf{r}_1\mathbf{r}_2; \bar{\mathbf{r}}_1\bar{\mathbf{r}}_2, t\bar{t}) \Sigma^{\rm ab}_0(\bar{\mathbf{r}}_1\bar{\mathbf{r}}_2; \mathbf{r}'_1\mathbf{r}'_2; \bar{t}t'). \tag{15}$$

With this result, Eq. (3) for $g^{\rm ab}$ can be rewritten as

$$g^{\rm ab}(\mathbf{r}_1\mathbf{r}_2;\mathbf{r}_1'\mathbf{r}_2';tt') = g^{\rm ab}_{\rm HF}(\mathbf{r}_1\mathbf{r}_2;\mathbf{r}_1'\mathbf{r}_2';tt') + g^{\rm ab}_{\rm corr}(\mathbf{r}_1\mathbf{r}_2;\mathbf{r}_1'\mathbf{r}_2';tt')\,, \tag{16}$$

where $g^{\rm ab}_{\rm corr}$ is completed by the Hartree-Fock contribution which, in the new notation, acquires the form

$$g^{\rm ab}_{\rm HF}(\mathbf{r}_1\mathbf{r}_2;\mathbf{r}_1'\mathbf{r}_2';tt') = G_0^{\rm ab}(\mathbf{r}_1\mathbf{r}_2;\mathbf{r}_1'\mathbf{r}_2';tt') \pm \delta_{\rm ab}G_0^{\rm ab}(\mathbf{r}_1\mathbf{r}_2;\mathbf{r}_2'\mathbf{r}_1';tt')\,. \tag{17}$$

According to Eq. (2), $g^{\rm ab}(tt')$ reads in terms of creation and annihilation operators (suppressing spatial arguments),

$$g^{\rm ab}(tt') = {\rm i}^2\langle \Psi^\dagger_{\rm a}(t')\Psi^\dagger_{\rm b}(t')\Psi_{\rm b}(t)\Psi_{\rm a}(t)\rangle\,. \tag{18}$$

It is important to notice that, due to its structure, this quantity is different from the double occupancy, which is directly accessible using the single-particle selfenergy, cf. [19],

$$d(1) \propto \int_{\mathcal{C}} {\rm d}3\,\Sigma(13)g(31^+)\,. \tag{19}$$

Now, to compute the pair distribution function from Eq. (16), we have to extract $g^{{\rm ab},<}$ from the Keldysh matrix and, in particular, the "<" component from the contour integral in Eq. (15). The solution of this problem is well known for one-particle functions, and for two-particle functions it has been solved in Ref. [34]. However, the latter results are not needed here. Indeed, although we are dealing with two-particle functions $G_0^{\rm ab}$ and $\Sigma_0^{\rm ab}$, in the present approximation, they have the same time-dependence (they depend just on t, t') as occurs in the case of single-particle functions in the integral term of the Keldysh Kadanoff-Baym equations. As a consequence, both these functions are 3×3 Keldysh matrices and, therefore, also the two-particle Green function $g^{\rm ab}$, Eq. (15), is a 3×3 Keldysh matrix.

The Keldysh components are classified by location of the two time-arguments on the contour $\mathcal{C}$. They include correlation functions, ">, <" with two real arguments, "⌈, ⌉" having one real and one imaginary time argument and the Matsubara component, $\alpha = {\rm M}$, where both arguments lie on the imaginary track. Below, we will need only the "<" component which determines the time-dependent nonequilibrium PDF and the Matsubara component which yields the PDF in thermodynamic equilibrium. We start the analysis with the former and consider the latter in Sec. 5.

Each of the nine components $G_0^{{\rm ab},\alpha}$ of the matrix $G_0^{\rm ab}$ is of the product form

$$G_0^{{\rm ab},\alpha}(\mathbf{r}_1\mathbf{r}_2;\mathbf{r}_1'\mathbf{r}_2';tt') \;=\; g^{{\rm a},\alpha}(\mathbf{r}_1t;\mathbf{r}_1't')g^{{\rm b},\alpha}(\mathbf{r}_2t;\mathbf{r}_2't')\,, \tag{20}$$

which allows for an easy matrix multiplication of $G_0^{\rm ab}$ and $\Sigma_0^{\rm ab}$ under the integral in Eq. (15). Since each component α has the same dependence on the coordinates as in Eq. (15), we will suppress the space arguments and integrations in the remainder of this section.

The result for the "<" component is well known (Langreth rules) and consists of an initial correlation and a collision term

$$g^{{\rm ab},<}_{\rm corr}(t,t') = g^{{\rm ab},<}_{\rm IC}(t,t') + g^{{\rm ab},<}_{\rm col}(t,t')\,, \tag{21}$$

which arise, respectively, from the imaginary and real part of the contour $\mathcal{C}$,

$$g^{ab,<}_{\rm IC}(t,t') \;=\; -{\rm i}\int_0^\beta {\rm d}\tau\, G_0^{{\rm ab},\rceil}(t,\tau)\Sigma_0^{{\rm ab},\lceil}(\tau,t')\,, \tag{22}$$

$$g^{ab,<}_{\rm col}(t,t') \;=\; \int_0^\infty {\rm d}\bar t\left\{G_0^{{\rm ab},<}(t,\bar t)\Sigma_0^{{\rm ab},{\rm A}}(\bar t,t') + G_0^{{\rm ab},R}(t,\bar t)\Sigma_0^{{\rm ab},<}(\bar t,t')\right\}\,. \tag{23}$$

Here, $G_0^{\mathrm{ab},<}$ denotes the product of two single-particle correlation functions $g^{\mathrm{a},<}, g^{\mathrm{b},<}$, both having two real time arguments, whereas the components $\rceil$ and $\lceil$ describe products of functions depending on one real and one complex (τ) time argument describing the propagation of initial correlations, for a detailed discussion, see, e.g., Refs. [11, 17]. Further, we introduced retarded and advanced functions defined by

$$a^{\mathrm{R/A}}(t,t') = \pm\Theta\left[\pm(t-t')\right]\left\{a^>(t,t') - a^<(t,t')\right\} . \tag{24}$$

3.2. One-time two-particle correlation function $g^{\mathrm{ab},<}(t)$

We now perform the final step on the way to the nonequilibrium PDF—taking the time-diagonal limit, $t = t'$, in the above equations for $g^{\mathrm{ab},<}$ which has the general structure

$$g^{\mathrm{ab},<}(t) = g_{\mathrm{HF}}^{\mathrm{ab},<}(t) + g_{\mathrm{IC}}^{\mathrm{ab},<}(t) + g_{\mathrm{col}}^{\mathrm{ab},<}(t) . \tag{25}$$

Expressing the time-diagonal part of the single-particle correlation function by the single-particle density matrix $\rho^{\mathrm{a}}(t) = \pm\mathrm{i}\, g^{\mathrm{a}}(t,t)$, where the upper (lower) sign refers to bosons (fermions), we obtain for the two-particle correlation function on the time-diagonal

$$G_0^{\mathrm{ab},<}(t,t) = -\rho^{\mathrm{a}}(t)\rho^{\mathrm{b}}(t) . \tag{26}$$

With this relation and Eq. (17), we first obtain the nonequilibrium Hartree-Fock pair distribution function

$$g_{\mathrm{HF}}^{\mathrm{ab},<}(\mathbf{r}_1\mathbf{r}_2;\mathbf{r}_1'\mathbf{r}_2';t) = -\rho^{\mathrm{a}}(\mathbf{r}_1\mathbf{r}_1',t)\rho^{\mathrm{b}}(\mathbf{r}_2\mathbf{r}_2',t) \mp \delta_{\mathrm{ab}}\rho^{\mathrm{a}}(\mathbf{r}_1\mathbf{r}_2',t)\rho^{\mathrm{b}}(\mathbf{r}_2\mathbf{r}_1',t) . \tag{27}$$

The second contribution to Eq. (25) is due to the initial correlations follows from Eq. (22) and is given by

$$\begin{aligned} g_{\mathrm{IC}}^{\mathrm{ab},<}(\mathbf{r}_1\mathbf{r}_2;\mathbf{r}_1'\mathbf{r}_2';t) &= -\mathrm{i} \int_0^\beta d\tau \int d^3\bar{r}_1 d^3\bar{r}_2 V^{\mathrm{ab}}(\bar{r}_{12}) g^{\mathrm{a},\rceil}(\mathbf{r}_1t;\bar{\mathbf{r}}_1\tau) g^{\mathrm{b},\rceil}(\mathbf{r}_2t;\bar{\mathbf{r}}_2\tau) \\ &\times\left\{ g^{\mathrm{a},\lceil}(\bar{\mathbf{r}}_1\tau,\mathbf{r}_1',t) g^{\mathrm{b},\lceil}(\bar{\mathbf{r}}_2\tau,\mathbf{r}_2';t) \pm \delta_{\mathrm{ab}} g^{\mathrm{a},\lceil}(\bar{\mathbf{r}}_1\tau,\mathbf{r}_2',t) g^{\mathrm{b},\lceil}(\bar{\mathbf{r}}_2\tau,\mathbf{r}_1';t)\right\} . \end{aligned} \tag{28}$$

Finally, the scattering contribution (23) contains two real-time integrals extending to t and t', respectively. Since now both times are equal, a partial cancellation is possible. Indeed, taking advantage of relation (24) allows to cancel all contributions with products of four "$>$" or four "$<$" functions, and we obtain

$$\begin{aligned} g_{\mathrm{col}}^{\mathrm{ab},<}(\mathbf{r}_1\mathbf{r}_2;\mathbf{r}_1'\mathbf{r}_2';t) &= \mathrm{i} \int_0^t d\bar{t} \int d^3\bar{r}_1 d^3\bar{r}_2 V^{\mathrm{ab}}(\bar{r}_{12}) \Big\{ g^{\mathrm{a},>}(\mathbf{r}_1t;\bar{\mathbf{r}}_1\bar{t}) g^{\mathrm{b},>}(\mathbf{r}_2t;\bar{\mathbf{r}}_2\bar{t}) \\ &\times\left[g^{\mathrm{a},<}(\bar{\mathbf{r}}_1\bar{t},\mathbf{r}_1',t) g^{\mathrm{b},<}(\bar{\mathbf{r}}_2\bar{t},\mathbf{r}_2';t) \pm \delta_{\mathrm{ab}} g^{\mathrm{a},<}(\bar{\mathbf{r}}_1\bar{t},\mathbf{r}_2',t) g^{\mathrm{b},<}(\bar{\mathbf{r}}_2\bar{t},\mathbf{r}_1';t)\right] \\ &-(>\leftrightarrow<)\Big\} . \end{aligned} \tag{29}$$

This expression is readily understood: the first term describes the correlation build-up due to scattering of a particle pair a, b out of state $|\bar{\mathbf{r}}_1\rangle\,|\bar{\mathbf{r}}_2\rangle$ into state $|\mathbf{r}_1\rangle\,|\mathbf{r}_2\rangle$, whereas the second contribution which is obtained by interchanging functions with the indices "$<$" and "$>$" describes the opposite process. The integral has the familiar non-Markovian structure ("memory") indicating that scattering processes from all times prior to the current one do contribute although

Journal of Physics: Conference Series **427** (2013) 012002 doi:10.1088/1742-6596/427/1/012002

their weight decreases since the single-particle Green function decay with increasing difference of their time arguments. Expression (29) has exactly the same structure as known from density operator theory [11]. The binary density operator in Born approximation is recovered, if the two-time correlation functions are eliminated using the generalized Kadanoff-Baym ansatz [39], see also Refs. [40],[41]. The present result is more general, because this reconstruction ansatz, which is only valid approximately and becomes increasingly inaccurate with increasing correlation effects, is avoided.

4. Nonequilibrium pair distribution function

The nonequilibrium pair distribution function $h^{\mathrm{ab}}(t)$ follows immediately from the two-particle correlation function according to Eq. (2) and the results (25), (27), (28) and (29). Explicit results depend on the choice of a basis. We start from the coordinate representation.

4.1. Coordinate representation

To obtain the standard pair distribution function, which is defined in configuration space, we set $\mathbf{r}_1 = \mathbf{r}_1'$ and $\mathbf{r}_2 = \mathbf{r}_2'$ in all expressions

$$\begin{aligned} h^{\mathrm{ab}}(\mathbf{r}_1, \mathbf{r}_2, t) =& \rho^{\mathrm{a}}(\mathbf{r}_1\mathbf{r}_1, t)\rho^{\mathrm{b}}(\mathbf{r}_2\mathbf{r}_2, t) \pm \delta_{\mathrm{ab}}\rho^{\mathrm{a}}(\mathbf{r}_1\mathbf{r}_2, t)\rho^{\mathrm{b}}(\mathbf{r}_2\mathbf{r}_1, t) \\ &- g_{\mathrm{IC}}^{\mathrm{ab},<}(\mathbf{r}_1\mathbf{r}_2; \mathbf{r}_1\mathbf{r}_2; t) - g_{\mathrm{col}}^{\mathrm{ab},<}(\mathbf{r}_1\mathbf{r}_2; \mathbf{r}_1\mathbf{r}_2; t) \,. \end{aligned} \tag{30}$$

In a spatially inhomogeneous system, it is advantageous to introduce center of mass and relative coordinates, $\mathbf{R} = (\mathbf{r}_1 + \mathbf{r}_2)/2$ and $\mathbf{r} = \mathbf{r}_1 - \mathbf{r}_2$, leading to the replacements $\mathbf{r}_{1,2} = \mathbf{R} \pm \mathbf{r}/2$, on the right hand side. Then $h^{\mathrm{ab}}(\mathbf{R}, \mathbf{r})$ is understood as the local probability to find a particle pair at distance $\mathbf{r}$ around space point $\mathbf{R}$.

If spatial inhomogeneities are not relevant or not of interest, a space integration leads to the space averaged nonequilibrium pair distribution

$$h^{\mathrm{ab}}(\mathbf{r}, t) = \int \mathrm{d}^d R \, h^{\mathrm{ab}}(\mathbf{R}, \mathbf{r}, t) \,, \tag{31}$$

where d is the dimensionality of the system. Note that h^{ab} is normalized to the total number of particles according to

$$\int \mathrm{d}^d h^{\mathrm{ab}}(\mathbf{r}, t) = N_{\mathrm{a}} \left\{ N_{\mathrm{b}} - \delta_{\mathrm{ab}} \right\} \,. \tag{32}$$

Finally, in some cases, just the dependence on the magnitude of the pair separation is of interest, which is obtained by an angle integration. The result is called radial distribution function, $h^{\mathrm{ab}}(r, t)$, and follows from Eq. (32) according to (we consider a two-dimensional system and change to polar coordinates)

$$h^{\mathrm{ab}}(r, t) = \int_0^{2\pi} \mathrm{d}\phi \, r h^{\mathrm{ab}}(r, \phi, t) \,, \tag{33}$$

with the normalization

$$\int_0^{\infty} \mathrm{d}r h^{\mathrm{ab}}(r, t) = N_{\mathrm{a}} \left\{ N_{\mathrm{b}} - \delta_{\mathrm{ab}} \right\} \,, \tag{34}$$

and similarly in 3D.

4.2. Arbitrary basis

Let us now consider the case of an arbitrary orthonormal stationary basis $\{\phi_i(\mathbf{r})\}$, which is of relevance, in particular, for spatially inhomogenenous systems. Then, the single-particle Green functions become matrices according to

$$g^{\mathrm{b},\alpha}(1,1') = \sum_{ij} \phi_i(\mathbf{r}_1) g^{\mathrm{b},\alpha}_{ij}(t_1,t_1')\phi^*_i(\mathbf{r}_1'), \tag{35}$$

which holds for any Keldysh component "α". Analogously, the two-particle single-time correlation function is represented by a four-dimensional matrix

$$g^{\mathrm{ab},<}(\mathbf{r}_1\mathbf{r}_2;\mathbf{r}_1'\mathbf{r}_2';t) = \sum_{ijkl} \phi_i(\mathbf{r}_1)\phi_j(\mathbf{r}_2) g^{\mathrm{ab},<}_{ijkl}(t)\phi^*_k(\mathbf{r}_1')\phi^*_l(\mathbf{r}_2'). \tag{36}$$

To compute the nonequilibrium PDF (25), we have to expand the Hartree-Fock, the initial correlation and the scattering part in this basis. For the Hartree-Fock contribution, we obtain, in analogy to (36),

$$g^{\mathrm{ab},<}_{\mathrm{HF}}(\mathbf{r}_1\mathbf{r}_2;\mathbf{r}_1'\mathbf{r}_2';t) = \sum_{ijkl} \phi_i(\mathbf{r}_1)\phi_j(\mathbf{r}_2) g^{\mathrm{ab},<}_{\mathrm{HF},ijkl}(t)\phi^*_k(\mathbf{r}_1')\phi^*_l(\mathbf{r}_2'), \tag{37}$$

$$g^{\mathrm{ab},<}_{\mathrm{HF},ijkl}(t) = g^{\mathrm{a},<}_{ik}(t) g^{\mathrm{b},<}_{jl}(t) \pm \delta_{\mathrm{ab}} g^{\mathrm{a},<}_{il}(t) g^{\mathrm{b},<}_{jk}(t). \tag{38}$$

Similarly, we obtain from Eq. (28) for the initial correlation contribution

$$g^{\mathrm{ab},<}_{\mathrm{IC}}(\mathbf{r}_1\mathbf{r}_2;\mathbf{r}_1'\mathbf{r}_2';t) = \sum_{ijkl} \phi_i(\mathbf{r}_1)\phi_j(\mathbf{r}_2) g^{\mathrm{ab},<}_{\mathrm{IC},ijkl}(t)\phi^*_k(\mathbf{r}_1')\phi^*_l(\mathbf{r}_2'), \tag{39}$$

$$\begin{aligned} g^{\mathrm{ab},<}_{\mathrm{IC},ijkl}(t) &= \int_0^\beta \mathrm{d}\tau \sum_{mnrs} V^{\mathrm{ab}}_{mnrs}\, g^{\mathrm{a},\rceil}_{m,i}(t,\tau) g^{\mathrm{b},\rceil}_{n,j}(t,\tau) \times \\ &\quad \left\{ g^{\mathrm{a},\lceil}_{r,k}(\tau,t) g^{\mathrm{b},\lceil}_{s,l}(\tau,t) \pm \delta_{\mathrm{ab}}\, g^{\mathrm{a},\lceil}_{r,l}(\tau,t) g^{\mathrm{b},\lceil}_{s,k}(\tau,t) \right\}, \end{aligned} \tag{40}$$

and from Eq. (29)

$$g^{\mathrm{ab},<}_{\mathrm{col}}(\mathbf{r}_1\mathbf{r}_2;\mathbf{r}_1'\mathbf{r}_2';t) = \sum_{ijkl} \phi_i(\mathbf{r}_1)\phi_j(\mathbf{r}_2) g^{\mathrm{ab},<}_{\mathrm{col},ijkl}(t)\phi^*_k(\mathbf{r}_1')\phi^*_l(\mathbf{r}_2'), \tag{41}$$

$$\begin{aligned} g^{\mathrm{ab},<}_{\mathrm{col},ijkl}(t) &= \mathrm{i}\int_0^t \mathrm{d}\bar{t} \sum_{mnrs} V^{\mathrm{ab}}_{mnrs} \Big\{ g^{\mathrm{a},>}_{i,m}(t,\bar{t}) g^{\mathrm{b},>}_{j,n}(t,\bar{t}) \times \\ &\quad \left[g^{\mathrm{a},<}_{k,r}(\bar{t},t) g^{\mathrm{b},<}_{l,s}(\bar{t},t) \pm \delta_{\mathrm{ab}}\, g^{\mathrm{a},<}_{l,r}(\bar{t},t) g^{\mathrm{b},<}_{k,s}(\bar{t},t) \right] \\ &\quad -(>\leftrightarrow< \text{ and } g_{1,2} \leftrightarrow g_{2,1}) \Big\}. \end{aligned} \tag{42}$$

Basis expansions for inhomogeneous quantum many-body systems have been successfully applied to electrons in quantum dots ("artificial atoms"), e.g., [35, 36, 37], and small atoms and molecules, e.g., [16, 38]. For these systems, the Keldysh/Kadanoff-Baym equations are solved for the matrix function $g^<_{ij}(t,t')$. Using these results and formulas (37),(39) and (41), the two-particle correlation function, $g^{\mathrm{ab},<}$ in configuration space, Eq. (36), can be reconstructed. Besides, also the matrix elements $g^{\mathrm{ab},<}_{ijkl}(t)$ themselves are of interest, as they carry extensive information on the many-body system. For example, the matrix components $g^{\mathrm{ab},<}_{ijij}(t)$ describe the correlation of two particles "a,b" occupying the orbitals i and j, respectively, at a given moment t.

5. Equilibrium pair distribution function

To compute the pair distribution in thermodynamic equilibrium from the Matsubara Green function, we return to Eq. (15) for the Keldysh matrix and extract the Matsubara component. According to Eq. (20), it is given by a product of one-particle Matsubara Green functions,

$$G_0^{\mathrm{ab,M}}(\mathbf{r}_1\mathbf{r}_2;\mathbf{r}_1'\mathbf{r}_2';\tau) = g^{\mathrm{a,M}}(\mathbf{r}_1\mathbf{r}_1',\tau)g^{\mathrm{b,M}}(\mathbf{r}_2\mathbf{r}_2',\tau)\,, \tag{43}$$

where $\tau = t - t'$. Similarly as in nonequilibrium, the two-particle Matsubara Green function consists of a Hartree-Fock and correlation part,

$$g^{\mathrm{ab,M}}(\tau) = g_{\mathrm{HF}}^{\mathrm{ab,M}}(\tau) + g_{\mathrm{col}}^{\mathrm{ab,M}}(\tau) \tag{44}$$

where the latter is obtained from Eq. (20) using the Langreth rules, in analogy to Eq. (22),

$$g_{\mathrm{col}}^{\mathrm{ab,M}}(\tau) = -\mathrm{i}\int_0^\beta \mathrm{d}\bar\tau\, G_0^{\mathrm{ab,M}}(\tau-\bar\tau)\Sigma_0^{\mathrm{ab,M}}(\bar\tau)\,. \tag{45}$$

Restoring the coordinate arguments and using the definitions of G_0^{ab} and Σ_0^{ab}, Eqs. (10) and (11), we obtain

$$g_{\mathrm{HF}}^{\mathrm{ab,M}}(\mathbf{r}_1\mathbf{r}_2;\mathbf{r}_1'\mathbf{r}_2') = -\rho^{\mathrm{a}}(\mathbf{r}_1\mathbf{r}_1')\rho^{\mathrm{b}}(\mathbf{r}_2\mathbf{r}_2') \mp \delta_{\mathrm{ab}}\rho^{\mathrm{a}}(\mathbf{r}_1\mathbf{r}_2')\rho^{\mathrm{b}}(\mathbf{r}_2\mathbf{r}_1')\,, \tag{46}$$

$$\begin{aligned} g_{\mathrm{col}}^{\mathrm{ab,M}}(\mathbf{r}_1\mathbf{r}_2;\mathbf{r}_1'\mathbf{r}_2',\tau) &= -\mathrm{i}\int_0^\beta \mathrm{d}\bar\tau \int \mathrm{d}^3\bar r_1 \mathrm{d}^3\bar r_2 V^{\mathrm{ab}}(\bar r_{12}) \times \\ &\Big\{g^{\mathrm{a,M}}(\mathbf{r}_1\bar{\mathbf{r}}_1,\tau-\bar\tau)g^{\mathrm{b,M}}(\mathbf{r}_2\bar{\mathbf{r}}_2,\tau-\bar\tau)\times \\ &\big[g^{\mathrm{a,M}}(\bar{\mathbf{r}}_1\mathbf{r}_1',\bar\tau)g^{\mathrm{b,M}}(\bar{\mathbf{r}}_2\mathbf{r}_2',\bar\tau) \pm \delta_{\mathrm{ab}}g^{\mathrm{a,M}}(\bar{\mathbf{r}}_1\mathbf{r}_2',\bar\tau)g^{\mathrm{b,M}}(\bar{\mathbf{r}}_2\mathbf{r}_1',\bar\tau)\big]\Big\}\,. \end{aligned} \tag{47}$$

From this, the equilibrium PDF is obtained by introducing center of mass and relative coordinates as was done in the nonequilibrium situation, cf. Sec. 4.1.

6. Numerical example: pair correlations in an electron-hole bilayer system

To illustrate the results obtained so far, we consider an example of a two-component system where strong correlations play a prominent role. The system of interest consists of two layers of zero thickness which contain an equal finite number N of negative (electrons) and positive (holes) charged particles with a finite layer spacing d. In each plane, the particles are confined by a harmonic potential of frequency Ω. Such electron-hole bilayers have been actively studied in recent years, e.g., [21, 22, 23], because they allow to study strongly correlated excitons, which may exhibit Bose condensation and superfluidity, as well as liquid-like and crystal-like behavior. For more details, we refer to Refs. [26, 29].

6.1. Model

The Hamiltonian of the quasi-two-dimensional electron-hole bilayer is given by

$$H = H_{\mathrm{e}} + H_{\mathrm{h}} + H_{\mathrm{eh}} \tag{48}$$

$$H_{\mathrm{e}} = \sum_{i=1}^{N_{\mathrm{e}}}\frac{1}{2}\left(-\Delta_{i,\mathrm{e}} + \mathbf{r}_{i,\mathrm{e}}^2\right) + \lambda\sum_{i<j=2}^{N_{\mathrm{e}}}\frac{1}{\sqrt{(\mathbf{r}_{i,\mathrm{e}}-\mathbf{r}_{j,\mathrm{e}})^2}}\,, \tag{49}$$

$$H_{\mathrm{h}} = \sum_{i=1}^{N_{\mathrm{h}}}\frac{1}{2}\left(-\frac{m_{\mathrm{e}}^*}{m_{\mathrm{h}}^*}\Delta_{i,\mathrm{h}} + \frac{m_{\mathrm{h}}^*}{m_{\mathrm{e}}^*}\mathbf{r}_{i,\mathrm{h}}^2\right) + \lambda\sum_{i<j=2}^{N_{\mathrm{h}}}\frac{1}{\sqrt{(\mathbf{r}_{i,\mathrm{h}}-\mathbf{r}_{j,\mathrm{h}})^2}}\,, \tag{50}$$

$$H_{\mathrm{eh}} = -\lambda\sum_{i=1}^{N_{\mathrm{e}}}\sum_{j=1}^{N_{\mathrm{h}}}\frac{1}{\sqrt{(\mathbf{r}_{i,\mathrm{e}}-\mathbf{r}_{j,\mathrm{h}})^2 + d^{*2}}}\,, \tag{51}$$

where we introduced dimensionless variables by rescaling length and energy by the units,

$$r_0 = \sqrt{\frac{\hbar}{m_e \Omega}}, \qquad E_0 = \hbar\Omega\,. \tag{52}$$

The term $m^*_{e/h}$ denotes the effective mass of electrons/holes and d^* is the effective distance between the layers. Further, we introduced the coupling parameter λ measuring the strength of the Coulomb interaction relative to the confinement energy

$$\lambda = \frac{r_0}{a_B}, \qquad a_B = \frac{\hbar^2}{m_e e^2}\,. \tag{53}$$

Here, a_B is the Bohr radius of a Hydrogen-like bound state—an exciton (we use the electron mass instead of the reduced mass).

The coupling parameter measures the strength of the Coulomb interaction among identical particles in each layer as well as the correlation between electron and holes. Generally, one may expect that for $\lambda \leq 1$, i.e. for very strong confinement, there is a strong wave function overlap, and the system approaches ideal gas like behavior. In the opposite case, $\lambda \gg 1$, the Coulomb interaction dominates, and particles will tend to become localized. Finally, variation of the layer separation gives an additional control of the many-particle state: for $d \to \infty$, both layers will be decoupled, containing independent electron and hole populations, whereas for decreasing d, Coulomb attraction plays an increasing role. This gives rise to formation of indirect excitons which behave (approximately) as bosons and exhibit dipole interaction, e.g., [26, 29].

6.2. Equilibrium PDF of the electron-hole bilayer—Comparison to Path Integral Monte-Carlo results

Preliminary results for the equilibrium PDF of the electron-hole bilayer according to Sec. 5 were obtained recently [43], however they still require further numerical tests. Therefore, to illustrate the physical content of the PDF, in this section, we show some results obtained from path integral Monte-Carlo (PIMC) calculations by *Böning et al.* [29]. Their approach to the description of the bilayer system is different to the one presented in Sec. 6, as they assume beforehand, that the layer separation d and the interaction strength λ in Eq. (48) between the electrons and holes, respectively, induce the formation of indirect excitons, i.e. quasiparticles, comprised of strongly bound but spatially separated electron-hole pairs. The excitons exhibit an interaction, which for large distances is of dipole type and for small distances approaches a soft Coulomb potential, for details see Ref. [29]. With these assumptions, *Böning et al.* compute the equilibrium PDF $h^{X_cX}(\mathbf{r})$ of one exciton X relative to a fixed exciton X_c in the center for different values of the exciton density n in a $ZnS_xSe_{1-x}/ZnSe$ quantum well with doping factor $x = 0.3$. The density is measured in units of $a_B^{*-2} = 1.06 \cdot 10^{17}\mathrm{m}^{-2}$, where $a_B^* = \hbar^2\epsilon/(e^2 m_e^*)$ is the electronic Bohr radius with the material constants $\epsilon = 8.7$ and $m_e^* = 0.15\,m_0$. The temperature is chosen to be $k_B T = 0.001\,\mathrm{Ha}^*$, where the energy unit is defined as $\mathrm{Ha}^* = e^2/(\epsilon a_B^*) = 53.93\,\mathrm{meV}$, resulting in a temperature of $T = 0.63\,\mathrm{K}$. In Fig. 6.2, $h^{X_cX}(\mathbf{r})$ is shown for different densities. One can see that, for low densities, the excitons in the system are melted, showing no localization or radial ordering. If the density is increased, an exciton crystal starts to form, due to pressure crystallization. For further increased density, this exciton crystal melts again, which is an effect of the Coulomb-like character of the exciton-exciton interaction at short distances, the excitons undergo quantum melting. The advantage of the PDF is clear from this figure: different phases of the system can be clearly distinguished, which is not possible on the basis of single-particle quantities such as the density.

PIMC is very efficient for computing the thermodynamic properties and also spectral properties [44] of bosons, such as excitons. However, at high densities, excitons break up and form an

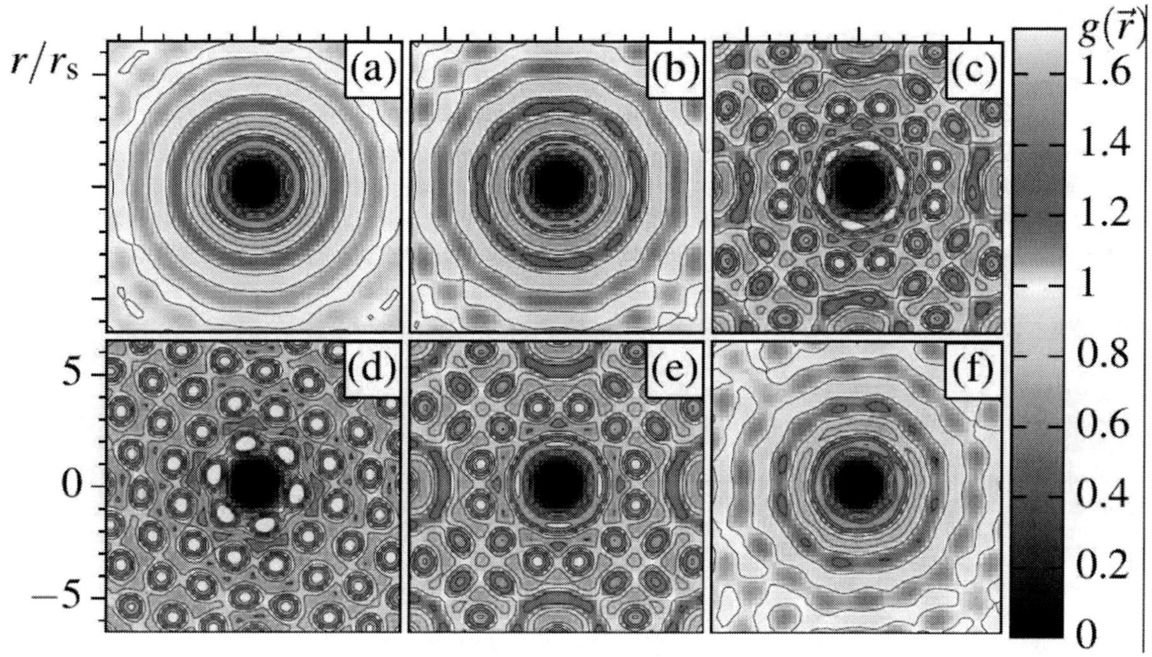

Figure 1. Radial Exciton-Exciton PDF $h^{X_cX}(\mathbf{r})$ in a $ZnS_{0.3}Se_{0.7}/ZnSe$ quantum well at $T = 0.63\,\mathrm{K}$ with layer separation $d = 40.83\,\mathrm{nm}$. Densities in units of $1.06 \cdot 10^{17}\mathrm{m}^{-2}$: (a) $0.84 \cdot 10^{-3}$, (b) $1.3 \cdot 10^{-3}$, (c) $1.7 \cdot 10^{-3}$, (d) $3.2 \cdot 10^{-3}$, (e) $3.6 \cdot 10^{-3}$, (f) $4.0 \cdot 10^{-3}$. The radial length is measured in units of the so called Brueckner parameter $r_s = 3.26 \cdot 10^8 \times a$, where a is the mean interparticle distance. The illustration is taken from Ref. [29].

electron-hole plasma (Mott effect) consisting of fermions. PIMC simulations of fermions at low temperatures are still hampered by the notorious sign problem [45]. In contrast, with nonequilibrium Green functions, this regime is easily accessible ,whereas limitations arise with increasing coupling strength. Therefore, PIMC and NEGF have complementary areas of applicability. Moreover, NEGF should allow one to access the time-evolution of the PDF, as demonstrated above.

7. Discussion

In this paper, we presented an approach to the pair distribution function of a quantum many-body system in the frame of the nonequilibrium real-time Green functions. This problem is complicated due to the fact that the standard approach used in NEGFs uses a formal decoupling of the Martin-Schwinger hierarchy on the level of the first equation: the two-particle Green function is eliminated by introduction of the single-particle selfenergy, cf. Eq. (1). With this elimination, direct access to the pair correlations is also lost. Thus, one has to reconstruct the pair correlations form the single-particle Green function within a chosen approximation for the selfenergy.

To solve this reconstruction problem, we started the analysis from the equation of motion of the two-particle Green function—the Bethe-Salpeter equation and simplified it systematically. Its solution for the two-particle Green function requires an initial value $g^{ab}(t_0)$, which is consistently determined from the (known) initial values of the single-particle Green function and the selfenergy, $g^{a,b}(t_0)$ and $\Sigma^{a,b}\left[g^{a,b}(t_0)\right]$, respectively, using Eq. (9). To further include more general arbitrary initial conditions for $g^{ab}(t_0)$, we refer the reader to, e.g., [5, 46].

Concerning the selfenergy, we concentrated on the case of the static second Born approximation, because for it a large number of numerical solutions of the KBE exist, for which it would be desirable to evaluate the PDF. It was shown that in second Born approximation a closed expression for the nonequilibrium PDF can be derived, which can be straightforwardly evaluated. The result involves combinations of four single-particle functions and is computationally expensive. Numerical results will be presented in a forthcoming paper.

Acknowledgments
We thank Th. Bornath for stimulating discussions. This work is supported in part by the Deutsche Forschungsgemeinschaft via SFB-TR24 and project BO1366/9.

References
[1] Danielewicz P 1984 *Ann. Phys.* (N.Y.) **152** 304
[2] Garny M and Müller M M 2009 *Phys. Rev.* D **80** 085001
[3] Haug H and Jauho A-P 1996 *Quantum Kinetics in Transport and Optics of Semiconductors*, Springer
[4] Bonitz M, Henning C and Block D 2010 *Reports Prog. Physics* **73** 066501
[5] Kremp D, Schlanges M and Kraeft W-D 2005 *Quantum Statistics of Nonideal Plasmas*, Springer
[6] Krausz F and Ivanov M 2009 *Rev. Mod. Phys.* **81** 163
[7] Krasovskii E E and Bonitz M 2007 *Phys. Rev. Lett.* **99** 247601
[8] Becker U 2008 *Nature Physics* **5** 649
[9] Bonitz M and Kremp D 1996 *Phys. Lett.* A **212** 83
[10] Kremp D, Bonitz M, Kraeft W D and Schlanges M 1997 *Ann. Phys.* (N.Y.) **258** 320
[11] Bonitz M 1998 *Quantum Kinetic Theory*, Teubner, Stuttgart, Leipzig
[12] Bauch S and Bonitz M 2008 *Phys. Rev.* A **78** 043403
[13] Bonitz M, Kremp D, Scott D C, Binder R, Kraeft W D and Köhler H S 1996 *J. Phys. Cond. Matt.* **8** 6057
[14] Bonitz M, Semkat D and Kremp D 1997 *Phys. Rev.* E **56** 1246
[15] Kwong N H and Bonitz M 2000 *Phys. Rev. Lett.* **84** 1768
[16] Dahlen N E and van Leeuwen R 2007 *Phys. Rev. Lett.* **98** 153004
[17] Stan A, Dahlen N E and van Leeuwen R 2009 *J. Chem. Phys.* **103** 176404
[18] von Friesen P, Verdozzi C and Almbladh C-O 2009 *Phys. Rev. Lett.* **103** 176404
[19] von Friesen P, Verdozzi C and Almbladh C-O 2011 *Europhys. Lett.* **95** 27005
[20] Bonitz M 1996 *Phys. Lett.* A **221** 85
[21] De Palo S, Rapisarda F and Senatore G 2002 *Phys. Rev. Lett.* **88** 206401
[22] Filinov A, Ludwig P, Golubnychiy V, Bonitz M and Lozovik Yu E 2003 *phys. stat. sol.* (c) **0** 1518
[23] Hartmann P, Donko Z and Kalman G 2005 *Europhys. Lett.* **72** 396
[24] Schleede J, Filinov A, Bonitz M and Fehske H 2012 *Contrib. Plasma Phys.*
[25] Filinov A, Ludwig P, Bonitz M and Lozovik Yu E 2009 *J. Phys.* A **42** 214016
[26] Ludwig P, Filinov A, Lozovik Yu E, Stolz H and Bonitz M 2007 *Contrib. Plasma Phys.* **47** 335
[27] Ludwig P, Balzer K, Filinov A, Stolz H and Bonitz M 2008 *New Journal of Physics* **10** 083031
[28] Kalman G, Hartmann P, Donko Z and Golden K 2007 *Phys. Rev. Lett.* **98** 236801
[29] Böning J, Filinov A and Bonitz M 2011 *Phys. Rev.* B **84** 075130
[30] Boercker D B and Dufty J W 1979 *Ann. Phys.* (N.Y.) **119** 43
[31] Semkat D, Kremp D and Bonitz M 1999 *Phys. Rev.* E **59** 1557
[32] Semkat D, Kremp D and Bonitz M 2000 *J. Math. Phys.* **42** 7458
[33] Semkat D, Bonitz M and Kremp D 2000 *Contrib. Plasma Phys.* **43** 321
[34] Bornath T, Kremp D and Schlanges M 1999 *Phys. Rev.* E **60** 6382
[35] Balzer K, Bonitz M, van Leeuwen R, Dahlen N E and Stan A 2009 *Phys. Rev* B **79** 245306
[36] Balzer K, Hermanns S and Bonitz M 2012 *EPL* **98** 67002
[37] Hermanns S, Balzer K and Bonitz M 2012 *Phys. Scr.* **T151** 014036
[38] Balzer K, Bauch S and Bonitz M 2010 *Phys. Rev* A **81** 022510
[39] Lipavski P, Spicka V and Velicky B 1986 *Phys. Rev.* B **34** 6933
[40] Hermanns S, Balzer K and Bonitz M 2012 *J. Phys. Conf. Series*, this issue
[41] Balzer K, Hermanns S and Bonitz M 2012 *J. Phys. Conf. Series*, this issue
[42] Rosenthal L 2009 *Green's functions approach to electron-hole bilayers* (in German), Diploma thesis, Kiel University
[43] Kobusch K 2012 *Nonequilibrium Green's function approach to the pair distribution function in the second Born approximation* (in German), Diploma thesis, Kiel University

[44] Filinov A, Prokof'ev N V and Bonitz M 2010 *Phys. Rev. Lett.* **105** 070401,
Filinov A and Bonitz M 2012 *Phys. Rev.* A **86** 043628
[45] Schoof T, Bonitz M, Filinov A, Hochstuhl D and Dufty J W 2011 *Contrib. Plasma Phys.* **51** 687-697
[46] van Leeuwen R and Stefanucci G 2011 *Phys. Rev* B **85** 115119

Journal of Physics: Conference Series **427** (2013) 012003 doi:10.1088/1742-6596/427/1/012003

Competition between the electronic and phonon–mediated scattering channels in the out–of–equilibrium carrier dynamics of semiconductors: an *ab–initio* approach

Andrea Marini

Istituto di Struttura della Materia of the National Research Council, Via Salaria Km 29.3, I-00016 Monterotondo Stazione, Italy.

E-mail: andrea.marini@cnr.it

Abstract. The carrier dynamics in bulk Silicon, a paradigmatic indirect gap semiconductor, is studied by using the Baym–Kadanoff equations. Both the electron–electron (e–e) and electron–phonon (e–p) self–energies are calculated fully *ab–initio* by using a semi–static out–of–equilibrium *GW* approximation in the e–e case and a Fan self–energy in the e–p case. By using the generalized Baym–Kadanoff ansatz the two–time evolution is replaced by the only dynamics on the macroscopic time axis. The enormous numerical difficulties connected with a real–time simulation of realistic systems is overcome by using a completed collision approximation that further simplifies the memory effects connected to the time evolution. The carrier dynamics is shown to reduce in such a way to have stringent connections to the well–known equilibrium electron–electron and electron–phonon self–energies. This link allows to use general arguments to motivate the relative balance between the e–e and e–p scattering channels on the basis of the carrier energies.

1. Introduction

Since the seminal works of two chemists, Norrish and Porter (recognized by the 1976 Nobel prize in Chemistry), laser technology has grown explosively until reaching in recent years laser durations of few attoseconds and power flux densities larger than 10^{18} W/cm^2 [1]. This led to the opening of the new era of femtosecond (fs) nanoscale physics, where combined non–linear and non–adiabatic phenomena [2] occur and where it is possible to monitor real–time the dynamics of photo–induced electronic excitations with unprecedented precision [3].

Ultrafast science at the nano-scale is clearing the path for nanostructure devices with efficient light emission spectra, high optical gain and controllable atomic deformations, thus paving the way to strategic applications in chemistry, biophysics and medicine. The design of nano-scale devices is, however, inevitably linked to a detailed knowledge of the structural and dynamical properties of these systems. Despite the massive number of available experimental results there are still scarce numerical and theoretical methods in use of the scientific community.

Nanostructures and biological systems are, indeed, formed by hundreds/thousands of atoms and their peculiar properties are related to their reduced dimensionality and extended surface. Any reliable theory is inevitably linked to a detailed knowledge of their structural and dynamical

Journal of Physics: Conference Series **427** (2013) 012003 doi:10.1088/1742-6596/427/1/012003

properties. Due to the complexity of these systems, however, state-of-the art methods are confined to rather simple models with empirical parameters. In this way the theory is deprived of its predictive aspiration, and of the possibility of inspiring new experiments and practical applications. This situation is unavoidably creating a gap between theory and experiment.

Density Functional Theory (DFT)[4] represents the most up–to–date, systematic and predictive approach to study the electronic, structural and optical properties of an impressive wide range of materials[5], including solids and nanostructures. DFT is, however, a ground–state theory that, as it will be shortly discussed in this paper, can only provide a suitable, parameter–free (and thus predictive) basis where the actual real–time simulations are carried on. All the class of methods developed starting from the DFT are labeled as *ab–initio* methods.

It is well known that extended systems have been mostly studied by using the many-body perturbation theory (MBPT) [6] and Time–Dependent Density–Functional–Theory (TDDFT)[7]. The different treatment of correlation and nonlinear effects marks the range of applicability of the two approaches. The real-time TDDFT makes possible to investigate nonlinear effects like second harmonic generation[8] or hyperpolarizabilities of molecular systems[9]. However the standard approaches used to approximate the exchange-correlation functional of TDDFT treat correlation effects only on a mean-field level. As a consequence, while finite systems—such as molecules—are well described, in the case of extended systems—such as periodic crystals and nano-structures—the real-time TDDFT does not capture the essential features of the optical absorption [10] even qualitatively. On the contrary MBPT allows to include correlation effects using controllable and systematic approximations for the self-energy Σ, that is a one-particle operator non-local in space and time.

2. The equilibrium and out–of–equilibrium dynamics in an *ab–initio* framework

In recent years, the MBPT approach has been merged with DFT by using the Kohn-Sham [7] Hamiltonian as zeroth-order term in the perturbative expansion of the interacting Green's functions. This approach is parameter free and completely *ab-initio*, [10] and in this paper will be addressed as *ab-initio*-MBPT (*Ai*-MBPT) to mark the difference with the conventional MBPT.

In practice the Kohn–Sham equation is first solved:

$$\left[-\frac{\nabla^2}{2} + V_{KS}\left[\{\mathbf{R}\}\right](\mathbf{r})\right]\phi_{n\mathbf{k}}(\mathbf{r}) = \epsilon_{n\mathbf{k}}\phi_{n\mathbf{k}}(\mathbf{r}), \tag{1}$$

with

$$V_{KS}\left[\{\mathbf{R}\}\right](\mathbf{r}) = V_{ion}\left[\{\mathbf{R}\}\right](\mathbf{r}) + \int d\mathbf{r}'\, v(\mathbf{r},\mathbf{r}')\, n(\mathbf{r}') + V_{xc}([n],\mathbf{r}). \tag{2}$$

V_{KS} is, thus, written in terms of the bare Coulomb interaction $v(\mathbf{r}_1,\mathbf{r}_2) \equiv |\mathbf{r}_1 - \mathbf{r}_2|^{-1}$, the electronic density $n(\mathbf{r})$, the exchange–correlation potential $V_{xc}([n],\mathbf{r})$ and the total ionic potential $V_{ion}\left[\{\mathbf{R}\}\right](\mathbf{r})$:

$$V_{ion}\left[\{\mathbf{R}\}\right](\mathbf{r}) = \sum_I V_{ion}(\mathbf{r} - \mathbf{R}_I), \tag{3}$$

with $\mathbf{R}_I$ the position of the generic atom I. All the simulations presented in this work have been carried on by using the Quantum Espresso[11] and the Yambo[12] by using pseudo–potential approach written in a plane–waves basis. In this approach $V_{ion}(\mathbf{r} - \mathbf{R}_I)$ is a pseudo–potential defined only for the valence electrons (8 in the case of bulk Silicon) in order to restrict the electronic dynamics to valence electron only. Therefore core–electrons are not considered in the present work.

Journal of Physics: Conference Series **427** (2013) 012003 doi:10.1088/1742-6596/427/1/012003

By plugging into Eq.3 the precise atomic structure of a given material, the solution of Eq.1 provides a basis for the electronic levels suitable to rewrite the Many–Body problem. Indeed any Green's function $G(1,2;\beta)$ can be easily rewritten as a matrix in the KS basis

$$G(1,2;\beta) = \sum_{nn'\mathbf{k}} \left[\mathbf{G}_{\mathbf{k}}(t_1,t_2;\beta)\right]_{nn'} \phi^*_{n\mathbf{k}}(\mathbf{r}_1)\,\phi_{n'\mathbf{k}}(\mathbf{r}_2), \tag{4}$$

where $1=(\mathbf{r}_1,t_1)$ and $2=(\mathbf{r}_2,t_2)$ represent global position, time and (eventually) spin indexes and $\beta = k_B T_{th}$ with k_B the Boltzmann constant and T_{ph} the equilibrium temperature.

However the *Ai*-MBPT is a very cumbersome technique that, based on a perturbative concept, increases its level of complexity with the order of the expansion. As an example, this makes the extension of this approach beyond the linear response regime quite complex, though there have been recently some applications of the *Ai*-MBPT in nonlinear optics [13–15].

Another stringent restriction of the *Ai*-MBPT is that it cannot be applied when non-equilibrium phenomena take place: for example it cannot be applied to study the light emission after an ultrafast laser pulse excitation. A generalization of MBPT to non-equilibrium situations has been proposed by Kadanoff and Baym[16–19]. In their seminal works the authors derived a set of equations for the real-time Green's functions, the Kadanoff-Baym equations (KBE's), that provide the basic tools of the non-equilibrium Green's Function theory and allowed essential advances in non equilibrium statistical mechanics.

In the equilibrium MBPT, due to the time translation invariance, the relevant variable used to calculate the Green's functions is the frequency ω. Instead, out of equilibrium, the time variables acquire a special role and much more attention is devoted to the their propagation properties. The time propagation avoids the explosive dependence, beyond the linear response, of the MBPT on high order Green's functions. Moreover the KBE's are non-perturbative in the external field therefore weak and strong fields can be treated on the same footing.

The key ingredient of the KBE's are the two times lesser/greater Green's functions[16–19] $G^{\lessgtr}_{n_1 n_2 \mathbf{k}}(t_1,t_2)$ and self–energies $\Sigma^{\lessgtr}_{n_1 n_2 \mathbf{k}}(t_1,t_2)$. If we consider the time evolution of $G^{<}$ only on the macroscopic axis ($t_1 = t_2 = T$) we get the starting form of the KBE's I will further analyze in the later sections of this paper:

$$i\frac{\partial}{\partial T}\mathbf{G}^{<}_{\mathbf{k}}(T,T;\beta) = \left[\mathbf{h}^{KS}_{\mathbf{k}} + \mathbf{V}^{Hartree}_{\mathbf{k}}\left[n(T;\beta)\right] + \right.$$
$$\left. \mathbf{\Sigma}^{s}_{\mathbf{k}}(T;\beta) + \mathbf{U}_{\mathbf{k}}(T;\beta), \mathbf{G}^{<}_{\mathbf{k}}(T,T;\beta)\right] + \mathbf{S}^{<}_{\mathbf{k}}(T;\beta). \tag{5}$$

In Eq.5 all quantities are matrices in the space of KS states (see Eq.4), e.g., $\Sigma^{s}_{n_1 n_2 \mathbf{k}}$, $\beta = (k_B T_{el})^{-1}$ where T_{el} is the phonon bath temperature, and I have already introduced the fundamental ingredients that will be discussed in the following of this paper:

- The equilibrium KS Hamiltonian, $\mathbf{h}^{KS}$.
- The Hartree potential $\mathbf{V}^{Hartree}$. The ionic and kinetic part is already embodied in $\mathbf{h}^{KS}$.
- The static self–energy $\mathbf{\Sigma}^{s}$. This is responsible for introducing static e–e correlations that, if properly defined [20] reduce the KBE, in the linear–response regime, to the well–known Bethe–Salpeter equation[6].
- The interaction with the external, arbitrary electric field. The only assumption here is the the field wavelength is long compared with the unit cell size (optical limit).
- The interaction kernel $\mathbf{S}^{<}$. This is the most important part of the equation that embodies dynamical e–e and e–p scatterings. It will be defined and carefully analyzed in Sec.4.

One of the first attempts to apply the KBE's for investigating optical properties of semiconductors was presented in the seminal paper of Schmitt-Rink and co-workers. [21] Later the KBE's were applied to study quantum wells, [22] laser excited semiconductors, [23] and luminescence [24]. However, only recently it was possible to simulate the Kadanoff-Baym dynamics in real-time. [25–29]

Regarding the out–of–equilibrium carrier dynamics there is a vast literature [18] where the KBE's have been solved using several kind of approximations. I do not want to give here an exhaustive list of all the works in this field, but a key aspect of most of them is that they rely on very strong approximations for what regards the underlying electronic structure and microscopic dielectric properties of the materials used. For example a large number of works use few (two or four) bands models like in Refs.[30–35]. In all these works the electronic levels are approximated as well as the screening of the electron–electron interaction.

In this work I want to go well beyond these approximations by solving the KBE's in a fully *ab–initio* framework. I will refer to this novel methodology as *ab–initio* Non–Equilibrium Green's function (*Ai*–NEGF) theory.

3. The equilibrium self–energies

In order to discuss the physics involved in the dynamics of carriers taken out–of–equilibrium it is useful to introduce the properties of the two self–energies that will be used to describe the e–e and e–p coupling. In the next two subsections, indeed, I will introduce the approximations used and the key definitions of the different ingredients needed to build the self–energies. I will later rewrite these equilibrium self–energies in the framework of the *Ai*–NEGF theory.

The e–e interaction will be treated in the so called GW approximation[36] where the self–energy is expanded at the first order in the dynamically screened interaction. This approximation has provided a priceless tool to evaluate accurately the electronic levels of a wide class of materials [10].

Since its first application to semiconductors[37] the GW self-energy has been shown to correctly reproduce quasi-particle energies and lifetimes for a wide range of materials[36]. Furthermore, by using the static limit of the GW self-energy as a scattering potential of the Bethe-Salpeter equation (BSE)[6], it is possible to calculate response functions including electron-hole interaction effects.

In contrast to the e–e case the e–p coupling has received only minor attention in the *ab–initio* community. And this is somehow astonishing if we consider that the e–p coupling is well known to play a key role in several physical phenomena. For example, beside inducing the well–known superconducting phase of several materials [38], it affects the renormalization of the electronic bands[39–43].

Despite the development of more powerful and efficient computational resources the actual calculation of the effects induced by the e–p coupling in realistic materials remains a challenging task. In addition to the numerical difficulties, it has been assumed, for a long time, that this interaction can yield only minor corrections (of the order of meV) to the electronic levels. As a consequence the majority of the *ab–initio* simulations of the electronic and optical properties of a wide class of materials are generally performed by keeping the atoms frozen in their crystallographic positions. More importantly it is well–known that phonons are atomic vibrations and, as a such, can be easily populated by increasing the temperature. This naive observation is *de-facto* used to associate the effect of the electron–phonon coupling to a temperature effect that vanishes as the temperature goes to zero. However this is not correct as the atoms posses an intrinsic spatial indetermination due to their quantum nature, that is independent on the temperature. These quantistic oscillations are taken in account by the e–p coupling when $T \to 0$ in the form of a zero–point–motion effect.

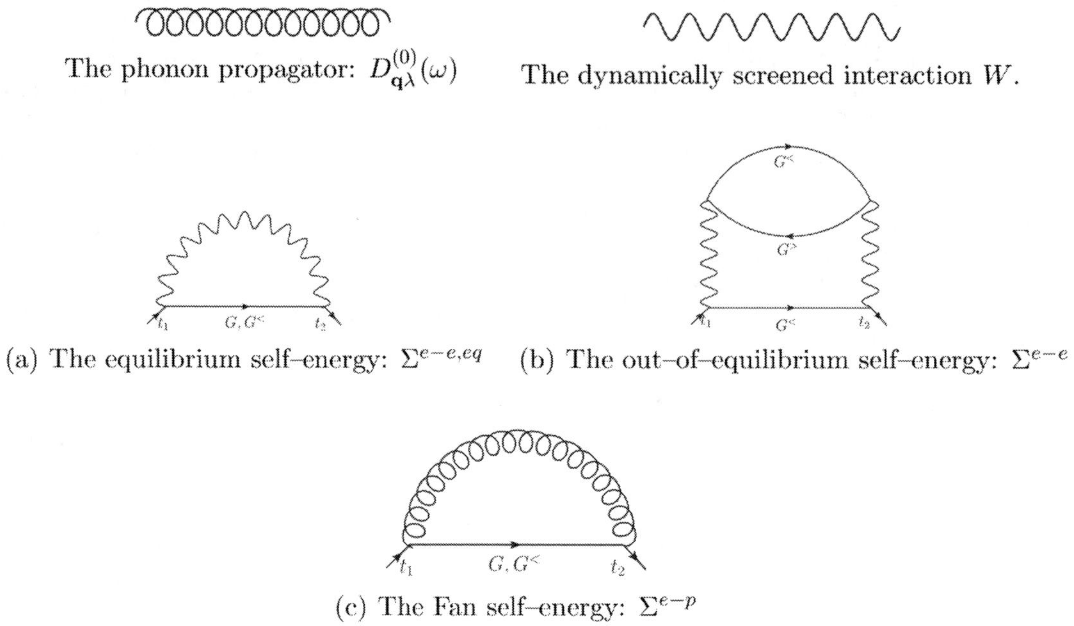

Figure 1. The self-energy diagrams at the equilibrium and out–of–the–equilibrium used to account for e–e and e–p coupling. The kind of Green's functions used in the diagrams are shown. Diagram (a) refers to the equilibrium case only, diagram (b) to the out–of–the–equilibrium case while diagram (c) can be used both at and out–of–the equilibrium. This is beacause in this work the phonons are assumed to be in thermal equilibrium with the external temperature. This means that no phonon dynamics is included. For the e–p case I consider the first order expansion in the atomic displacements corresponding to the Fan self–energy (diagram (c)). In the e–e case, instead, I first consider an equilibrium GW self-energy (diagram (a)) and the corresponding out–of–equilibrium version (diagram (b)) . In the diagram (c) the independent particle polarization corresponds to the dynamical part of the diagram. Indeed the two vertical wiggled lines correspond to a statically screened microscopical e–e interaction (see Eq.(33)).

3.1. The GW approximation for the electronic self–energy

The Feynman diagram corresponding to the GW approximation (GWA) in the equilibrium is showed in the panel (b) of Fig.1. It corresponds to the lowest order contribution to the self–energy expressed as a Taylor expansion in the screened interaction $W(1,2)$:

$$\Sigma^{eq}_{e-e}(1,2) = iW\left(1^+,2\right) G_1(1,2)\,. \tag{6}$$

In the present case we will use a spinless formulation where spin indexes are summed out. In Eq.(6) $1 = (\mathbf{r}_1, t_1)$ and $2 = (\mathbf{r}_2, t_2)$ represent global position and time. In all diagrams showed in Fig.1 the time lies on the real–time axis. The out–of–equilibrium version of the GWA will be introduced in the next section.

Figure 2. The equation for the screened interaction W within the RPA.

In the GWA the screened interaction is written as:

$$W(1,2) = v(1,2) + \int d34\, v(1,3)\, \tilde{\chi}(3,4)\, W(4,2), \tag{7}$$

with $v(1,2) \equiv v(\mathbf{r}_1,\mathbf{r}_2)\,\delta(t_1 - t_2)$ is the bare Coulomb interaction and $\tilde{\chi}$ is the irreducible electronic response function [6]. An approximation for W follows by choosing $\tilde{\chi}$ to be calculated in the independent–particle approximation which corresponds, for W, to the random–phase–approximation (RPA), where

$$\tilde{\chi}(1,2) \approx \chi_0(1,2) = -iG_0(1,2)\, G_0(2,1). \tag{8}$$

The actual evaluation of the space dependence appearing in Eq.8 is treated in reciprocal space, by summing on the lattice vectors $\mathbf{G}, \mathbf{G}'$ and integrating over the Brillouin Zone (BZ):

$$\chi(\mathbf{r}_1,\mathbf{r}_2;t) = \int_{BZ} \frac{d\mathbf{q}}{(2\pi)^3} \sum_{\mathbf{G},\mathbf{G}'} e^{i(\mathbf{q}+\mathbf{G})\cdot\mathbf{r}_1} \chi_{\mathbf{G},\mathbf{G}'}(\mathbf{q},t)\, e^{-i(\mathbf{q}+\mathbf{G}')\cdot\mathbf{r}_2}. \tag{9}$$

It follows that, within the Random–Phase approximation (RPA):

$$\chi_{\mathbf{G}\,\mathbf{G}'}(\mathbf{q},\omega) = \chi^0_{\mathbf{G}\,\mathbf{G}'}(\mathbf{q},\omega) + \sum_{\mathbf{G}''} \chi^0_{\mathbf{G}\,\mathbf{G}''}(\mathbf{q},\omega) \frac{4\pi}{|\mathbf{q}+\mathbf{G}''|^2} \chi_{\mathbf{G}''\,\mathbf{G}'}(\mathbf{q},\omega). \tag{10}$$

Eq.10 is used together with the definition of the single–particle Green's function

$$G_0(\mathbf{r}_1,\mathbf{r}_2;\omega) = 2\sum_n \sum_{\mathbf{k}\in BZ} \phi_{n\mathbf{k}}(\mathbf{r}_1)\, \phi^*_{n\mathbf{k}}(\mathbf{r}_2)\, G_{n\mathbf{k}}(\omega), \tag{11}$$

with

$$G_{n\mathbf{k}}(\omega) = \frac{f_{n\mathbf{k}}}{\omega - \epsilon_{n\mathbf{k}} - i\delta} + \frac{1 - f_{n\mathbf{k}}}{\omega - \epsilon_{n\mathbf{k}} + i\delta}; \tag{12}$$

to write the final expression of the GW self-energy . In the present work we are mainly interested in the carrier dynamics and, in order to keep the notation as compact as possible, we restrict our attention on the imaginary part only of Σ^{eq}_{e-e}. This, indeed, defines the quasi–particle linewidth as

$$\Gamma^{e-e,eq}_{n\mathbf{k}} = \Im\left[\Sigma^{e-e,eq}_{n\mathbf{k}}(\epsilon_{n\mathbf{k}})\right], \tag{13}$$

where we have used the simplest on–the–mass–shell approximation[36]. By doing simple algebraic operations it is possible to rewrite $\Gamma^{e-e,eq}$ as:

$$\Gamma^{e-e,eq}_{n\mathbf{k}} = \frac{2\pi}{N\Omega} \sum_m \sum_{\mathbf{q}} \sum_{\mathbf{G},\mathbf{G}'} \Big\{ \tilde{\rho}_{nm}(\mathbf{k},\mathbf{q},\mathbf{G}) \left[\tilde{\rho}_{nm}(\mathbf{k},\mathbf{q},\mathbf{G}')\right]^* Im\left[W_{\mathbf{G},\mathbf{G}'}(\mathbf{q},\epsilon_{n\mathbf{k}} - \epsilon_{m\mathbf{k}-\mathbf{q}})\right]$$
$$\left[\underbrace{(2 - f_{m\mathbf{k}-\mathbf{q}})\,\theta(\epsilon_{n\mathbf{k}} - \epsilon_{m\mathbf{k}-\mathbf{q}})}_{\textbf{electron scattering}} - \underbrace{f_{m\mathbf{k}-\mathbf{q}}\theta(\epsilon_{m\mathbf{k}-\mathbf{q}} - \epsilon_{n\mathbf{k}})}_{\textbf{hole scattering}} \right] \Big\}. \tag{14}$$

The term (*electron scattering*) represents an electron with energy $\epsilon_{n\mathbf{k}}$ that decays into an empty state with lower energy $\epsilon_{m\mathbf{k}-\mathbf{q}}$ dissipating the corresponding energy difference in positive energy e–h pairs (represented by $W_{\mathbf{G},\mathbf{G}'}(\mathbf{q},\epsilon_{n\mathbf{k}}-\epsilon_{m\mathbf{k}-\mathbf{q}})$). The (*hole scattering*) channel is equivalent to the (*electron scattering*) with the only difference that the hole jumps into a full state with higher energy.

In Eq.14 we have defined

$$\tilde{\rho}_{nm}(\mathbf{k},\mathbf{q},\mathbf{G}) = \langle n\mathbf{k}|e^{i(\mathbf{q}+\mathbf{G})\cdot\mathbf{r}_1}|m\mathbf{k}-\mathbf{q}\rangle = \int d\mathbf{r}\, u^*_{n\mathbf{k}}(\mathbf{r})\, u_{m\mathbf{k}-\mathbf{q}}(\mathbf{r})\, e^{i(\mathbf{q}+\mathbf{G})\cdot\mathbf{r}}, \tag{15}$$

where $u_{n\mathbf{k}}(\mathbf{r})$ is the periodic part of the Bloch function. We have also introduced the dynamically screened interaction $W_{\mathbf{G},\mathbf{G}'}(\mathbf{q},\omega)$ defined as

$$W_{\mathbf{G},\mathbf{G}'}(\mathbf{q},\omega) = \frac{(4\pi)}{|\mathbf{q}+\mathbf{G}|^2}\delta_{\mathbf{G},\mathbf{G}'} + \frac{(4\pi)}{|\mathbf{q}+\mathbf{G}|^2}\chi_{\mathbf{G},\mathbf{G}'}(\mathbf{q},\omega)\frac{(4\pi)}{|\mathbf{q}+\mathbf{G}'|^2}. \tag{16}$$

3.2. The electron–phonon Fan self–energy

The phonon contribution to the equilibrium self-energy comes from the oscillations of the atoms around their equilibrium positions. The vibrational states of the system are described, fully *ab–initio*, by using the well–known extension of DFT, Density Functional Perturbation Theory (DFPT)[44, 45] where the change in the ionic potential due to a generic atomic displacement is taken in account, self–consistently, with the solution of the KS equation. In the DFPT framework the bare ionic potential is screened by the electronic polarization and the interaction Hamiltonian between electrons and phonons $\hat{H}_{el-at}$ is given by

$$\hat{H}_{el-at} = \int_{crystal} d\mathbf{r}\, \hat{\rho}(\mathbf{r})\, \hat{V}_{scf}[\{\mathbf{R}\}](\mathbf{r}). \tag{17}$$

In Eq.(17) the self–consistent ionic potential, calculated at the first order in the atomic displacements, is given by[44, 45]

$$\hat{V}_{scf}[\{\mathbf{R}\}](\mathbf{r}) = \int d\mathbf{r}_1 \left[\delta(\mathbf{r}-\mathbf{r}_1) + \int d\mathbf{r}_2 f_{Hxc}(\mathbf{r},\mathbf{r}_2)\, \underline{\chi}(\mathbf{r}_2,\mathbf{r}_1)\right] V_{ion}[\{\mathbf{R}\}](\mathbf{r}_1), \tag{18}$$

where $\underline{\chi}$ is the TDDFT response function calculated using the f_{Hxc} kernel. In this work I have used the Local Density Approximation for V_{xc} and f_{Hxc}. If we now consider a configuration of lattice displacements $\mathbf{u}_{Is}$, H can be expressed as a Taylor expansion

$$\hat{H}_{el-at} - \overline{\hat{H}_{el-at}} = \sum_{I\alpha} \overline{\frac{\partial \hat{V}_{scf}[\{\mathbf{R}\}](\mathbf{r})}{\partial R_{I\alpha}}}\, \hat{u}_{I\alpha}, \tag{19}$$

where α is the Cartesian coordinate. Any overlined operator $\overline{\hat{O}}$ is evaluated with the atoms frozen in their crystallographic equilibrium positions. The link with the perturbative expansion is readily done by transforming Eq. (19) from the space of the lattice displacements to the space of the canonical lattice vibrations by means of the identity[46]:

$$\hat{u}_{I\alpha} = \sum_{\mathbf{q}\lambda} (2N_q M_I \omega_{\mathbf{q}\lambda})^{-1/2}\, \xi_\alpha(\mathbf{q}\lambda|I)\, e^{i\mathbf{q}\cdot(\mathbf{R}_I)} \left(\hat{b}^\dagger_{-\mathbf{q}\lambda} + \hat{b}_{\mathbf{q}\lambda}\right), \tag{20}$$

where N is the number of cells (or, equivalently the number of q–points) used in the simulation and M_I is the atomic mass of the I–th atom in the unit cell. $\xi_\alpha(\mathbf{q}\lambda|I)$ is the phonon polarization vector and $\hat{b}^\dagger_{-\mathbf{q}\lambda}$ and $\hat{b}_{\mathbf{q}\lambda}$ are the bosonic creation and annihilation operators.

By inserting Eq. (20) into Eq. (19) we get

$$\hat{H} - \overline{\hat{H}} = \frac{1}{\sqrt{N_q}} \sum_{\mathbf{k}nn'\mathbf{q}\lambda} g^{\mathbf{q}\lambda}_{n'n\mathbf{k}} c^{\dagger}_{n'\mathbf{k}+\mathbf{q}} c_{n\mathbf{k}} \left(\hat{b}^{\dagger}_{-\mathbf{q}\lambda} + \hat{b}_{\mathbf{q}\lambda}\right) \tag{21}$$

In Eq. (21) we have introduced the ($g^{\mathbf{q}\lambda}_{n'n\mathbf{k}}$) electron–phonon matrix elements. This is obtained by rewriting V_{scf} and by making explicit its dependence on the atomic positions:

$$\hat{V}_{scf}\left[\{\mathbf{R}\}\right](\mathbf{r}) = \sum_{Is} \hat{V}^{(s)}_{scf}(\mathbf{r} - \mathbf{R}_{Is}). \tag{22}$$

By using Eq.22 I get an explicit form for the electron–phonon matrix elements

$$g^{\mathbf{q}\lambda}_{n'n\mathbf{k}} = \sum_{s\alpha} (2M_s\omega_{\mathbf{q}\lambda})^{-1/2} e^{i\mathbf{q}\cdot\tau_s} \langle n'\mathbf{k}+\mathbf{q}|\frac{\partial \hat{V}^{(s)}_{scf}(\mathbf{r})}{\partial R_{s\alpha}}|n\mathbf{k}\rangle \xi_{\alpha}(\mathbf{q}\lambda|s). \tag{23}$$

We have used the short form $R_{s\alpha} = R_{Is\alpha}|_{I=0}$.

Note that, in principle, the interaction Hamiltonian, Eq.(21), can be also used to obtain a diagrammatic representation of the phonon self–energy. If the electronic and phononic problems is solved at the same time, however, the exchange–correlation effects introduced within the DFPT would lead to severe double counting problems [47]. In the present case this problem is avoided by using bare DFPT phonon modes without taking in account additional Many–Body corrections.

The actual calculation of the Fan diagram is straightforward. The Fan diagram is similar to the GW one where the screened electronic interaction W is replaced by a phonon propagator of wavevector $\mathbf{q}$ and branch λ[48]. By applying the finite temperature diagrammatic rules it is possible to define the Fan self-energy operator $\Sigma^{e-p}_{n\mathbf{k}}(\omega)$, recovering the expression originally evaluated by Fan[49] for the phonon mediated quasi–particle linewidth

$$\Gamma^{e-p,eq}_{n\mathbf{k}}(\beta) = \Im\left[\Sigma^{e-p,eq}_{n\mathbf{k}}(\epsilon_{n\mathbf{k}}, \beta)\right], \tag{24}$$

with

$$\begin{aligned}\Gamma^{e-p,eq}_{n\mathbf{k}}(\beta) = \Gamma^{e-p,eq}_{n\mathbf{k}}\Big|_{e} + \Gamma^{e-p,eq}_{n\mathbf{k}}\Big|_{h} + \Gamma^{e-p,eq}_{n\mathbf{k}}(\beta)\Big|_{ph} = \\ \frac{\pi}{N}\sum_{n'\mathbf{q}\lambda} | g^{\mathbf{q}\lambda}_{n'n\mathbf{k}} |^2 \left(1 - f_{n'\mathbf{k}-\mathbf{q}}\right) \delta\left(\epsilon_{n'\mathbf{k}-\mathbf{q}} - \epsilon_{n\mathbf{k}} + \omega_{\mathbf{q}\lambda}\right) + \\ \frac{\pi}{N}\sum_{n'\mathbf{q}\lambda} | g^{\mathbf{q}\lambda}_{n'n\mathbf{k}} |^2 f_{n'\mathbf{k}-\mathbf{q}} \delta\left(\epsilon_{n'\mathbf{k}-\mathbf{q}} - \epsilon_{n\mathbf{k}} - \omega_{\mathbf{q}\lambda}\right) + \\ \frac{\pi}{N}\sum_{n'\mathbf{q}\lambda} | g^{\mathbf{q}\lambda}_{n'n\mathbf{k}} |^2 N_{\mathbf{q}\lambda}(T) \left[\delta\left(\epsilon_{n'\mathbf{k}-\mathbf{q}} - \epsilon_{n\mathbf{k}} + \omega_{\mathbf{q}\lambda}\right) + \delta\left(\epsilon_{n'\mathbf{k}-\mathbf{q}} - \epsilon_{n\mathbf{k}} - \omega_{\mathbf{q}\lambda}\right)\right],\end{aligned} \tag{25}$$

with $N_{\mathbf{q}\lambda}(T)$ is the Bose function distribution of the phonon mode $(\mathbf{q}, \lambda)$ at temperature T. The e–p lifetime is splitted in Eq.25 in three different contributions corresponding to the scattering of an electron/hole $\Gamma^{e-p,eq}_{n\mathbf{k}}\Big|_{e/h}$ with the phonon bath and an explicitly temperature term, $\Gamma^{e-p,eq}_{n\mathbf{k}}(\beta)\Big|_{ph}$. The Fan self-energy can be also derived by using the framework of the Heine–Allen–Cardona (HAC) theory [39, 40, 50]. The HAC approach is based on the static Rayleigh-Schrödinger perturbation theory done using the $\hat{u}_{Is\alpha}$ displacement operators as scalar variables.

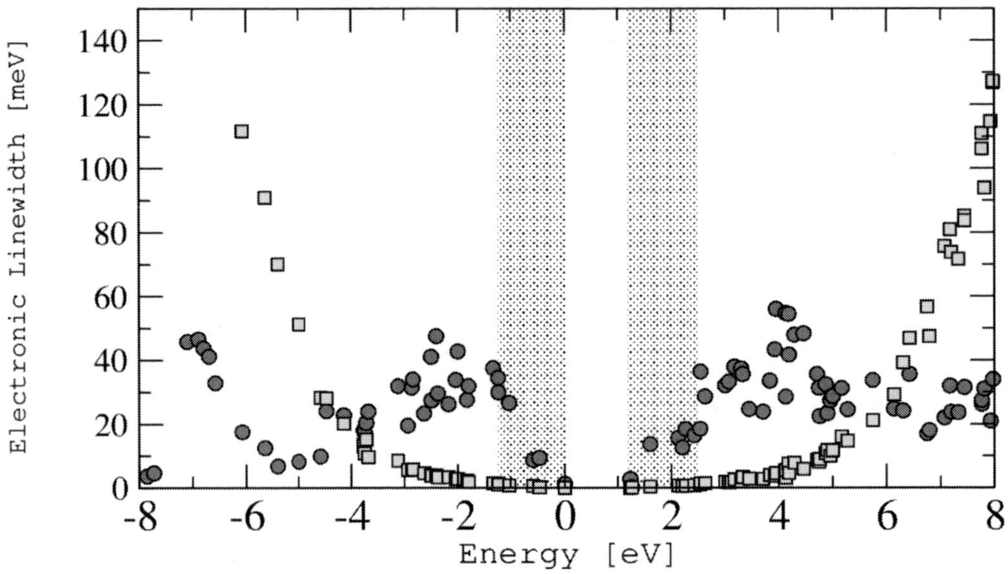

Figure 3. Quasiparticle linewidths of bulk Silicon calculated by using the *GW* approximation for the e–e scattering (green boxes) and the Fan approximation for the e–p scattering (red circles). The two gray areas denote the energy regions (as large as twice the electronic indirect gap $2E_g \approx 1.2\,\text{eV}$) where, by simple energy conservation arguments, the electronic linewidth are zero by definition. In these energy conditions the e–p contribution is stronger and the corresponding linewidths are larger then the e–e ones. The quadratic energy dependence of the e–e linewidths inverts this trend at about five times the indirect gap.

3.3. Electron–phonon versus electron–electron contribution to the equilibrium electronic lifetimes

From Eq.14 and Eq.25 we can deduce a simple physical property of the equilibrium self–energies. By simple energetic conditions it is clear that in a system with a direct or indirect gap E_g the e–e linewidths $\Gamma^{e-e,eq}_{n\mathbf{k}}$ are exactly zero whenever $\epsilon_{n\mathbf{k}} < \epsilon_{CBM} + 2E_g$ or $\epsilon_{n\mathbf{k}} > \epsilon_{VBM} - 2E_g$, with CBM referring to the condunction band minumum and VBM to the valence band maximum. These two energy conditions define the energy ranges represented by gray areas in Fig.3.

On the contrary the e–p linewidths, thanks to the continuum of phonon frequencies, have no zero energy regions. It follows that in the case of low energy electrons or hole the e–p linewidths largely dominate on the e–e part. Only above (or below) the energy forbidden regions the e–e linewidths start growing quadratically, following the Fermi liquid expected behavior, and cross the e–p part when $\epsilon_{n\mathbf{k}} = -4\,\text{eV}$ (valence bands) and $6\,\text{eV}$.

4. An out–of–equilibrium formulation of the carrier dynamics including e–e and e–p scatterings

Now that the equilibrium self–energies have been introduced and defined we can move in the out–of–equilibrium regime by giving a precise form to the interaction kernel $\mathbf{S}_{\mathbf{k}}(T)$. First of all this is formally defined [16–19] as

$$\mathbf{S}^{<}_{\mathbf{k}}(T;\beta) = \int_{-\infty}^{T} d\tau\, \Big[\boldsymbol{\Sigma}^{>}_{\mathbf{k}}(T,\tau;\beta)\,\mathbf{G}^{<}_{\mathbf{k}}(\tau,T;\beta) + \mathbf{G}^{<}_{\mathbf{k}}(T,\tau;\beta)\,\boldsymbol{\Sigma}^{>}_{\mathbf{k}}(\tau,T;\beta) - \boldsymbol{\Sigma}^{<}_{\mathbf{k}}(T,\tau;\beta)\,\mathbf{G}^{>}_{\mathbf{k}}(\tau,T;\beta) - \mathbf{G}^{>}_{\mathbf{k}}(T,\tau;\beta)\,\boldsymbol{\Sigma}^{<}_{\mathbf{k}}(\tau,T;\beta)\Big]. \quad (26)$$

The key point in Eq.26 is that the definition of $\mathbf{S}$ is not closed as it involves two–times functions and both the greater and lesser Green's functions. This means that one should consider the two–time evolution of the KBE together with the equation of motion for the $\mathbf{G}_{\mathbf{k}}^{\lessgtr}$ and, as we will shortly see, for the $\mathbf{G}_{\mathbf{k}}^{(r/a)}$ Green's functions.

As a two–time evolution of the KBE is, at the present stage, computationally not feasible for the real–time dynamics of complex solids and nanostructures we want to explore a different path based on several key ansatz and approximations.

For the e–e case the extension of the GWA, Eq.7, to the out–of–equilibrium regime is well–known [17]. The diagrammatic form is represented in the frame (b) of Fig.1. What I want to do here is the rewrite this extension in a *ab–initio* framework introducing some key approximations. First of all I split $\mathbf{\Sigma}$ and $\mathbf{S}$ in the e–e and e–p parts:

$$\mathbf{\Sigma}_{\mathbf{k}}^{\lessgtr}(t_1,t_2;\beta) = \mathbf{\Sigma}_{\mathbf{k}}^{(e-e)\lessgtr}(t_1,t_2;\beta) + \mathbf{\Sigma}_{\mathbf{k}}^{(e-p)\lessgtr}(t_1,t_2;\beta)\,, \tag{27}$$

$$\mathbf{S}_{\mathbf{k}}^{\lessgtr}(T;\beta) = \mathbf{S}_{\mathbf{k}}^{(e-e)\lessgtr}(T;\beta) + \mathbf{S}_{\mathbf{k}}^{(e-p)\lessgtr}(T;\beta)\,. \tag{28}$$

The GWA for $\mathbf{\Sigma}^{(e-e)\lessgtr}$ out–of–equilibrium reads:

$$\Sigma_{n_1n_2\mathbf{k}}^{(e-e)\lessgtr}(t_1,t_2;\beta) = \frac{-i}{2N\Omega}\sum_{m_1,m_2}\sum_{\mathbf{G}_1,\mathbf{G}_2}\sum_{\mathbf{q}}\{\tilde{\rho}_{n_1m_1}(\mathbf{k},\mathbf{q},\mathbf{G}_1)\left[\tilde{\rho}_{n_2m_2}(\mathbf{k},\mathbf{q},\mathbf{G}_2)\right]^* \times W_{\mathbf{G}_1\mathbf{G}_2}^{\lessgtr}(\mathbf{q},t_1,t_2;\beta)\, G_{m_1m_2\mathbf{k}-\mathbf{q}}^{\lessgtr}(t_1,t_2;\beta)\}, \tag{29}$$

with the screened interaction $W^{\lessgtr}$ written as

$$W_{\mathbf{G}_1,\mathbf{G}_2}^{\lessgtr}(\mathbf{q},t_1,t_2) = \int_{-\infty}^{t_1} d\tau_1 \int_{-\infty}^{t_2} d\tau_2 \sum_{\mathbf{G}_3,\mathbf{G}_4} W_{\mathbf{G}_1,\mathbf{G}_3}^{(r)}(\mathbf{q},t_1,\tau_1;\beta) \times L_{\mathbf{G}_3,\mathbf{G}_4}^{\lessgtr}(\mathbf{q},\tau_1,\tau_2;\beta)\, W_{\mathbf{G}_4,\mathbf{G}_2}^{(a)}(\mathbf{q},\tau_2,t_2;\beta)\,. \tag{30}$$

In Eq.30 $W^{(r/a)}$ are the retarded/advanced screened interactions and $L^{\lessgtr}$ is the independent particle polarization diagram represented by the closed loop in the diagram (b) of Fig.1:

$$L_{\mathbf{G}_1,\mathbf{G}_2}^{\lessgtr}(\mathbf{q},t_1,t_2;\beta) = \frac{-i}{2N\Omega}\sum_{m_2,m_3}\sum_{m_2',m_3'}\left[\tilde{\rho}_{m_3m_2'}(\mathbf{p},\mathbf{q},\mathbf{G}_1)\right]^* \tilde{\rho}_{m_3'm_2}(\mathbf{p},\mathbf{q},\mathbf{G}_2) \times G_{m_3m_3'\mathbf{p}}^{\lessgtr}(t_1,t_2;\beta)\, G_{m_2'm_2\mathbf{p}-\mathbf{q}}^{\gtrless}(t_2,t_1;\beta)\,. \tag{31}$$

Before analyzing in detail the temporal dependence of $\mathbf{\Sigma}^{(e-e)\lessgtr}$ it is instructive to introduce the out–of–equilibrium e–p self–energy. The extension out–of–equilibrium of the FAN self–energy is straightforward[17]:

$$\Sigma_{n_1n_2\mathbf{k}}^{(e-p)\lessgtr}(t_1,t_2;\beta) = \frac{-i}{N}\sum_{m_1,m_2}\sum_{\mathbf{q},\lambda}\left(g_{n_1m_1\mathbf{k}}^{\mathbf{q}\lambda}\right)^* g_{n_2m_2\mathbf{k}}^{\mathbf{q}\lambda} D_{\lambda}^{\lessgtr}(\mathbf{q},t_1,t_2;\beta)\, G_{m_1m_2\mathbf{k}-\mathbf{q}}^{\lessgtr}(t_1,t_2;\beta)\}, \tag{32}$$

with D_λ the phonon propagator that I will shortly define.

Now I introduce the complete set of approximations I will use to transform Eq.29 in a simpler set of equations of motion to describe, *ab–initio*, the carrier dynamics in realistic materials. These approximation are guided by simple physical arguments introduced with the aim of limiting the computation effort. This is a crucial requirement to make the simulations feasible. At the same time those approximations represent a first step in the device of a sound and complete framework for the *Ai*–NEGF approach and their validity will be carefully analyzed in the further development of this field.

The approximations are:

(i) The screened interaction entering Eq.30 are assumed to be static. I introduce this semi–static approximation (SSA) of the out–of–equilibrium version of the GWA being inspired by the well–known Coulomb-Hole plus Screened-Exchange approximation (COHSEX)[10]. The SSA reads:

$$W^{(r/a)}_{\mathbf{G}_1,\mathbf{G}_2}(\mathbf{q}, t_1, t_2; \beta) \approx W_{\mathbf{G}_1,\mathbf{G}_2}(\mathbf{q}) . \tag{33}$$

This approximation is well motivated in semiconductors and insulators as an electronic gap of 1 eV corresponds to a time scale of 0.66 fs. This a very short time compared to the time scale of the simulations presented in the present work.

(ii) The simulation is done at a fixed temperature β that is assumed to not change during the simulation. This approximation corresponds to take the phonon population at the equilibrium and assuming that their energy distribution does never deviate from a Bose distribution. In practice this means that:

$$D^{<}_{\lambda}(\mathbf{q}, t_1, t_2; \beta) \approx (-i) \sum_{\pm} N^{\pm}_{\mathbf{q}\lambda}(\beta) e^{\pm i\omega_{\mathbf{q}\lambda}(t_1 - t_2)} . \tag{34}$$

Note that $D^{>}_{\lambda}(\mathbf{q}, t_1, t_2; \beta) = -\left[D^{<}_{\lambda}(\mathbf{q}, t_1, t_2; \beta)\right]^*$.

(iii) I use the Generalized Baym–Kadanoff ansatz (GBKA) [17–19].

(iv) I neglect the off–diagonal matrix elements of the $\mathbf{G}^{\lessgtr}_{\mathbf{k}}$ and $\mathbf{\Sigma}^{\lessgtr}_{\mathbf{k}}$ matrices

$$G^{<}_{nn'\mathbf{k}}(T, T; \beta) \approx i\delta_{nn'} f_{n\mathbf{k}}(T; \beta) , \tag{35}$$

$$G^{>}_{nn'\mathbf{k}}(T, T; \beta) \approx i\delta_{nn'} \left[f_{n\mathbf{k}}(T; \beta) - 1\right] , \tag{35'}$$

$$\Sigma^{\lessgtr}_{nn'\mathbf{k}}(T, T; \beta) \approx \delta_{nn'} \Sigma_{n\mathbf{k}}(T; \beta) . \tag{35''}$$

As it will be showed in the following these approximations will make possible to perform the actual simulations of complex materials at a reasonable computational cost.

4.1. The collision operator in the Generalized Baym–Kadanoff ansatz

The GBKA provides the most important tool to disentangle the complex two–times evolution by rewriting:

$$G^{\lessgtr}_{n\mathbf{k}}(t_1, t_2; \beta) \approx (i) \left[G^{(r)}_{n\mathbf{k}}(t_1 - t_2; \beta) G^{\lessgtr}_{n\mathbf{k}}(t_2, t_2; \beta) - G^{\lessgtr}_{n\mathbf{k}}(t_1, t_1; \beta) G^{(a)}_{n\mathbf{k}}(t_1 - t_2; \beta)\right] , \tag{36}$$

that, in the lesser case, acquires a simple form written in terms of the time–dependent electronic occupations:

$$G^{<}_{n\mathbf{k}}(t_1, t_2; \beta) \approx -\left[G^{(r)}_{n\mathbf{k}}(t_1 - t_2; \beta) f_{n\mathbf{k}}(t_2; \beta) - f_{n\mathbf{k}}(t_1; \beta) G^{(a)}_{n\mathbf{k}}(t_1 - t_2; \beta)\right] . \tag{37}$$

A similar identity holds for the greater Green's function. By using Eq.36 the KBE's for the electronic occupations $f_{n\mathbf{k}}(T, \beta)$ acquire a closed form if a given analytic form for the advanced/retarded Green's functions is given. But before discussing the specific form of the (r/a) Green's functions I use Eq.36 to rewrite the two contributions to $\mathbf{S}$.

The algebraic passages to rewrite $\mathbf{S}^{e-p,<}$ by using the GBKA can be found, for example, in Ref.[51] in the simpler case of a Jellium model with e–p interaction treated using a Fröhlich interaction $g_{nn\mathbf{k}}^{\mathbf{q}\lambda} \approx g_{|\mathbf{q}|}$ and a single optical phonon mode.

After some involved but conceptually simple algebra it can be shown that

$$S_{n\mathbf{k}}^{e-p,<}(T;\beta) = N^{-1}\sum_{\mathbf{q}\lambda}\sum_{s=\pm} e^{is\omega_{\mathbf{q}\lambda}T}\left\{ S_{n\mathbf{k}}^{\mathbf{q}\lambda s}(T;\beta) + \left[S_{n\mathbf{k}}^{\mathbf{q}\lambda -s}(T;\beta)\right]^{*}\right\}, \tag{38}$$

with

$$\begin{aligned} S_{n\mathbf{k}}^{\mathbf{q}\lambda s}(T;\beta) = (-i)\int_{-\infty}^{T} d\tau\, e^{-is\omega_{\mathbf{q}\lambda}\tau} \sum_{m} |g_{nm\mathbf{k}}^{\mathbf{q}\lambda}|^2 G_{m\mathbf{k}-\mathbf{q}}^{(r)}(T-\tau;\beta)\times \\ \left[N_{\mathbf{q}\lambda}^{-s}(\beta)\left(2-f_{m\mathbf{k}-\mathbf{q}}(\tau;\beta)\right) f_{n\mathbf{k}}(\tau;\beta) - N_{\mathbf{q}\lambda}^{s}(\beta) f_{m\mathbf{k}-\mathbf{q}}(\tau;\beta)\left(2-f_{n\mathbf{k}}(\tau;\beta)\right)\right]\times \\ \left[G_{n\mathbf{k}}^{(a)}(\tau;\beta)\right]^{*}. \end{aligned} \tag{39}$$

As it will be clear in the next section the sum of $S_{n\mathbf{k}}^{\mathbf{q}\lambda s}$ and of its adjoin in Eq.38 is crucial to keep the time–evolution unitary in such a way that the number of electrons is conserved. The actual time decay of the photo–excited carriers will, then, be connected to the actual memory of the system linked with the τ integral appearing in Eq.38 rather that with the imaginary part of the self–energy, as commonly done in equilibrium theories. This detail marks the basic physical difference between the equilibrium QP linewidths and the out–of–equilibrium QP decay time, as it will be clear from the discussion in Sec.5.

In the case of the e–e scattering the application of the GBKA is more involved as it must be applied to three Green's functions instead of one, like in the e–p case. Nevertheless, once a simple form for $G^{(r/a)}$ will be given the physical interpretation of the collision operators will be equally simple. By following a procedure similar to the e–p case we find that

$$\begin{aligned} S_{n\mathbf{k}}^{e-e,<}(T;\beta) = \frac{1}{2(N\Omega)^2}\sum_{\mathbf{qp}}\sum_{m_1m_2m_3}\left|W_{\substack{nm_1\mathbf{k}\\ m_3m_2\mathbf{p}}}^{\mathbf{q}}\right|^2\int_{-\infty}^{T} d\tau \Big\{ \\ \Big[G_{m_1\mathbf{k}-\mathbf{q}}^{(r)}(T-\tau;\beta)\,G_{m_3\mathbf{p}}^{(r)}(T-\tau;\beta)\,G_{m_2\mathbf{p}-\mathbf{q}}^{(a)}(\tau-T;\beta)\,G_{n\mathbf{k}}^{(a)}(\tau-T;\beta) + \\ G_{m_1\mathbf{k}-\mathbf{q}}^{(a)}(\tau-T;\beta)\,G_{m_3\mathbf{p}}^{(a)}(\tau-T;\beta)\,G_{m_2\mathbf{p}-\mathbf{q}}^{(r)}(T-\tau;\beta)\,G_{n\mathbf{k}}^{(r)}(T-\tau;\beta)\Big]\times \\ \left[\left(2-f_{m_1\mathbf{k}-\mathbf{q}}(\tau;\beta)\right)\left(2-f_{m_3\mathbf{p}}(\tau;\beta)\right) f_{m_2\mathbf{p}-\mathbf{q}}(\tau;\beta)\, f_{n\mathbf{k}}(\tau;\beta) - \right. \\ \left. f_{m_1\mathbf{k}-\mathbf{q}}(\tau;\beta)\, f_{m_3\mathbf{p}}(\tau;\beta)\left(2-f_{m_2\mathbf{p}-\mathbf{q}}(\tau;\beta)\right)\left(2-f_{n\mathbf{k}}(\tau;\beta)\right)\right]\Big\}. \end{aligned} \tag{40}$$

Eq.38–40 represent the expressions for the time–dependent kernel $\mathbf{S}$ that I will further simplify in the next section. By using this expression for $\mathbf{S}$ the final equation of motion for the electronic occupation reads:

$$\frac{\partial}{\partial T} f_{n\mathbf{k}}(T;\beta) = \left.\frac{\partial}{\partial T} f_{n\mathbf{k}}(T;\beta)\right|_{coh} - S_{n\mathbf{k}}^{<}(T;\beta). \tag{41}$$

with

$$\left.\frac{\partial}{\partial T} f_{n\mathbf{k}}(T;\beta)\right|_{coh} = -\left[\mathbf{h}_{\mathbf{k}}^{KS} + \mathbf{V}_{\mathbf{k}}^{Hartree}[n(T;\beta)] + \mathbf{\Sigma}_{\mathbf{k}}^{s}(T;\beta) + \mathbf{U}_{\mathbf{k}}(T;\beta), \mathbf{G}_{\mathbf{k}}^{<}(T,T;\beta)\right]_{nn}. \tag{41'}$$

4.2. The Completed Collision Approximation

Thanks to the GKBA the KBE's can be rewritten as a set of non–linear differential equations for the occupation functions. Nevertheless the whole history of these functions should be integrated numerically in Eqs.39–40. In order to further simplify the theory I introduce two more approximations to the list presented in Sec.2:

(v) The $G^{(r/a)}$ functions are assumed to have a QP form (exponential):

$$G^{(r)}_{n\mathbf{k}}(T;\beta) \approx -ie^{-iE_{n\mathbf{k}}(\beta)T-\Gamma_n k(\beta)T}, \tag{42}$$

with $G^{(a)}_{n\mathbf{k}}(T;\beta) = \left[G^{(r)}_{n\mathbf{k}}(-T;\beta)\right]^*$. This means that we neglect any deviation from the pure exponential behavior induced by the short–time dynamics[52] and the e–e and e–p correlations[53]. In eq.42 the QP energies $E_{n\mathbf{k}}$ and widths $\Gamma_{n\mathbf{k}}$ are calculated *ab-initio* by means of the equilibrium GWA and Fan self–energies. The explicit expressions for $\Gamma_{n\mathbf{k}}$ are given by Eq.14 and Eq.25.

(vi) I adopt the so called Completed Collision Approximation (CCA)[51] that makes possible to solve the time–integration appearing in Eqs.39–40 analytically by assuming that the occupation functions vary only slowly on the time scale defined by $G^{(ra)}_{n\mathbf{k}}(T-\tau;\beta)$. This time scale is given by $\Gamma_{n\mathbf{k}}(\beta)^{-1}$ and, as shown in Fig.3, it is of the order of $10-20$ fs. Under this assumption the f functions can be taken out of the integrals appearing in Eqs.39–40 and the rest of the integral can be done analytically by using Eq.42.

Thanks to the CCA Eq.39 can be rewritten as:

$$\begin{aligned} S^{\mathbf{q}\lambda s}_{n\mathbf{k}}(T;\beta) \approx (-i)\sum_m |g^{\mathbf{q}\lambda}_{nm\mathbf{k}}|^2 \Big[& N^{-s}_{\mathbf{q}\lambda}(\beta)\left(2-f_{m\mathbf{k}-\mathbf{q}}(T;\beta)\right) f_{n\mathbf{k}}(T;\beta) - \\ & N^{s}_{\mathbf{q}\lambda}(\beta)\, f_{m\mathbf{k}-\mathbf{q}}(T;\beta)\left(2-f_{n\mathbf{k}}(T;\beta)\right)\Big] \times \\ & \left\{\left[\int_{-\infty}^{T} d\tau\, e^{-is\omega_{\mathbf{q}\lambda}\tau} G^{(r)}_{m\mathbf{k}-\mathbf{q}}(T-\tau;\beta)\left[G^{(a)}_{n\mathbf{k}}(\tau;\beta)\right]^*\right]\right\}. \end{aligned} \tag{43}$$

By using Eq.42 the τ integral reads:

$$\int_{-\infty}^{T} d\tau\, e^{-is\omega_{\mathbf{q}\lambda}\tau} G^{(r)}_{m\mathbf{k}-\mathbf{q}}(T-\tau;\beta)\left[G^{(a)}_{n\mathbf{k}}(\tau;\beta)\right]^* = \frac{e^{-is\omega_{\mathbf{q}\lambda}T}}{E_{m\mathbf{k}-\mathbf{q}}(\beta)-E^*_{n\mathbf{k}}(\beta)-is\omega_{\mathbf{q}\lambda}}. \tag{44}$$

Thanks to Eq.44 the $S^{\mathbf{q}\lambda s}_{n\mathbf{k}}(T;\beta)$ functions can be rewritten in terms of three simple lifetimes: $\gamma^{(e-p,h)}_{n\mathbf{k}}(T;\beta)$, $\gamma^{(e-p,e)}_{n\mathbf{k}}(T;\beta)$ and $\gamma^{(e-p,ph)}_{n\mathbf{k}}(T;\beta)$:

$$S^{\mathbf{q}\lambda s}_{n\mathbf{k}}(T;\beta)\Big|_{CCA} = -\gamma^{(e-p,h)}_{n\mathbf{k}}(T;\beta)\left[1-f_{n\mathbf{k}}(T;\beta)\right] + \gamma^{(e-p,e)}_{n\mathbf{k}}(T;\beta)\, f_{n\mathbf{k}}(T;\beta) + \gamma^{(e-p,ph)}_{n\mathbf{k}}(T;\beta). \tag{45}$$

In order to define and interpret the physical properties of these lifetimes I introduce two functions: P^{abs} and P^{emit}.

$$P^{(abs)}_{nn'\mathbf{kq}}(\beta) = \frac{|\Im E_{n\mathbf{k}}(\beta)| + |\Im E_{n\mathbf{k}-\mathbf{q}}(\beta)|}{\left[\Re\left(E_{n\mathbf{k}-\mathbf{q}}(\beta)-E_{n\mathbf{k}}(\beta)\right)-\omega_{\mathbf{q}\lambda}\right]^2 + \left[|\Im E_{n\mathbf{k}}(\beta)| + |\Im E_{n\mathbf{k}-\mathbf{q}}(\beta)|\right]^2}, \tag{46}$$

and

$$P^{(emit)}_{nn'\mathbf{kq}}(\beta) = \frac{|\Im E_{n\mathbf{k}}(\beta)| + |\Im E_{n\mathbf{k}-\mathbf{q}}(\beta)|}{\left[\Re\left(E_{n\mathbf{k}-\mathbf{q}}(\beta)-E_{n\mathbf{k}}(\beta)\right)+\omega_{\mathbf{q}\lambda}\right]^2 + \left[|\Im E_{n\mathbf{k}}(\beta)| + |\Im E_{n\mathbf{k}-\mathbf{q}}(\beta)|\right]^2}. \tag{47}$$

These two functions represent the energy constrain in the emission (P^{emit}) or absorption (P^{abs}) of a phonon smeared out due to the QP approximation for the retarded Green's function that. This, indeed, yields a Lorentzian indetermination in the energy conservation. By using these two simple functions the three e–p induced lifetimes can be easily rewritten as

$$\gamma_{n\mathbf{k}}^{(e-p,h)}(T;\beta) = \sum_{n'\mathbf{q}\lambda} \frac{\left|g_{nn'\mathbf{k}}^{\mathbf{q}\lambda}\right|^2}{N} P_{nn'\mathbf{kq}}^{(abs)}(\beta)\, f_{n'\mathbf{k}-\mathbf{q}}(T;\beta), \tag{48}$$

$$\gamma_{n\mathbf{k}}^{(e-p,e)}(T;\beta) = \sum_{n'\mathbf{q}\lambda} \frac{\left|g_{nn'\mathbf{k}}^{\mathbf{q}\lambda}\right|^2}{N} P_{nn'\mathbf{kq}}^{(emit)}(\beta)\left(2 - f_{n'\mathbf{k}-\mathbf{q}}(T;\beta)\right), \tag{48'}$$

$$\gamma_{n\mathbf{k}}^{(e-p,ph)}(T;\beta) = \sum_{n'\mathbf{q}\lambda} \frac{\left|g_{nn'\mathbf{k}}^{\mathbf{q}\lambda}\right|^2}{N} N_{\mathbf{q}\lambda}(\beta)\left[f_{n\mathbf{k}}(T;\beta) - f_{n'\mathbf{k}-\mathbf{q}}(T;\beta)\right]\left[P_{nn'\mathbf{kq}}^{(abs)}(\beta) + P_{nn'\mathbf{kq}}^{(emit)}(\beta)\right]. \tag{48''}$$

The dynamics induced by the e–p kernel on the electronic occupations can be easily visualized by using Eq.45 to simplify Eq.5 for the diagonal matrix elements of $G^<$. If I do not consider the coherent part of the KBE equation I get

$$\frac{\partial}{\partial T} f_{n\mathbf{k}}(\beta;T) = \gamma_{n\mathbf{k}}^{(e-p,h)}(T;\beta)\left[2 - f_{n\mathbf{k}}(T;\beta)\right] - \gamma_{n\mathbf{k}}^{(e-p,e)}(T;\beta)\, f_{n\mathbf{k}}(T;\beta) - \gamma_{n\mathbf{k}}^{(e-p,ph)}(T;\beta). \tag{49}$$

Eq.49 makes the interpretation of the carrier dynamics induced by the e–p coupling straightforward. We start indeed noticing that, simply because of their definition, $\gamma_{n\mathbf{k}}^{(e-p,h)}$ and $\gamma_{n\mathbf{k}}^{(e-p,e)}$ are always positive. In contrast $\gamma_{n\mathbf{k}}^{(e-p,ph)}$ can be positive or negative.

To see why this property is important let's consider two limiting cases when $f_{n\mathbf{k}}(\beta;T=0)=0$ or $f_{n\mathbf{k}}(\beta;T=0)=1$. In the first case, from Eq.49 it follows that, if we consider the zero temperature case,

$$f_{n\mathbf{k}}(\beta\to\infty;T=0) = 0 \Rightarrow \frac{\partial}{\partial T} f_{n\mathbf{k}}(\beta\to\infty;T=0) > 0, \tag{50}$$

$$f_{n\mathbf{k}}(\beta\to\infty;T=0) = 2 \Rightarrow \frac{\partial}{\partial T} f_{n\mathbf{k}}(\beta\to\infty;T=0) < 0. \tag{50'}$$

This simply means that an initially completely filled level will be emptied by the e–p interaction while an initially empty level will be filled. The $\gamma_{n\mathbf{k}}^{(e-p,ph)}$ is non zero only at finite temperature and works to build up a finite temperature quasi–equilibrium among the carrier occupations. The $\gamma_{n\mathbf{k}}^{(e-p,e/h)}$ lifetimes represent a single scattering of the state $|n\mathbf{k}\rangle$ to the state $|n\mathbf{k}-\mathbf{q}\rangle$ with emission/absorption of a phonon mode $|\mathbf{q}\lambda\rangle$. This process is similar to the one studied in the equilibrium case and, indeed, if we consider the Lorentzian function appearing in P^{abs} and P^{emit} as approximated delta functions we see that the analytic form of $\gamma_{n\mathbf{k}}^{(e-p,e/h)}$ are proportional to two terms appearing in Eq.25. Nevertheless, as we will clearly see in the case of bulk silicon, the dynamics induced by the KBE equation is a cascade process where many phonons are emitted or adsorbed and the final decay time of the electronic levels will be very different from the equilibrium case.

By using approximations (vi) and (vii) the e–e kernel, Eq.40, can be worked out in a way similar to the e–p case. Although the math is more involved the final outcome is quite similar to the e–p case. Indeed it turns out that

$$S^{e-e,<}_{n\mathbf{k}}(T;\beta)\Big|_{CCA} = -\xi^{(e-e,h)}_{n\mathbf{k}}(T;\beta)\left[2-f_{n\mathbf{k}}(T;\beta)\right] + \xi^{(e-e,e)}_{n\mathbf{k}}(T;\beta)\, f_{n\mathbf{k}}(T;\beta)\,. \quad (51)$$

The two additional lifetimes $\xi^{(e-e,e/h)}_{n\mathbf{k}}$ are defined to be:

$$\xi^{e-e,e}_{n\mathbf{k}}(T;\beta) = 4\,(N\Omega)^{-2}\sum_{\mathbf{qp}}\sum_{m_1m_2m_3}\left|W^{\mathbf{q}}_{\substack{nm_1\mathbf{k}\\ m_3m_2\mathbf{p}}}\right|^2 R^{\mathbf{q}}_{\substack{nm_1\mathbf{k}\\ m_3m_2\mathbf{p}}}\left(2-f_{m_1\mathbf{k}-\mathbf{q}}(T;\beta)\right)\left(2-f_{m_3\mathbf{p}}(T;\beta)\right) f_{m_2\mathbf{p}-\mathbf{q}}(T;\beta)\,, \quad (52)$$

and

$$\xi^{e-e,h}_{n\mathbf{k}}(T;\beta) = 4\,(N\Omega)^{-2}\sum_{\mathbf{qp}}\sum_{m_1m_2m_3}\left|W^{\mathbf{q}}_{\substack{nm_1\mathbf{k}\\ m_3m_2\mathbf{p}}}\right|^2 R^{\mathbf{q}}_{\substack{nm_1\mathbf{k}\\ m_3m_2\mathbf{p}}}\, f_{m_1\mathbf{k}-\mathbf{q}}(T;\beta)\, f_{m_3\mathbf{p}}(T;\beta)\left(2-f_{m_2\mathbf{p}-\mathbf{q}}(T;\beta)\right). \quad (52')$$

We have introduce the function R defined as:

$$R^{\mathbf{q}}_{\substack{nm_1\mathbf{k}\\ m_3m_2\mathbf{p}}} = \left[|\Im E_{m_1\mathbf{k}-\mathbf{q}}(\beta)| + |\Im E_{m_3\mathbf{p}}(\beta)| + |\Im E_{m_2\mathbf{p}-\mathbf{q}}(\beta)| + |\Im E_{n\mathbf{k}}(\beta)|\right] \left\{\left[\Re\left(E_{m_1\mathbf{k}-\mathbf{q}}(\beta) + E_{m_3\mathbf{p}}(\beta) - E_{m_2\mathbf{p}-\mathbf{q}}(\beta) - E_{n\mathbf{k}}(\beta)\right)\right]^2 + \left[|\Im E_{m_1\mathbf{k}-\mathbf{q}}(\beta)| + |\Im E_{m_3\mathbf{p}}(\beta)| + |\Im E_{m_2\mathbf{p}-\mathbf{q}}(\beta)| + |\Im E_{n\mathbf{k}}(\beta)|\right]^2\right\}^{-1}. \quad (53)$$

The physical interpretation of the e–e assisted carrier dynamics is very similar to the e–p case. The only difference is that, instead of phonons, the carrier decay is due to the creation of e–h pairs. The basic transition that contributes to the e–e scattering is a four bodies process: the initial electron (or hole) scatters in an e–h pair plus another electron (or hole). If we distinguish between the bath electrons and the optically pumped ones (the carriers) we see that these four bodies can, quite generally, be distributed in both group of states.

Nevertheless in the case of semiconductors and insulators it is meaningful to distinguish between intra–carrier and extra–carrier channels. In the first case the e–h pair is created among the optically pumped carriers while in the second case the pair is created across the gap involving the bath electrons. Of course these two channels, in general, contribute at the same time to the overall dynamics. But, as we will see in Sec.5 their relative strength depends on the carrier energy.

In conclusion the total rate equation governing the carrier dynamics that will be used in the Sec.5 is given by Eq.41 with $\mathbf{S}^<$ defined by Eqs.45 and 51. I coded this equation in the Yambo code[12], in the same numerical framework where the equilibrium GW and Fan self–energies are coded. The existing structure of Yambo has already produced several publications and it provides, therefore, the most appropriate environment to perform out–of–equilibrium simulations in a controllable manner. The e–p matrix elements and phonon modes are calculated, instead, with the Quantum Espresso package[11].

4.3. A detailed balance condition to ensure that the number of electrons is conserved
In the case of e–e scattering the actual possibility of creating high energy electron–hole (e–h) pairs can modify, during the real–time dynamics, the number of carriers that fill the conduction

bands with respect to the pairs initially photo–excited. This process is particularly evident for carriers created with energy larger than $2E_g$ below the VBM, or $2E_g$ above the CBM.

In the e–p case, instead, the maximum available energy that any carrier can dissipate during the relaxation is the Debye energy that, in the systems I consider in this work, is much smaller than the gap. In this case, indeed, it is important that the theory preserves the correct number of photo–excited electrons during the whole simulation.

I use this example to discuss a methodological aspect embodied in the use of a fully *ab–initio* framework. Indeed the e–p matrix elements $g^{\mathbf{q}\lambda}_{n'n\mathbf{k}}$, defined in Eq.23, can be written as

$$g^{\mathbf{q}\lambda}_{n'n\mathbf{k}} \equiv g^{\mathbf{k}-\mathbf{p}\lambda}_{n'n\mathbf{k}} = \sum_{s\alpha} \left(2M_s\omega_{\mathbf{k}-\mathbf{p}\lambda}\right)^{-1/2} e^{i(\mathbf{k}-\mathbf{p})\cdot\tau_s} \langle n'\mathbf{p}|\frac{\partial \hat{V}^{(s)}_{scf}(\mathbf{r})}{\partial R_{s\alpha}}|n\mathbf{k}\rangle_0 \xi_\alpha(\mathbf{k}-\mathbf{p},\lambda|s). \quad (54)$$

We can, now, consider the KBE for the two isolate states $|n\mathbf{k}\rangle$ and $|m\mathbf{p}\rangle$ that are scattered one into the other by means of the $S^{e-p,<}_{n\mathbf{k}}(T;\beta)$. For simplicity I only consider the dynamics induced by the e–p kernel at zero temperature and for a single phonon band μ. This dynamics can be rewritten in terms of four lifetimes: $\gamma^{(e/h)}_{n\mathbf{k}}, \gamma^{(e/h)}_{m\mathbf{p}}$:

$$\gamma^{(h)}_{n\mathbf{k}}(T) = \frac{\left|g^{\mathbf{k}-\mathbf{p},\mu}_{nm\mathbf{k}}\right|^2}{N} P^{(abs)}_{nm\mathbf{k}(\mathbf{k}-\mathbf{p})\mu} f_{m\mathbf{p}}(T), \quad (55)$$

$$\gamma^{(e)}_{n\mathbf{k}}(T) = \frac{\left|g^{\mathbf{k}-\mathbf{p},\mu}_{nm\mathbf{k}}\right|^2}{N} P^{(emit)}_{nm\mathbf{k}(\mathbf{k}-\mathbf{p})\mu} \left(2 - f_{m\mathbf{p}}(T)\right), \quad (55'')$$

$$\gamma^{(h)}_{m\mathbf{p}}(T) = \frac{\left|g^{\mathbf{p}-\mathbf{k},\mu}_{mn\mathbf{k}}\right|^2}{N} P^{(abs)}_{mn\mathbf{p}(\mathbf{p}-\mathbf{k})\mu} f_{n\mathbf{k}}(T), \quad (55'')$$

$$\gamma^{(e)}_{m\mathbf{p}}(T) = \frac{\left|g^{\mathbf{p}-\mathbf{k},\mu}_{mn\mathbf{k}}\right|^2}{N} P^{(emit)}_{mn\mathbf{p}(\mathbf{p}-\mathbf{k})\mu} \left(2 - f_{n\mathbf{k}}(T)\right). \quad (55''')$$

Now we notice that, thanks to simple symmetry arguments it can be shown that

$$g^{\mathbf{p}-\mathbf{k},\mu}_{mn\mathbf{k}} = \left[g^{\mathbf{k}-\mathbf{p},\mu}_{nm\mathbf{k}}\right]^*. \quad (56)$$

Moreover from Eqs.46–47 it follows that $P^{(emit)}_{nm\mathbf{k}(\mathbf{k}-\mathbf{p})\mu} = P^{(abs)}_{mn\mathbf{p}(\mathbf{p}-\mathbf{k})\mu}$ and, consequently, from Eqs.55 we see that:

$$\frac{\partial}{\partial T} f_{n\mathbf{k}}(T) = \frac{\left|g^{\mathbf{k}-\mathbf{p},\mu}_{nm\mathbf{k}}\right|^2}{N} P^{(emit)}_{nm\mathbf{k}(\mathbf{k}-\mathbf{p})\mu} \left(2 - f_{m\mathbf{p}}(T)\right)\left(2 - f_{n\mathbf{k}}(T)\right) - \frac{\left|g^{\mathbf{k}-\mathbf{p},\mu}_{nm\mathbf{k}}\right|^2}{N} P^{(abs)}_{nm\mathbf{k}(\mathbf{k}-\mathbf{p})\mu} f_{m\mathbf{p}}(T) f_{n\mathbf{k}}(T), \quad (57)$$

and

$$\frac{\partial}{\partial T} f_{m\mathbf{p}}(T) = \frac{\left|g^{\mathbf{k}-\mathbf{p},\mu}_{nm\mathbf{k}}\right|^2}{N} P^{(abs)}_{nm\mathbf{k}(\mathbf{k}-\mathbf{p})\mu} \left(2 - f_{m\mathbf{p}}(T)\right)\left(2 - f_{n\mathbf{k}}(T)\right) - \frac{\left|g^{\mathbf{k}-\mathbf{p},\mu}_{nm\mathbf{k}}\right|^2}{N} P^{(emit)}_{nm\mathbf{k}(\mathbf{k}-\mathbf{p})\mu} f_{m\mathbf{p}}(T) f_{n\mathbf{k}}(T). \quad (58)$$

Journal of Physics: Conference Series **427** (2013) 012003 doi:10.1088/1742-6596/427/1/012003

By summing Eq.57 and Eq.58 one gets

$$\frac{\partial}{\partial T}\left(2 - f_{m\mathbf{p}}(T) - f_{n\mathbf{k}}(T)\right) = \\ -\frac{\left|g_{nm\mathbf{k}}^{\mathbf{k}-\mathbf{p},\mu}\right|^2}{N}\left[P^{(abs)}_{nm\mathbf{k}(\mathbf{k}-\mathbf{p})\mu} + P^{(emit)}_{nm\mathbf{k}(\mathbf{k}-\mathbf{p})\mu}\right]\left(2 - f_{m\mathbf{p}}(T) - f_{n\mathbf{k}}(T)\right). \quad (59)$$

But from the boundary condition $(2 - f_{m\mathbf{p}}(T=0) - f_{n\mathbf{k}}(T=0)) = 0$ it follows $\frac{\partial}{\partial T}(2 - f_{m\mathbf{p}}(T) - f_{n\mathbf{k}}(T)) = 0$. The total number of electrons in the two levels made scattering by the e–p interaction is conserved, as it should be.

Eq.56 thus represents a sort of detailed balance condition that ensures that the fraction of charge lost by a given state $|n\mathbf{k}\rangle$ is exactly compensated by the charge added to the $|m\mathbf{p}\rangle$ state. In total there is no nonphysical charge lost or increase. Of course the level of accuracy of this detailed balance condition is ensured by the numerical precision under which Eq.56 is satisfied. This precision must be carefully checked and, eventually imposed, as on long–time simulations (of the order of picoseconds) tiny deviations can induce important deviations of the total number of electrons conservation.

5. An application in a paradigmatic solid: bulk silicon

After all the math and approximations introduced in the previous sections it is instructive to apply all the machinery composed by the KBE's written in the *ab–initio* framework to perform realistic simulations. The paradigmatic case I consider is bulk silicon. This a quite simple material that, nevertheless, has an interesting property: the minimal gap is indirect.

From the point of view of the carrier dynamics this is an important aspect. Indeed the optical gap is located at the Γ point and it is larger ($\sim 3.2\,eV$) of the indirect gap ($\sim 1.2\,eV$). Thus if an electron is optically pumped in the conduction states this will populate the energy region surrounding the CBM, that is the Γ_{15} state, and the relaxation is expected to occur along the ΓX line (see the upper frame of Fig.4) towards the X_1 point.

This has been, indeed, experimentally measured[54] by using Time–Resolved Two–Photon Photoemission (2PPE) and the time–decay of a level measured along the ΓX has been found to be around 40 fs.

I do not want, however, to interpret and reproduce the 2PPE experiment but to use bulk silicon to illustrate the relative balance of the e–e and e–p channels in the carrier dynamics on the basis of the initial carrier energy and density. I will also briefly discuss the effect of the scattering with acoustic phonons keeping in mind the large number of simulations done using simple optical–only phonon models.

The simulation is performed by defining an *ad–hoc* initial condition for the occupations and let the system evolve by means of Eq.5. In the first series of calculations I take

$$f_{n\mathbf{k}}(t=0;\beta) = 2 \quad \text{when} \quad |n\mathbf{k}\rangle = |\Gamma_{15}\rangle. \quad (60)$$

Two electrons (to account for the two spin channels) are moved by hand in two out of the three states that are energetically degenerate with the Γ_{15} stat. This corresponds to an initial carried density of $3.2 \times 10^{21}\,cm^{-3}$. As the Γ_{15} state is at the border of the $2E_g$ threshold its extra–carriers contribution to the e–e lifetime is very small. This is because the energy of the state is not enough to create e–h pairs across the minimum gap. The real–time occupations of three paradigmatic states: $|\Gamma_{15}\rangle$, $|L_1\rangle$ and $|X_1\rangle$ are showed in Fig.4. In the $\mathbf{k}$–grid I have used the X_1 state is the nearest to the CBM.

In Fig.4 the functions $f_{\Gamma_{15}}(T;\beta)$ (red curves), $f_{L_1}(T;\beta)$ (blue curves) and $f_{X_1}(T;\beta)$ (green curves) are showed comparing the dynamics with only the e–p scattering (continuous lines)

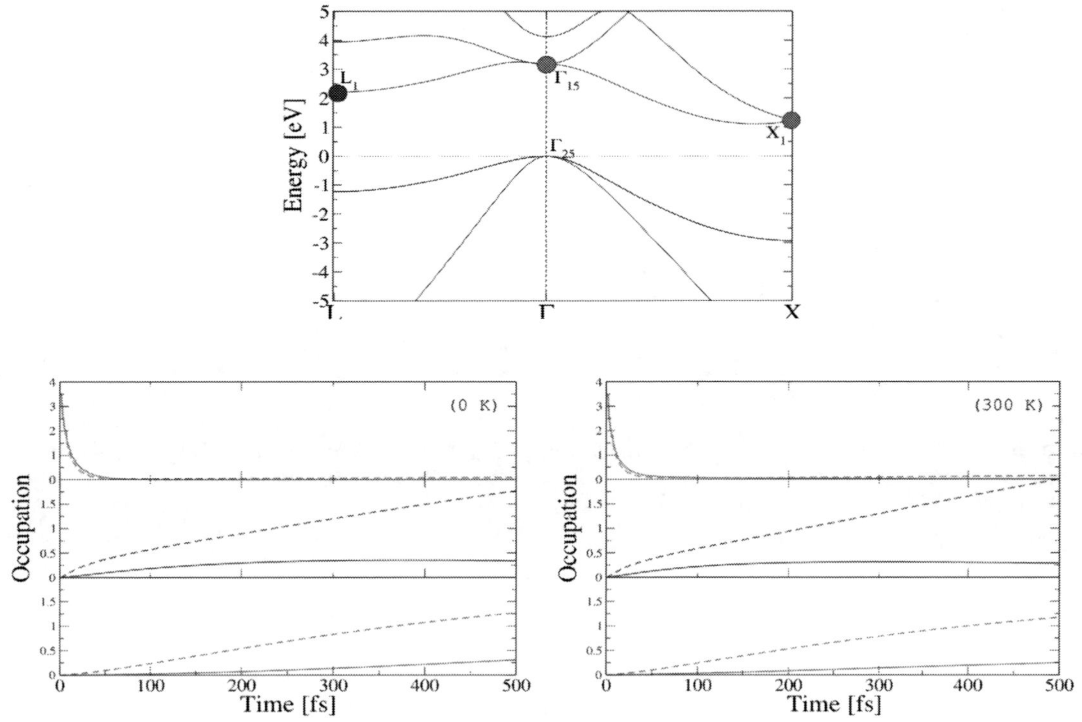

Figure 4. Time dependent occupations (lower frames) $f_{n\mathbf{k}}(T;\beta)$ as a function of the T for two external fixed temperature ($0\,K$ left frame and $300\,K$ right frame) for three selected states in the band–structure of Silicon (upper frame). At $T=0$ only the Γ_{15} is artificially populated with 4 electrons needed to fill 2 of the 3 energy degenerated states. This explains why the y–axis goes from 0 to 4. In the lower frames the dynamics is done including only e–p scattering (continuous lines) and e–e plus e–p scattering (dashed lines). In this case, as shown in Fig.3, the energy of the $|\Gamma_{15}\rangle$ state is below the $2E_g$ threshold. This means, indeed, that at $T=0$ the e–p scattering is largely dominant over the e–p one. But, as the time increases and the initial charge is fragmented in lower energy states the intra–carriers e–e scattering quickly increases leading to a much faster population of the X_1 state.

and with the e–p and the e–e scatterings (dashed lines) included. We immediately see some clear trends. The decay if the Γ_{15} is very fast, with an exponential time–dependence with a lifetime of the order of $12\,fs$. This value is very similar to the value of $16\,eV$ obtained by calculating $\Gamma^{e-p,eq}_{\Gamma_{15}}(\beta\to\infty)$ showing that the e–e contribution is practically zero. As the time increases the initial charge is fragmented and lower state are populated. At this point the intra–carriers contribution to the e–e is rapidly increased and the contribution to the filling is largely underestimated.

The message is thus clear. In the very short time the dynamics of carriers with energy below the $2E_g$ threshold is e–p dominated. Nevertheless in this regime the use of the CCA is highly questionable and a more careful full–memory treatment is required. As time increases also in this energy range the intra–carriers contribution to the carrier dynamics makes the e–e scattering dominant compared to the e–p channel.

An important aspect that must be taken in account in the analysis of the time–dependencies showed in Fig.4 is that the overall filling of the X_1 and L_1 states is enhanced when the e–e scattering is included also because of the promotion of additional electrons from the valence to the conduction by the creation of e–p pairs across the gap. This process, although energetically impossible for the Γ_{15} state, is made possible by the energy indetermination caused by the use

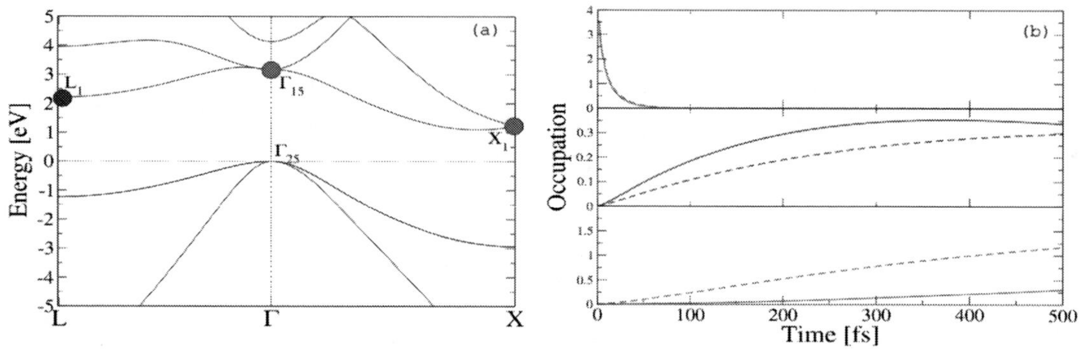

Figure 5. Time dependent occupations (right frame) $f_{n\mathbf{k}}(T;\beta)$ as a function of the T for an external vanishing temperature for three selected states in the band–structure of Silicon (left frame). At $T = 0$ only the Γ_{15} is artificially populated with 4 electrons needed to fill 2 of the 3 energy degenerated states. In the right frame the dynamics is done only including the e–p scattering but including only optical phonon modes (continuous lines) or all modes (dashed lines). We see that the effect of the acoustic modes is not at all negligible, especially for the X_1 state filling.

of the QP approximation for the retarded Green's function. Indeed thanks to the Lorentzian line–shape, the R functions (see Eq.53) can be non–zero for arbitrary values of the energy levels involved in the e–e scattering. This is, in my opinion, a stringent limitation imposed by the QP approximation that can be easily avoided by using more refined approximations or by calculating the equilibrium spectral functions numerically in the GW and Fan approximations.

By looking at the two lower frames of Fig.4 we see that the increase of the temperature from 0 to 300 K does not appreciably change the occupation dynamics in the first 500 fs. This is reasonable as the e–e lifetimes are not explicitly temperature sensitive and, as far as the e–p channel is considered, the Silicon Debye temperature is as large as 645 K. Thus at room temperature only the low–energy acoustic modes are populated. Although those contribute for half of the e–p lifetime the overall effect on the dynamics is marginal.

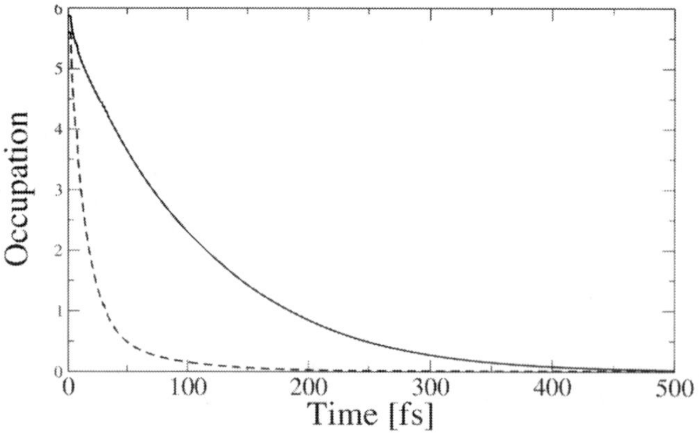

Figure 6. Time dependent occupation as a function of the T for the set of electronic states with energy 5.76 eV, well above the $2E_g$ threshold energy. The simulation is performed with the e–p scattering and including (dashed line) and not including (continuous line) the e–e contribution. For this high energy level the equilibrium e–e and e–p lifetimes are similar and both contribute to the decay of the state. This contrasts the Γ_{15} case where the e–p scattering dominates in the short–time dynamics.

An important issue that can be fully exploited by the present calculations is, indeed, the role played by the acoustic phonon modes that, in model calculations, are often neglected. As shown in Fig.5 the acoustic modes, even at zero temperature, contributes for more than half of the filling of the X_1 state. Their contribution to the Γ_{15} state is, instead, minor and it slowly increase as the energy of the state approaches the CBM. This phenomena is connected with the physics of both the e–e and e–p induced relaxation of the carriers. This is indeed a cascade process where the initial Γ_{15} electrons jump towards other states by the emission of, initially, a single phonon and a single e–h pair. This elementary process is very fast, of the order of even faster of 10 fs. But in order to reach the X_1 state the electron undergoes several of these elementary processes and the overall relaxation is a much slower process.

The last aspect I want to analyze by using bulk Silicon as paradigmatic system is the relative influence of the e–e and e–p scattering channels in the short–time regime for carriers pumped well above the $2E_g$ threshold. To this end I change the boundary condition give by Eq.60 to be

$$f_{n\mathbf{k}}(t=0;\beta) = 2 \quad \text{when} \quad E_{n\mathbf{k}} = 5.76\, eV \tag{61}$$

This corresponds to an initial carried density of $1.9 \times 10^{22}\, cm^{-3}$, an order of magnitude larger then in the Γ_{15} case. From Fig.3 we see that the equilibrium QP e–e and e–p lifetimes are very similar. And, indeed, the decay of the $5.76\, eV$ state is very much affected by the e–e scattering that, in this case, at short times is dominated by the extra–carriers channel.

6. Conclusions

In conclusion I have presented an *ab–initio* approach to the solution of the Baym–Kadanoff equations in realistic materials. The present approach is based on two major approximations: the generalized Baym–Kadanoff ansatz and the completed collision approximation. These makes the calculation feasible, as shown in the case of bulk silicon and provide a consistent and solid theoretical and numerical framework that can be applied to any kind of material that can be studied using the standard *ab–initio* Many–Body Perturbation Theory techniques. Indeed, a major point of the present approach is that all the presented theoretical and numerical tools has been coded in an existing long–term, stable and highly productive code: the Yambo code[12]. By using bulk silicon as a paradigmatic material I show that, in semiconductors and insulators, the presence of an electronic gap makes the e–p channel dominant over the e–e one in the short–time regime ($T \lesssim 100$ fs) and for low energy carriers injected or optically pumped above the conduction band minimum but below the $2E_g$ threshold. This is due to the fact that one of the most important kinematic channels in the e–e scattering is the one that involves e–h pairs created across the energy gap, and those pairs are energetically forbidden when the carriers are pumped with energy that is $2E_g$ below the CBM. It is important to stress that on the basis of the relative balance of the e–e and e–p scattering channels presented in this work it is not possible be draw general conclusions for all kind of materials. If on one side I am convinced that, even if the CCA weakens the results obtained in the very–short time scale, the arguments based on the energy of the initial carrier to motivate a potential bigger importance of the e–p channel is reliable for any kind of semiconductor or insulator. Of course the same argument does not apply to metals. On the other hand silicon is not a system known for its strong e–p coupling. There are systems like bulk Diamond[42], carbon nano–structures[43] or super–conductors[38] where the e–p scattering is enormously strong and this can strengthen the e–p channel over the e–e one.

7. Acknowledgements

Financial support was provided by the *Futuro in Ricerca* grant No. RBFR12SW0J of the Italian Ministry of Education, University and Research.

Journal of Physics: Conference Series **427** (2013) 012003 doi:10.1088/1742-6596/427/1/012003

References

[1] Krausz F and Ivanov M 2009 *Rev. Mod. Phys.* **81** 163–234

[2] Rossi F and Kuhn T 2002 *Rev. Mod. Phys.* **74** 895–950

[3] Goulielmakis E, Loh Z H, Wirth A, Santra R, Rohringer N, Yakovlev V S, Zherebtsov S, Pfeifer T, Azzeer A M, Kling M F, Leone S R and Krausz F 2010 *Nature* **466** 739

[4] Dreizler R M and Gross E K U 1990 *Density Functional Theory* (Springer-Verlag)

[5] Kohn W 1999 *Rev. Mod. Phys.* **71** 1253–1266

[6] Strinati G 1988 *Riv. Nuovo Cimento* **11** 1

[7] Marques M, Maitra N, Nogueira F, Gross E and Rubio A (eds) 2012 *Fundamentals of Time-Dependent Density Functional Theory* (Springer)

[8] Takimoto Y, Vila F D and Rehr J J 2007 *J. Chem. Phys.* **127** 154114

[9] Castro A, Appel H, Oliveira M, Rozzi C A, Andrade X, Lorenzen F, Marques M A L, Gross E K U and Rubio A 2006 *Phys. Status Solidi B* **243** 2465–2488

[10] Onida G, Reining L and Rubio A 2002 *Rev. Mod. Phys.* **74** 601–659

[11] Giannozzi P, Baroni S, Bonini N, Calandra M, Car R, Cavazzoni C, Davide C, Chiarotti G L, Cococcioni M, Dabo I, Corso A D, de Gironcoli S, Fabris S, Fratesi G, Gebauer R, Gerstmann U, Gougoussis C, Kokalj A, Lazzeri M, Martin-Samos L, Marzari N, Mauri F, Mazzarello R, Paolini S, Pasquarello A, Paulatto L, Sbraccia C, Scandolo S, Sclauzero G, Seitsonen A P, Smogunov A, Umari P and Wentzcovitch R M 2009 *J. Phys.: Cond. Matt.* **21** 395502

[12] Marini A, Hogan C, Grüning M and Varsano D 2009 *Comput. Phys. Commun.* **180** 1392–1403

[13] Chang E K, Shirley E L and Levine Z H 2001 *Phys. Rev. B* **65** 035205

[14] Leitsmann R, Schmidt W G, Hahn P H and Bechstedt F 2005 *Phys. Rev. B* **71** 195209

[15] Luppi E, Hübener H and Véniard V 2010 *Phys. Rev. B* **82** 235201

[16] Kadanoff L and Baym G 1962 *Quantum statistical mechanics* (W.A. benjamin)

[17] Haug H and Jauho A P 2008 *Quantum Kinetics in Transport and Optics of Semiconductors*

[18] Schafer W and Wegener M 2002 *Semiconductor optics and transport phenomena* (Springer)

[19] Bonitz M 1998 *Quantum Kinetic Theory* (B.G.Teubner Stuttgart. Leipzig)

[20] Attaccalite C, Grüning M and Marini A 2011 *Phys. Rev. B* **84** 245110

[21] Schmitt-Rink S, Chemla D S and Haug H 1988 *Phys. Rev. B* **37** 941–955

[22] Pereira M F and Henneberger K 1998 *Phys. Rev. B* **58** 2064–2076

[23] Henneberger K and Haug H 1988 *Phys. Rev. B* **38** 9759–9770

[24] Hannewald K, Glutsch S and Bechstedt F 2001 *Phys. Rev. Lett.* **86** 2451–2454

[25] Khler H, Kwong N and Yousif H A 1999 *Comput. Phys. Commun.* **123** 123 – 142

[26] Puig von Friesen M P, Verdozzi C and Almbladh C O 2009 *Phys. Rev. Lett.* **103** 176404

[27] Kwong N H and Bonitz M 2000 *Phys. Rev. Lett.* **84** 1768–1771

[28] Dahlen N E and van Leeuwen R 2007 *Phys. Rev. Lett.* **98** 153004

[29] Pal G, Pavlyukh Y, Hbner W and Schneider H C 2011 *EPJ B* **79** 327–334

[30] Gartner P, Bányai L and Haug H 2000 *Phys. Rev. B* **62** 7116–7120

[31] Sun B Y, Zhou Y and Wu M W 2012 *Phys. Rev. B* **85** 125413

[32] Vu Q T, Bányai L, Haug H, Camescasse F X, Likforman J P and Alexandrou A 1999 *Phys. Rev. B* **59** 2760–2767

[33] Jahnke F and Koch S W 1995 *Phys. Rev. A* **52** 1712–1727

[34] Bányai L, Vu Q T, Mieck B and Haug H 1998 *Phys. Rev. Lett.* **81** 882–885

[35] El Sayed K, Kenrow J A and Stanton C J 1998 *Phys. Rev. B* **57** 12369–12377

[36] Aulbur W G, Jonsson L and Wilkins J W 1999 *Solid State Physics (edited by H. Ehrenreich and F. Spaepen), Academic press* **54** 1 – 218

[37] Strinati G, Mattausch H J and Hanke W 1980 *Phys. Rev. Lett.* **45** 290–294

[38] Mitsuhashi R, Suzuki Y, Yamanari Y, Mitamura H, Kambe T, Ikeda N, Okamoto H, Fujiwara A, Yamaji M, Kawasaki N, Maniwa Y and Kubozono Y 2010 *Nature* **464** 76

[39] Allen P B and Cardona M 1983 *Phys. Rev. B* **27** 4760–4769

[40] Cardona M 2006 *Sci. Technol. Adv. Mater.* **7** S60–S66

[41] Marini A 2008 *Phys. Rev. Lett.* **101** 106405

[42] Giustino F, Louie S G and Cohen M L 2010 *Phys. Rev. Lett.* **105** 265501

[43] Cannuccia E and Marini A 2011 *Phys. Rev. Lett.* **107** 255501

[44] Baroni S, de Gironcoli S, Dal Corso A and Giannozzi P 2001 *Rev. Mod. Phys.* **73** 515–562

[45] Gonze X 1995 *Phys. Rev. A* **52** 1096–1114

[46] Mattuck R 1976 *A guide to Feynman diagrams in the Many-Body problem* (McGraw-Hill, New York)

[47] van Leeuwen R 2004 *Phys. Rev. B* **69** 115110

[48] Mahan G 1998 *Many-Particle Physics* ((New York: Plenum))

[49] Fan H 1950 *Phys. Rev.* **78** 808

[50] Allen P and Heine V 1976 *J. Phys. C* **9** 2305

[51] Haug H 1992 *Phys. Status Solidi B* **173** 139

[52] Bonitz M, Semkat D and Haug H 1999 *EPJ B - Condensed Matter and Complex Systems* **9** 309–314

[53] Haug H and Bnyai L 1996 *Solid State Commun.* **100** 303 – 306

[54] Ichibayashi T, Tanaka S, Kanasaki J, Tanimura K and Fauster T 2011 *Phys. Rev. B* **84** 235210

Progress in Nonequilibrium Green's Functions V (PNGF V) IOP Publishing
Journal of Physics: Conference Series **427** (2013) 012004 doi:10.1088/1742-6596/427/1/012004

Configuration-picture description of quantum-dot carrier scattering for the interaction with LO phonons

Alexander Steinhoff,[1] **Paul Gartner,**[1,2] **and Frank Jahnke**[1]
[1] Institute for Theoretical Physics, University of Bremen, 28334 Bremen, Germany
E-mail: www.itp.uni-bremen.de
[2] National Institute of Materials Physics, Bucharest-Magurele, Romania

Abstract.
A theory of scattering processes between the electronic configurations of a single quantum dot (QD) in interaction with LO phonons is presented. Polaronic renormalizations of the QD carriers are taken into account using a Green's function approach and play an essential role in the calculation of the scattering rates. Moreover, the rates are configuration-dependent, so that additional carriers in the QD, that do not take part in a transition, influence its efficiency. The QD carrier correlations are fully included by using a von Neumann-Lindblad formalism in the configuration basis. Numerical results for scattering rates and QD carrier dynamics are shown to illustrate the theory.

1. Introduction

Semiconductor quantum dots (QDs) provide a three-dimensional carrier confinement with atom-like discrete density of states. In the weak excitation regime, optical properties are determined by multi-exciton states. Single QD spectroscopy allows to follow the temporal dynamics of excitation and cascaded emission from various multi-exciton configurations [1]. This leads to the question to what extent the QD carrier kinetics can be described in terms of one-particle occupation probabilities, that are averaged over the QD ensemble and/or over repeated measurements, or whether the system needs to be analyzed in terms of many-body configurations. Related to this is the importance of carrier correlations and their inclusion in the theoretical models.

Due to the finite number of confined carriers in a QD, we are able to address the many-body dynamics of the QD carriers in terms of the von Neumann equation. This provides a description of carrier scattering processes in which the QD carriers enter via multi-exciton configurations. For Coulomb scattering processes, the QD carriers can be coupled to carriers in delocalized states via Lindblad terms in the Born-Markov

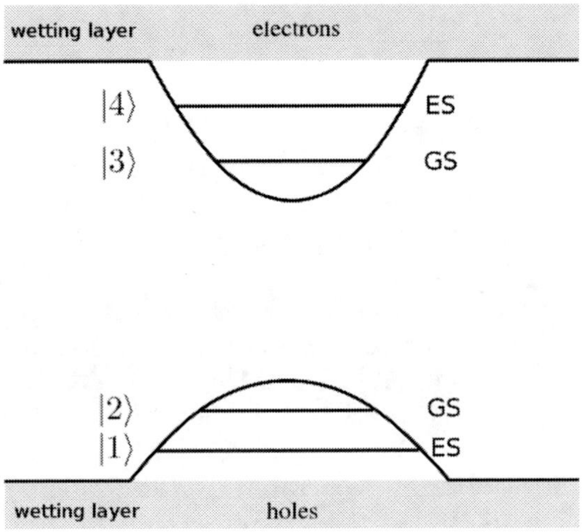

Figure 1: Schematic drawing of confined energy levels in the quantum-dot (QD) on wetting-layer (WL) system. The quasi-continuum of WL states (grey area) has larger interband transition energies than the discrete QD states for holes ($|1\rangle$, $|2\rangle$) and electrons ($|3\rangle$, $|4\rangle$).

approximation [2]. In this paper, we consider carrier scattering due to interaction with LO phonons, which can not be treated in the Born approximation. We overcome this problem by including polaronic quasiparticle properties of the QD carrier configurations, which are obtained from solutions for the retarded Green's functions.

Numerical results for the thus obtained rates of LO-phonon-assisted processes and for the dynamics of the QD system due to these processes, as obtained from the von Neumann-Lindblad (vNL) equation, are presented.

2. Model system

We consider a QD-WL system in the limit of a small QD density with the QD containing a confined ground state (GS) and first excited state (ES) in each band. To simplify the theoretical description, we assume that any excitation of carriers in the WL or the QD is spin-polarized and that spin-flip processes can be neglected on the time scales discussed in this paper. Hence only one spin-subsystem will be considered. Moreover, we assume that further degeneracies of the excited state are sufficiently lifted due to an asymmetric QD shape, which for instance results from strain effects. Thus, we end up with the simplest model system that contains all features we would like to illustrate, including two confined states for holes ($|1\rangle$, $|2\rangle$) and electrons ($|3\rangle$, $|4\rangle$), as shown in Fig. 1. The theory presented in the next section, however, is more general and allows for an arbitrary number of single-particle states in the QD.

Journal of Physics: Conference Series **427** (2013) 012004 doi:10.1088/1742-6596/427/1/012004

3. Theory

3.1. Configuration picture

Carrier kinetics is concerned with the time evolution of the populations f_i of QD single-particle states, given by the expectation values of the number operators $\hat{n}_i$. However, the populations carry only a partial information about the state of the QD, whose full description is provided by its density operator $\rho(t)$, acting in the many-particle Hilbert space spanned by all configurations $|I\rangle = |n_1 n_2 n_3 n_4\rangle\,, \quad n_i \in \{0, 1\}$. The configurations contain information about the simultaneous occupancies of each single-particle state in the QD. In the process of scattering of a QD carrier, the presence and distribution of "spectator" carriers, which do not take part in the transition, influence the scattering event through the Pauli principle, which limits the availability of unoccupied states. This way the exclusion principle induces "Pauli correlations" which modify the carrier dynamics. Additional correlations are generated by the Coulomb interaction between the QD carriers. All these effects are taken into account exactly in the configuration basis of the QD.

On the other hand, a description only in terms of single-particle expectation values, e.g. by a Boltzmann equation for f_i, represents an approximate treatment, as every carrier in the QD reacts only to an independently averaged "mean-field" distribution of collision partners. As we have discussed in Ref. [2], this corresponds to a Hartree-Fock-like factorization of many-particle expectation values in the equations of motion for the QD single-particle state occupancies, which amounts to neglecting certain correlations between the QD carriers. We also showed that, as expected, there are situations where the QD as a system with a small number of single-particle state populations can not be described properly in the single-particle picture. For these reasons, in what follows we address the QD-carrier dynamics using the configuration picture.

3.2. System-reservoir interaction

The QD system is described by the system Hamiltonian

$$H_S = H_0 + H_{\text{Coul}}^{\text{QD}}, \tag{1}$$

consisting of a "free" part

$$H_0 = \sum_i \hat{n}_i\, \varepsilon_i^0, \tag{2}$$

with the confinement energies ε_i^0, and an interacting part

$$H_{\text{Coul}}^{\text{QD}} = \frac{1}{2}\sum_{ijkl} V_{ijkl}\, a_i^\dagger a_j^\dagger a_k a_l\,, \tag{3}$$

which accounts for Coulomb interaction between the QD carriers and leads to shifts of the configuration energies as well as configuration mixing. The picture is simplified by noticing [3] that the most important terms in $H_{\text{Coul}}^{\text{QD}}$ are those which are diagonal in the configuration basis. In other words, the dominant Coulomb effects correspond to (direct

Journal of Physics: Conference Series **427** (2013) 012004 doi:10.1088/1742-6596/427/1/012004

and exchange) energy renormalizations of the configurations, while configuration mixing is negligible.

$$H_{\mathrm{Coul}}^{\mathrm{QD}} \approx H_{\mathrm{Coul}}^{\mathrm{diag}}, \quad \left(H_0 + H_{\mathrm{Coul}}^{\mathrm{diag}}\right)|I\rangle = \varepsilon_I\,|I\rangle\,. \tag{4}$$

As an open system, the QD interacts with carriers in the WL and with phonons. Carriers can be captured from the WL into the QD or released back with certain rates. Thus, the WL is part of the QD environment as a fermionic bath with a given temperature and chemical potential. Similarly, the phonon system constitutes a bosonic energy reservoir, taken at the same temperature. While in Ref. [2] we considered the system dynamics only due to Coulomb interaction with the WL reservoir, we will focus here on carrier-phonon interaction.

To obtain the evolution of the reduced density matrix $\rho(t)$ of the QD system (S) in contact with the reservoirs (R), we consider the evolution generated by the total Hamiltonian

$$H = H_S + H_R + H_{SR} \tag{5}$$

in the Born-Markov limit [4]. H_R is the reservoir Hamiltonian, describing the quasi-free WL carriers and phonons with energies $\varepsilon_{\boldsymbol{k}}$ and $\hbar\omega_{\boldsymbol{q}}$, respectively:

$$H_R = \sum_{\boldsymbol{k}} \varepsilon_{\boldsymbol{k}} a_{\boldsymbol{k}}^\dagger a_{\boldsymbol{k}} + \sum_{\boldsymbol{q}} \hbar\omega_{\boldsymbol{q}} b_{\boldsymbol{q}}^\dagger b_{\boldsymbol{q}}\,. \tag{6}$$

In polar semiconductors the dominant electron-phonon interaction is provided by the LO phonons. Therefore we limit our discussion to carrier-LO-phonon scattering, and accordingly we assume, as it is customary, a dispersionless spectrum $\omega_{\boldsymbol{q}} = \omega_{LO}$. The interaction between system and reservoir is described by the Hamiltonian

$$H_{SR} = \sum_{i,j,\boldsymbol{q}} g_{\boldsymbol{q}}^{i,j} a_i^\dagger a_j \left(b_{\boldsymbol{q}} + b_{-\boldsymbol{q}}^\dagger\right), \tag{7}$$

with the matrix elements $g_{\boldsymbol{q}}^{i,j}$ as given by the Fröhlich coupling [5].

Here we have to distinguish between two cases. First, when both electronic states i, j correspond to QD states, the term describes the $|j\rangle \to |i\rangle$ scattering inside the QD assisted by the emission or absorption of a phonon. This is a redistribution of carrier population, like e.g. the relaxation of an electron from the excited state into the ground state of the QD. The second case involves a QD state and a WL one, and corresponds to the capture from or reemission into the WL of a carrier, again assisted by phonons.

Expressing the interaction, Eq. (7), into the configuration basis is straightforward but requires some care. We begin by discussing the first case. The operator $a_i^\dagger a_j$ produces transitions between QD-configuration differing only by the occupancies of states i and j. Such transitions are of the form $|J\rangle \to |I\rangle$, in which $|J\rangle = a_j^\dagger\,|\Phi\rangle$ and $|I\rangle = a_i^\dagger\,|\Phi\rangle$, with $|\Phi\rangle$ a reference state which specifies the "spectator" carriers and in which, of course, the states i, j are left empty. Such a process is present in H_{SR} by a term $s_{IJ}\Gamma_{IJ}$, where $s_{IJ} = |I\rangle\langle J|$ represents the system and $\Gamma_{IJ} = \sum_{\boldsymbol{q}} g_{\boldsymbol{q}}^{i,j}(b_{\boldsymbol{q}} + b_{-\boldsymbol{q}}^{\dagger})$ represents the reservoir. A similar treatment holds for the terms in which the number of QD carriers is not conserved. For instance, in the case of a carrier capture $|I\rangle = a_i^{\dagger}|\Phi\rangle$, as before, but now $|J\rangle = |\Phi\rangle$ and the annihilation operator belongs to the reservoir factor $\Gamma_{IJ} = \sum_{\boldsymbol{q},\boldsymbol{k}} g_{\boldsymbol{q}}^{i,j} a_{\boldsymbol{k}}(b_{\boldsymbol{q}} + b_{-\boldsymbol{q}}^{\dagger})$.

Adding up all the terms leads to the Hamiltonian

$$H_{SR} = \sum_{IJ} s_{IJ}\Gamma_{IJ}\,, \tag{8}$$

whose Born-Markov treatment can proceed in the standard way [4]. The result is a vNL equation of the form

$$\frac{\partial}{\partial t}\rho = -\frac{i}{\hbar}[H_S,\rho] + \sum_{I,J}\frac{\gamma_{IJ}}{2}\left[2s_{IJ}\rho s_{IJ}^{\dagger} - s_{IJ}^{\dagger}s_{IJ}\rho - \rho s_{IJ}^{\dagger}s_{IJ}\right]. \tag{9}$$

The commutator part represents the quantum-mechanical evolution of the system, whereas the second part describes, by means of the so-called Lindblad terms, the irreversible, dissipative kinetics due to the contact with the reservoir. Our task is to calculate the $|J\rangle \to |I\rangle$ transition rates

$$\gamma_{IJ} = \frac{2}{\hbar^2}\mathrm{Re}\int_{-\infty}^{t} dt' e^{i\omega_{IJ}(t-t')}\langle\Gamma_{JI}(t)\Gamma_{IJ}(t')\rangle_R\,, \tag{10}$$

where $\hbar\omega_{IJ} = \hbar(\omega_J - \omega_I)$ is the difference of the corresponding configuration energies, and the time dependence of the operators corresponds to the interaction picture with respect to $H_S + H_R$.

As an example consider again the case in which both i and j are QD states. Then Eq. (10) reads

$$\gamma_{IJ} = \frac{2}{\hbar^2}\mathrm{Re}\int_{-\infty}^{t} dt' e^{i(\tilde{\omega}_j-\tilde{\omega}_i)(t-t')}\sum_{\boldsymbol{q}}|g_{\boldsymbol{q}}^{i,j}|^2\Big\{(1+N_{LO})e^{-i\omega_{LO}(t-t')} + N_{LO}e^{i\omega_{LO}(t-t')}\Big\}, \tag{11}$$

in which N_{LO} is the phonon number at the given temperature and $\hbar\tilde{\omega}_i = \varepsilon_i^{\Phi}$, $\hbar\tilde{\omega}_j = \varepsilon_j^{\Phi}$ are the carrier energies of the two states involved, calculated *in the presence* of the "spectator" carriers as specified in the reference configuration $|\Phi\rangle$. It becomes clear that, in the present model, the expressions for the transition rates are very similar to the single-particle case, the only difference being that the energies involved are Coulomb renormalized by the presence of the other carriers. Similarly, the rate of the capture from the WL into the QD state i is given by

$$\gamma_{IJ} = \frac{2}{\hbar^2}\mathrm{Re}\int_{-\infty}^{t} dt' e^{-i\tilde{\omega}_i(t-t')}\sum_{\boldsymbol{qk}} f_{\boldsymbol{k}} e^{i\omega_{\boldsymbol{k}}(t-t')}|g_{\boldsymbol{q}}^{i,\boldsymbol{k}}|^2\Big\{(1+N_{LO})e^{-i\omega_{LO}(t-t')} + N_{LO}e^{i\omega_{LO}(t-t')}\Big\}, \tag{12}$$

with $f_{\boldsymbol{k}}$ the thermal equilibrium occupancy of the WL state with energy $\hbar\omega_{\boldsymbol{k}}$, from which the capture took place.

The time integrations in the Eqs. (11) and (12) lead to strict energy-conserving δ-functions, requiring a perfect compensation by the reservoir of the energy $\hbar\omega_{JI}$ exchanged in the system transition. This is a consequence of the low-order perturbative treatment of the system-reservoir interaction. While this is not a limitation for transitions enabled by Coulomb scattering processes, the constant LO-phonon frequency makes LO-phonon-assisted transitions between the energetically discrete system configurations in general impossible. This is a well-known problem, referred to in the literature as the "phonon bottleneck" [6–9]. It was shown [10–12] that by taking into account higher order effects, non-vanishing transition rates can be obtained. A systematic way of doing this is provided by Green's function (GF) techniques.

In what follows we will use transition rates obtained by GF methods in the Lindblad terms of Eq. (9). In this way not only the bottleneck is circumvented but also the correlation effects appearing in the kinetics are correctly accounted for.

As stated above, except for the presence of the "spectator" carriers, the formalism describing each transition is similar to single-particle scattering. What is not taken into account is the fact that the phonons are not only providing scattering but are also changing the bare electrons into polarons. In order to include polaronic effects, the quantum kinetic approach [11] replaces systematically the simple exponentials, oscillating with the one-particle frequency $\widetilde{\omega}_i = \varepsilon_i^{\Phi}/\hbar$, by the polaronic retarded GFs, since the latter represent the carrier dressed with the LO-phonon interaction (polarons) in the same way as the exponentials correspond to the bare-particle GF. As a consequence, in the transition rates Eqs. (11) and (12) we change the exponentials according to

$$\frac{1}{i\hbar}e^{-i\widetilde{\omega}_i(t-t')} \quad \Longrightarrow \quad G_i^r(t-t'). \tag{13}$$

Of course, the corresponding retarded GF should be calculated in the presence of the same reference state $|\Phi\rangle$. This problem is addressed in more detail below.

3.3. Spectral problem

As indicated in the previous section, renormalizations, that are due to the interaction of the carriers with the phonon reservoir, are included by calculating the time evolution of the retarded GFs under the influence of a polaron self-energy. Moreover, Hartree-Fock shifts, that do not emerge from the interaction of QD carriers, but also from the interaction with WL carriers, shall be taken into account. The retarded GFs are defined as expectation values with respect to a certain reference configuration $|\Phi\rangle$:

$$\begin{aligned}
G_i^r(t-t') &= \theta(t-t')\Big[G_i^>(t-t') - G_i^<(t-t')\Big], \\
G_i^>(t-t') &= -\frac{i}{\hbar}\langle\Phi|\, a_i(t)a_i^\dagger(t')\,|\Phi\rangle\,, \\
G_i^<(t-t') &= \frac{i}{\hbar}\langle\Phi|\, a_i^\dagger(t')a_i(t)\,|\Phi\rangle\,.
\end{aligned} \tag{14}$$

Thus the spectrum of a single-particle state i is probed by creating an electron (a hole) in the reference configuration at one time, annihilating it at another time and projecting the result back on the reference configuration. To calculate all possible transition rates in the above discussed scheme, we have to solve the Dyson equation for retarded GFs of the QD-WL system with a given self-energy for each reference configuration, which means all configurations but that of a full QD.
The retarded GFs obey the spectral Kadanoff-Baym equation [13, 14]

$$\left(i\hbar\frac{\partial}{\partial t}-\tilde{\varepsilon}_i\right)G_i^r(t)=\delta(t)+\int_0^t dt'\Sigma_i^r(t-t')G_i^r(t') \tag{15}$$

with the self-energy

$$\Sigma_i^r(t-t')=\delta(t-t')\Sigma_i^{HF}+\theta(t-t')\Big[\Sigma_i^>(t-t')-\Sigma_i^<(t-t')\Big]. \tag{16}$$

Here, Σ_i^{HF} contains instantaneous screened Hartree-Fock contributions due to Coulomb interaction between all carriers in the QD-WL system and the propagators $\Sigma_i^>$ and $\Sigma_i^<$ are taken in RPA [13]:

$$\begin{aligned}\Sigma_i^>(t-t')&=\sum_j(1-f_j)G_j^r(t-t')D_{i,j}^>(t-t'),\\ \Sigma_i^<(t-t')&=-\sum_j f_jG_j^r(t-t')D_{i,j}^<(t-t'),\end{aligned} \tag{17}$$

with the phonon propagators in thermal equilibrium

$$\begin{aligned}D_{i,j}^{\gtrless}(t-t')&=\sum_{\boldsymbol{q}}|g_{\boldsymbol{q}}^{i,j}|^2 d^{\gtrless}(t-t'),\\ d^{\gtrless}(t-t')&=\Big\{(1+N_{LO})e^{\mp i\omega_{LO}(t-t')}+N_{LO}e^{\pm i\omega_{LO}(t-t')}\Big\}.\end{aligned} \tag{18}$$

The occupancies of WL states j are assumed to be the equilibrium values. If i is a QD single-particle state, the QD state occupancies in the self-energy depend on the reference state, in which the spectral problem is solved. On the other hand, if i is a WL single-particle state, the QD state occupancies are in principle given by the average over the occupancies of all QD configurations, which in general depends on time. Assuming that the influence of the QD ensemble on the reservoir is negligible, we completely neglect renormalizations of WL states due to QD state population. Thus the spectral problem can be solved for each reference configuration beforehand, independently of the carrier kinetics, and then used for the calculation of modified Lindblad rates. As the RPA self-energy (17) itself contains the full retarded GFs again, the Kadanoff-Baym equation (15) is solved self-consistently.

3.4. Scattering rates

In Section 3.2, we provided expressions for transition rates due to LO-phonon-assisted carrier capture and relaxation, as they result from the Born-Markov treatment of system-reservoir interaction. If we perform the replacement (13) in Eq. (12), we obtain the following expression for the rate of carrier capture into a QD state i from the WL:

$$\gamma_{IJ} = 2\mathrm{Re} \int_{-\infty}^{t} dt' \sum_{\boldsymbol{k}} G_i^r(t-t')(G_{\boldsymbol{k}}^r(t-t'))^* f_{\boldsymbol{k}} D_{\boldsymbol{k},i}^{>}(t-t'). \tag{19}$$

For the reverse process, $G^r(t-t')$ has to be replaced by $(G^r(t-t'))^*$ and $f_{\boldsymbol{k}}$ by $(1-f_{\boldsymbol{k}})$. Replacing the exponential factors in Eq. (11) yields the following expression for the rate of relaxation from a QD state j into another QD state i:

$$\gamma_{IJ} = 2\mathrm{Re} \int_{-\infty}^{t} dt' G_i^r(t-t')(G_j^r(t-t'))^* D_{i,j}^{>}(t-t'). \tag{20}$$

4. Numerical results

In this section, we present numerical results that are obtained from the solution of the spectral problem formulated in Eqs. (15)-(17) and the subsequent evaluation of the vNL equation (9) with scattering rates that are modified according to Eqs. (19) and (20).

4.1. Model parameters

For the numerical results in this paper, we consider an InGaAs system consisting of a flat lens-shaped QD on a WL. We assume effective electron and hole masses $m_e = 0.067\, m_0$ and $m_h = 0.15\, m_0$ [15], respectively, and the static dielectric constant $\varepsilon = 12.5$ as well as the high-frequency dielectric constant $\varepsilon_\infty = 10.9$ and the LO-phonon energy $\omega_{LO} = 36\,\mathrm{meV}$. A WL thickness of 2.2 nm and additional 2.1 nm QD height are used, leading to large subband energy spacing in z-direction, so that only the energetically lowest confined state in this direction is considered. The finite height of the confinement potential for electrons and holes is taken to be 350 and 170 meV, respectively, so that equal z-confinement wave functions for electrons and holes can be adopted. Furthermore, we assume equal QD in-plane wave functions for electrons and holes, where the in-plane confinement potential is modeled as a harmonic oscillator potential with oscillator length $l = 5.4\,\mathrm{nm}$. Using these assumptions as well as an envelope approximation, the wave functions of the combined QD-WL system and with them the Coulomb matrix elements can be constructed following the procedure described in [16], where a similar system geometry is adopted. The ground state energies for electrons (holes) are taken to be 80 meV (30 meV) and the excited state energies are taken to be 40 meV (15 meV) below the WL continuum.

4.2. Spectral functions

The first step of the numerical procedure is to solve the polaron problem in each reference configuration $|\Phi\rangle$ to obtain the time evolution of the polaronic retarded GFs of all QD and WL single-particle states. Although the GFs are used in their time-dependent form to calculate scattering rates, the spectral properties of the single-particle states can

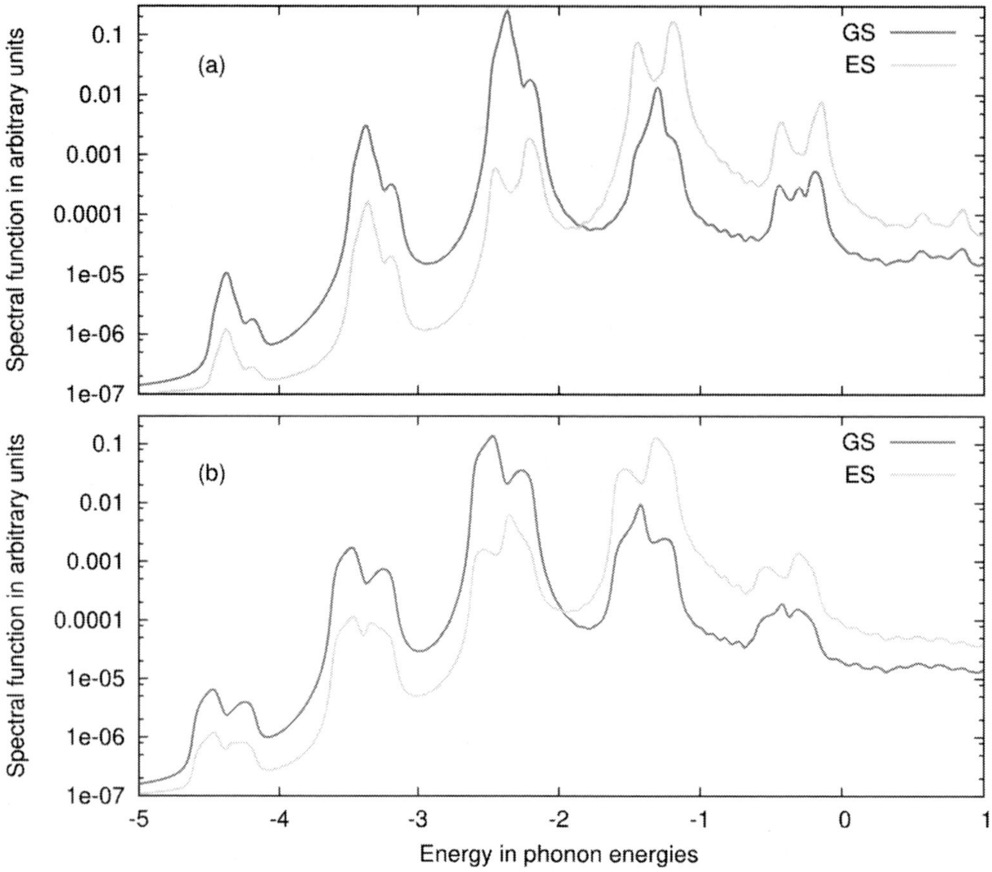

Figure 2: Spectral functions of the electronic ground state (GS) and excited state (ES) with (*a*) the empty QD ($|0X\rangle$) and (*b*) the ES exciton ($|1X_{ES}\rangle$) as reference state $|\Phi\rangle$ for a WL-carrier density of $n_{\mathrm{WL}} = 5 \cdot 10^{11}\,\mathrm{cm}^{-2}$ and a temperature $T = 300\,\mathrm{K}$.

be illustrated by means of their spectral functions, which are obtained as the Fourier transform of the retarded GFs:

$$\hat{G}_i(\omega) = -\frac{1}{\pi}\mathrm{Im}G_i^r(\omega). \tag{21}$$

As we consider only retarded GFs in thermal equilibirium, the spectral functions are not time-dependent. On the other hand, their shape clearly depends on the reference state in which the spectral problem is solved, as different QD state occupancies enter in the self-energy (17). In Fig. 2 we show spectral functions of the QD electron states for two different reference configurations. In both cases, the spectral functions exhibit the characteristic features of phonon replica and hybridisation that were discussed in literature [11]. Nevertheless, the actual form of these features is modified by the presence of additional carriers in the QD. The latter also cause Hartree-Fock shifts, which are reflected in the spectral positions of the main peaks but are small due to efficient screening by the WL carriers. These differences in the spectral properties of the QD states are reflected by the fact that the rate of a certain scattering process depends on the reference configuration it takes place in.

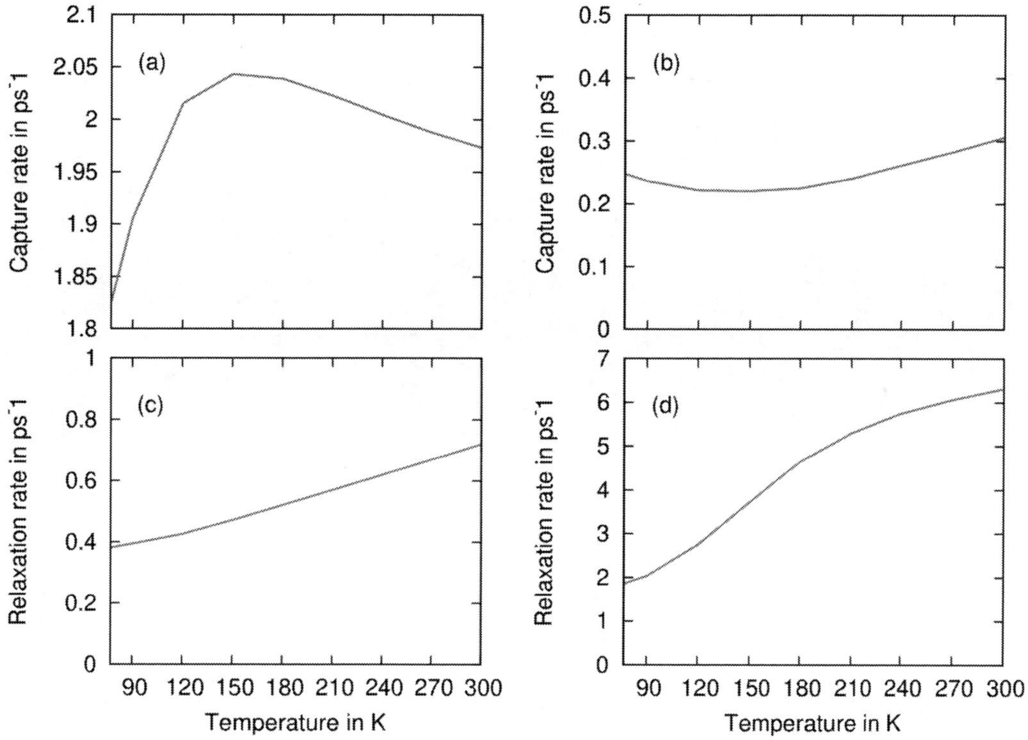

Figure 3: Scattering rates vs temperature, calculated with the empty QD as reference state $|\Phi\rangle$ for a WL-carrier density of $n_{\rm WL} = 5 \cdot 10^{11}\,\rm cm^{-2}$. (a): Capture rate into the hole excited state as given by Eq. (19), (b): Capture rate into the electron excited state as given by Eq. (19). (c): Relaxation rate from the hole excited state to the ground state as given by Eq (20). (d): Relaxation rate from the electron excited state to the ground state as given by Eq (20). Note the different ordinate scales.

4.3. Scattering rates

In this section, we present numerical results for the temperature dependence of selected phonon-assisted scattering rates, that are important in our model system, for a WL-carrier density of $n_{\rm WL} = 5 \cdot 10^{11}\,\rm cm^{-2}$. For the results shown in Fig. 3, we used retarded GFs that were calculated with the empty QD as reference state. As discussed in the previous section, a different reference state will in general modify the rates, but they are not likely to drastically change their overall behaviour.

It is shown that the capture rate for holes into the excited state is much larger than that for electron capture, as the hole states are energetically much closer to the WL continuum wherefrom the carriers are captured. Moreover, the electron capture rate depends stronger on temperature. The reason is, that in our model system, the electronic excited state is more than a phonon energy away from the WL continuum, so that electron capture relies on the spectral broadening of single-particle states, which is reduced at low

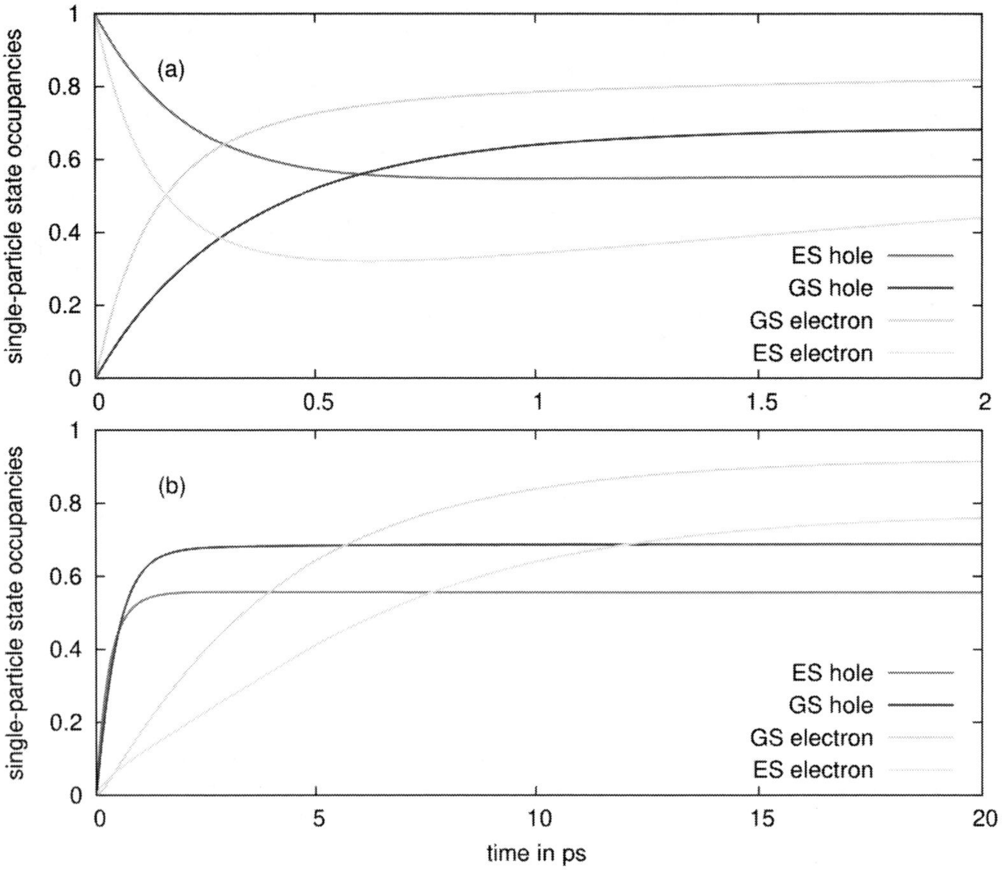

Figure 4: Time evolution of single-particle state occupancies in the QD at a temperature $T = 300\,\mathrm{K}$ in the presence of a WL-carrier density of $n_{\mathrm{WL}} = 5 \cdot 10^{11}\,\mathrm{cm}^{-2}$ for different initial conditions: (a) Excited state exciton. (b) Empty QD. Note the different time scales.

temperatures. Furthermore, the probability of phonon emission as an assisting process decreases with decreasing temperature. Both rates exhibit non-monotonous behaviour, which can be assigned to the fact that not only polaronic properties play a role here, but also other effects, that do not necessarily act in the same direction. These are the temperature dependence of the WL carrier population at the band edge, from where carriers are captured, as well as that of Hartree-Fock shifts. The relaxation rates of both electrons and holes decrease significantly with the temperature, as they do not depend on WL-carrier, but only on phonon populations as well as on spectral broadening. However, the electron relaxation is much faster than the hole relaxation, which is due to the fact that the energetic separation of electronic states in our model is much closer to the phonon energy than that of hole states.

4.4. Carrier dynamics

Finally, we use the renormalization-modified scattering rates to calculate the time evolution of the QD system given by the vNL equation. To demonstrate the efficiencies of different types of scattering processes, we consider two different initial situations. As a first example, the QD is prepared to be in a state of unity probability for the ES exciton, which is described as $\rho_0 = |1X_{ES}\rangle \langle 1X_{ES}| = |1001\rangle \langle 1001|$ in the configuration picture. The results are shown in Fig. 4(a). As expected, the occupancies of the excited electron and hole states decrease on a very short time scale due to fast carrier relaxation, whereas the GS is filled. While the hole ES reaches its thermal equilibrium occupancy after a ps, the electron ES is refilled on a much longer timescale due to slow carrier capture from the WL. As a second example, we use the empty QD as initial state, which is given by $\rho_0 = |0X\rangle \langle 0X| = |0000\rangle \langle 0000|$ in the configuration picture. The corresponding results are shown in Fig. 4(b). We see that while holes are captured into the QD and relax into the GS within a ps, the electron relaxation is strongly inhibited by the slow electron capture from the WL. As a consequence, the thermalization of the electrons is ten times slower than that of the holes. Note that the electron occupancies shown in Fig. 4(a) and (b) have not reached their thermal values after 2 ps and 20 ps, respectively. Nevertheless, in both examples, under von Neumann-Lindblad evolution the system relaxes to a thermal state, which is uniquely defined by the Lindblad rates and thus independent of the specific initial density matrix of the system. For low temperatures, one can expect a slower thermalization of carriers especially in the first situation, where relaxation is the dominant process. On the other hand, a low WL-carrier density will not change much in the first situation, while thermalization of an initially empty QD will become extremely slow.

5. Conclusions

We have presented a method to combine the description of a single QD in the configuration picture in terms of the vNL equation with many-particle renormalizations, that are calculated in a GF approach. Thus we fully take into account correlations between QD carriers and at the same time include polaronic properties, that arise due to their interaction with a phononic bath. This enables us to obtain finite rates for LO-phonon-assisted scattering processes between QD configurations, which is not possible in the usual Born-Markov treatment of system-reservoir interaction due to the discreteness of the QD spectrum. Moreover, the rates are configuration-dependent, so that additional carriers in the QD, that do not take part in a transition, influence its efficiency. We have shown numerical results for the temperature dependence of selected scattering rates as well as for the time evolution of a QD system due to phonon-assisted scattering processes for different initial conditions. We found a strong temperature dependence of carrier relaxation rates, while carrier capture rates are more or less constant. Moreover, it shows that the capture from the WL continuum represents a decelerating factor for the thermalization of electrons, which is much faster after an initial excitation of the QD.

The numerical results illustrate how the "phonon bottleneck" is efficiently circumvented by using a GF approach to include higher-order effects in the treatment of system-reservoir coupling.

Acknowledgments

This work was supported by the Deutsche Forschungsgemeinschaft.

References

[1] Laucht A, Kaniber A, Mohtashami A, Hauke N, Bichler M and Finley J J 2010 *Phys. Rev. B* **81** 241302

[2] Steinhoff A, Gartner P, Florian M and Jahnke F 2012 *Phys. Rev. B* **85** 205144

[3] Baer N, Gies C, Wiersig J and Jahnke F 2006 *Eur. Phys. J. B* **71** 411

[4] Carmichael H J 1999 *Statistical Methods in Quantum Optics 1* (Berlin Heidelberg New York: Springer)

[5] Mahan G D 2000 *Many-Particle Physics* 3rd ed (New York, Boston, Dordrecht: Kluwer Academic/Plenum Publishers)

[6] Benisty H, Sotomayor-Torrès C M and Weisbuch C 1991 *Phys. Rev. B* **44**(19) 10945–10948

[7] Urayama J, Norris T B, Singh J and Bhattacharya P 2001 *Phys. Rev. Lett.* **86** 4930

[8] Minnaert A W E, Silov A Y, van der Vleuten W, Haverkort J E M and Wolter J H 2001 *Phys. Rev. B* **63**(7) 075303

[9] Xu S, Mikhailovsky A A, Hollingsworth J A and Klimov V I 2002 *Phys. Rev. B* **65**(4) 045319

[10] Inoshita T and Sakaki H 1992 *Phys. Rev. B* **46**(11) 7260–7263

[11] Seebeck J, Nielsen T R, Gartner P and Jahnke F 2005 *Phys. Rev. B* **71** 125327

[12] Král K and Khás Z 1998 *Phys. Rev. B* **57**(4) R2061–R2064

[13] Haug H and Jauho A P 1998 *Quantum Kinetics in Transport and Optics of Semiconductors* (Berlin: Springer)

[14] Danielewicz P 1984 *Ann. Phys.* **152** 239

[15] Wimbauer T, Oettinger K, Efros A L, Meyer B K and Brugger H 1994 *Phys. Rev. B* **50(12)** 8889

[16] Nielsen T R, Gartner P and Jahnke F 2004 *Phys. Rev. B* **69** 235314

Progress in Nonequilibrium Green's Functions V (PNGF V) IOP Publishing
Journal of Physics: Conference Series **427** (2013) 012005 doi:10.1088/1742-6596/427/1/012005

Dielectric breakdown of Mott insulators – doublon production and doublon heating

Martin Eckstein[1] **and Philipp Werner**[2]

[1] Max Planck Research Department for Structural Dynamics, University of Hamburg, CFEL, Hamburg, Germany
[2] Department of Physics, University of Fribourg, 1700 Fribourg, Switzerland

E-mail: `martin.eckstein@mpds.cfel.de`

Abstract. Using dynamical mean-field theory and the non-crossing approximation as impurity solver, we study the response of a Mott insulator to strong dc electric fields. The breakdown of the Mott insulating state is triggered by field-induced creation of doublon-hole pairs. In a previous investigation, Ref. [1], it was found that the system approaches a long-lived quasi-steady state in which the current is time-independent although the number of carriers constantly increases. Here we investigate and clarify the nature of this state, which exists only because thermalization is slow in the Hubbard model at strong coupling. The current is time-independent because doublons and holes have an infinite temperature distribution. Evidence for this fact is obtained from spectral functions and by comparing the electric current with the field-induced doublon-hole creation rate. Implications to real experiments, in systems with energy dissipation, are discussed.

1. Introduction

The study of nonequilibrium phenomena in complex strongly correlated systems may reveal unexpected physics and eventually lead to new ways of tuning and controlling material properties on ultrafast timescales. The Mott insulator, a system with partially filled bands in which electrons are localized due to the strong Coulomb repulsion, is an ideal example of a correlated many-body state, and Mott insulators are common among transition metal oxides or organic charge-transfer salts [2]. A possible way to drive these systems out of equilibrium is to expose them to a strong (dc) electric field F. The resulting dc response, which is non-perturbative in the field and thus cannot be explained by response functions of the equilibrium state, is known as the dielectric breakdown. A considerable amount of work in the previous few years has been devoted to understand this conceptually simple, and yet theoretically challenging phenomenon. Experimentally, nonlinear transport in correlated insulators has been studied in both oxides [3, 4] and in organic materials [5, 6]. One observes a strong non-linearity in the current-field (j-F) characteristics, with a negative differential resistivity between weak current and large current regimes. The "strong-field physics" of correlated materials can also be addressed in experiments with ultra-cold atoms in optical lattices, e.g., as an elementary probe of the Mott phase [7].

In an intuitive picture, the Mott insulating ground state decays through the production of doublon-hole pairs in the presence of an electric field: In a scalar potential gauge, the field implies a potential energy difference F between two neighboring lattice sites in the field-direction (choosing units with lattice spacing $a = 1$ and electron charge $e = 1$). In the half-filled Mott insulator, with unit occupancy at every site, an electron can thus gain the energy U from the field which is needed to form a doublon (doubly occupied site) and leave behind a hole, by tunneling over a distance $\ell_U = U/F$. In this picture, the tunneling process alone contributes a current of the order

$$\Gamma_{\rm dh}(t) = \ell_U \dot{d}(t), \tag{1}$$

where $\dot{d}$ is the rate of doublon-hole (dh) pair production. In addition, doublons and holes are expected to carry a current after they have been created. We will refer to $\Gamma_{\rm dh}$ as the dh-creation current. Although its definition may seem ad-hoc here, $\Gamma_{\rm dh}$ will turn out to be very useful in the analysis of the data below.

For $F \ll 1$ the tunneling is over many lattice sites and thus occurs at an exponentially small rate. An exponential scaling with a threshold field $F_{\rm th}$ for the electric current,

$$j \propto F \exp\Big(-\frac{F_{\rm th}}{F}\Big), \tag{2}$$

and for observables related to the doublon-production rate (with different powers of F in the pre-factor) is indeed found quite generically, for the fermionic Hubbard model in one dimension [8, 9, 10, 11, 12, 13, 14], in infinite dimensions [1], and also for the bosonic Mott insulator [15]. A lot of work aimed at understanding the dependence of $F_{\rm th}$ on the charge gap Δ_c. Motivated by the dielectric breakdown of band insulators [9], which can be explained by the Landau-Zener mechanism [16], Oka *et al.* proposed a description of the Mott breakdown in the one-dimensional Hubbard model in terms of Landau-Zener tunneling between many-body eigenstates [8, 9, 12, 13, 17]. Both numerical work [8, 9] using exact diagonalization and density-matrix renormalization group (DMRG), and an analytical approach involving the ground state and the first excited state of the Bethe ansatz solution [12, 13] then suggest a relation $F_{\rm th} \sim \Delta_c^2$ for small U. Another analytical approach is the solution for the case of one spin-$\downarrow$ electron in a spin-$\uparrow$-polarized background, for which one finds $F_{\rm th} \propto \Delta_c^{3/2}$ [14]. DMRG results suggest that this behavior holds down to smaller polarization (the Mott insulator being the unpolarized case). Furthermore, if the voltage drop is applied over one lattice site rather than linearly, one finds $F_{\rm th} \propto \Delta_c$ [11]. In the infinite-dimensional case [1], $F_{\rm th}$ increases with Δ_c, but the regime close to the metal insulator transition ($\Delta_c \to 0$) has so far not been studied.

In the present paper we do not focus on the value of $F_{\rm th}$ close to the metal-insulator transition, but we want to clarify another fundamental question that arises in connection with studies of the dielectric breakdown: Most of the theories mentioned above involve isolated systems. After the field is switched on, one observes the emergence of a quasi-steady state in which doublons are produced at a more or less time-independent rate, and dc properties of the system are obtained from this quasi-steady state. In particular, also the electric current becomes time-independent [1, 10, 18], which seems quite peculiar if doublons and holes, whose number is constantly increasing, are interpreted as "charge carriers". The main purpose of the present paper is to clarify the nature of this quasi-steady state: We demonstrate that its existence is related to a failure of the quasi-equilibrium description, i.e., a lack of thermalization (Sec. 3.2). We furthermore clarify the relation of the electric current and the doublon production rate, and provide an explanation why the doublons and holes which are generated by the field appear to be immobile and do not contribute to the current (Sec. 5). This also leads to a better understanding of how temperature (Sec. 5) or generic energy dissipation mechanisms (Sec. 6) influence the dielectric breakdown, which is essential for understanding (and designing) experiments that can probe the non-perturbative effects of Eq. (2) (Sec. 7). Our analysis builds on the results of Ref. [1], in which the dielectric breakdown was studied in the limit of large dimensions [19], using dynamical mean-field theory (DMFT) [20].

Journal of Physics: Conference Series **427** (2013) 012005 doi:10.1088/1742-6596/427/1/012005

2. Model and methods

Throughout this paper we study the paramagnetic Mott insulating phase in the half-filled Hubbard model on a d-dimensional cubic lattice,

$$H = \sum_{\langle ij\rangle\sigma} V_{ij}(t)\, c_{i\sigma}^{\dagger} c_{j\sigma} + U \sum_{i} (n_{i\uparrow} - \tfrac{1}{2})(n_{i\downarrow} - \tfrac{1}{2}). \tag{3}$$

Here $c_{i\sigma}^{\dagger}$ ($c_{i\sigma}$) is the creation (annihilation) operator for an electron with spin σ at lattice site $\boldsymbol{R}_i$, U denotes the Coulomb repulsion, and V_{ij} is the matrix element for hopping between nearest neighbor sites i and j. We initially prepare the system in thermal equilibrium at temperature $T = 1/\beta$, and apply a homogeneous electric field $\boldsymbol{F}(t)$ for time $t > 0$. For convenience, $\boldsymbol{F}$ is always pointing along the body diagonal $\hat{\boldsymbol{\eta}} = (1 \dots 1)^t$, which simplifies the DMFT self-consistency (see below) [21]. The field is turned on to a value F within a switching time t_0 with a given ramp profile $r(x)$,

$$\boldsymbol{F}(t) = \hat{\boldsymbol{\eta}}\, F\, r(t/t_0), \tag{4}$$

$$r(x) = \begin{cases} \frac{1}{2} - \frac{3}{4}\cos(\pi x) + \frac{1}{4}\cos(\pi x)^3 & \text{for } 0 \le x \le 1 \\ 1 & \text{for } x \ge . \end{cases} \tag{5}$$

The smooth turn-on of F damps transient currents, but does not influence the long-time behavior [1]. To incorporate the field into Eq. (3) we use a gauge with pure vector potential $\boldsymbol{A}(t)$, i.e., $\boldsymbol{F}(t) = -\partial_t \boldsymbol{A}(t)/c$, for which the hopping matrix elements acquire a Peierls phase,

$$V_{ij}(t) = V_{ij}^{0} e^{ie(\boldsymbol{R}_j - \boldsymbol{R}_i)\boldsymbol{A}(t)/\hbar c}. \tag{6}$$

(For a discussion of various gauges within a tight-binding model, see Ref. [22].) In the limit $d = \infty$ [19], with rescaled hopping $V_{ij}^{0} = t^*/2\sqrt{d}$, the problem can be solved exactly using DMFT [20] in its nonequilibrium (Keldysh) formulation [23, 24]. The rescaled nearest neighbor hopping t^* is used as the unit of energy, such that the density of states is given by $\rho(\epsilon) \propto \exp(-\epsilon^2)$. Time and field are measured in units of $\hbar/t^*$ and t^*/ea, respectively ($\hbar = 1$, $e = 1$, $a = 1$).

The DMFT single-site problem is solved by means of the self-consistent hybridization expansion [25]. The entire set of equations for the geometry considered here has been discussed in Ref. [26], and details of the numerical implementation of the Keldysh equations are given in Ref. [27]. In the present work we use the lowest order of the strong-coupling solver, or non-crossing approximation (NCA) [28], which allows us to address certain issues that require long simulation times. In order to assess the validity of this approximation, we compare these results in Sec. 3.1 to the results of Ref. [1], which were obtained using the second order, or one-crossing approximation (OCA) [29].

Observables

From the DMFT solution we directly evaluate [27] expectation values of the double occupancy $d = \langle \frac{1}{L} \sum_i \hat{n}_{i\uparrow}\hat{n}_{i\downarrow} \rangle$ and the kinetic energy $E_{\text{kin}} = \langle \frac{1}{L} \sum_{\boldsymbol{k}\sigma} \epsilon_{\boldsymbol{k}+\boldsymbol{A}(t)} \hat{n}_{\boldsymbol{k}\sigma} \rangle$, where the band energy $\epsilon_{\boldsymbol{k}}$ is the Fourier transform of V_{ij}^{0}. The current is given by $\boldsymbol{j} = \langle \frac{1}{L} \sum_{\boldsymbol{k}\sigma} \boldsymbol{v}_{\boldsymbol{k}+\boldsymbol{A}(t)} \hat{n}_{\boldsymbol{k}\sigma} \rangle$, with the band velocity $\boldsymbol{v}_{\boldsymbol{k}} = \partial_{\boldsymbol{k}} \epsilon_{\boldsymbol{k}}$. One can then verify the general fact that the total internal energy

$$E_{\text{tot}} = E_{\text{kin}}(t) + U d(t) \tag{7}$$

changes according to the equation

$$\dot{E}_{\text{tot}}(t) = \boldsymbol{j}(t) \boldsymbol{F}(t). \tag{8}$$

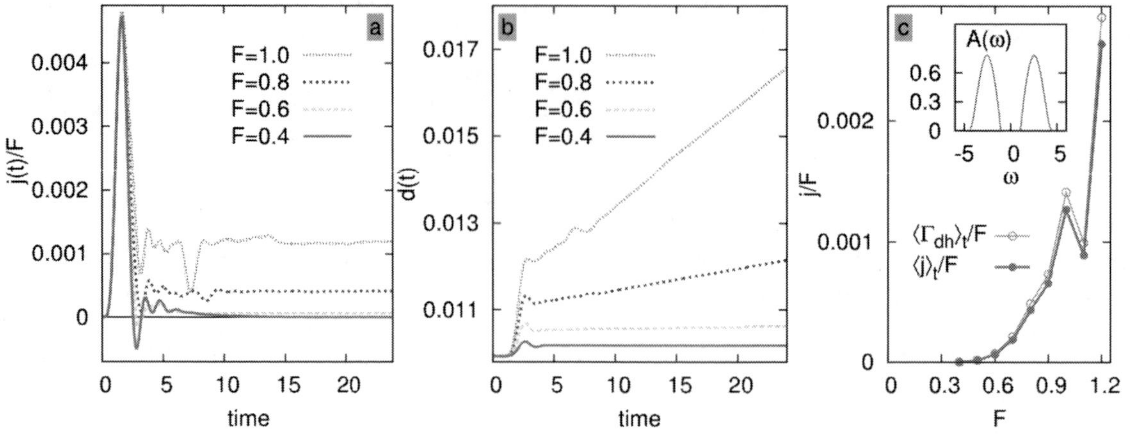

Figure 1. (a) Time-dependent current $j(t)$ for $U = 5$ and $\beta = 5$ and various electric fields F. (b) Double occupancy for the same parameters. (c) Current $\langle j \rangle_t$, averaged over times $40 < t < 50$. The curve $\langle \Gamma_{\text{dh}} \rangle_t$ (open symbols) shows corresponding data for the doublon current Eq. (1), see Sec. 5. *Inset:* Spectral function for the equilibrium state at $U = 5$ and $\beta = 5$.

Because the self-consistent strong-coupling solver is conserving in the sense of Kadanoff and Baym, Eq. (8) is satisfied also for the approximate solution.

In addition to the static observables, we define a spectral function from the Fourier transform of the retarded local Green function $G^{\text{ret}}(t,t') = -i\Theta(t-t')\langle\{c(t), c^\dagger(t')\}\rangle$,

$$A(\omega,t) = -\frac{1}{\pi}\text{Im}\int_0^{s_{\max}} ds\, e^{i\omega s}\, G^{\text{ret}}(t+s,t). \tag{9}$$

For a general non-equilibrium situation, $A(\omega,t)$ is not necessarily positive. However, when the t-dependence of $A(\omega,t)$ can be neglected, $A(\omega,t)$ is positive, and it can be related to photoemission and inverse photoemission spectra in the usual way [30]. In a quasi-steady state, as we will encounter below, these properties remain as long as the t-dependence of $A(\omega,t)$ is very slow compared to the inverse of the scale $\Delta\omega$ on which $A(\omega,t)$ changes as a function of ω. Technically, a finite cutoff $s_{\max}$ limits the frequency resolution. In analogy to Eq. (9) we will also look at the density of occupied states (corresponding to the photoemission spectrum),

$$A^<(\omega,t) = \frac{1}{\pi}\text{Im}\int_0^{s_{\max}} ds\, e^{i\omega s}\, G^<(t+s,t), \tag{10}$$

which is the Fourier transform of the lesser Green function $G^<(t,t') = i\langle c^\dagger(t')c(t)\rangle$. The occupation function

$$N(\omega,t) = A^<(\omega,t)/A(\omega,t) \tag{11}$$

gives the Fermi function $f(\omega)$ in equilibrium.

3. The dielectric breakdown current

3.1. The current

In this section we briefly recapitulate the main results of Ref. [1]. We use the lowest order of the strong-coupling expansion (NCA) to compute several quantities related to the dielectric breakdown, and compare the outcome to OCA results from Ref. [1]. Figure 1a shows the

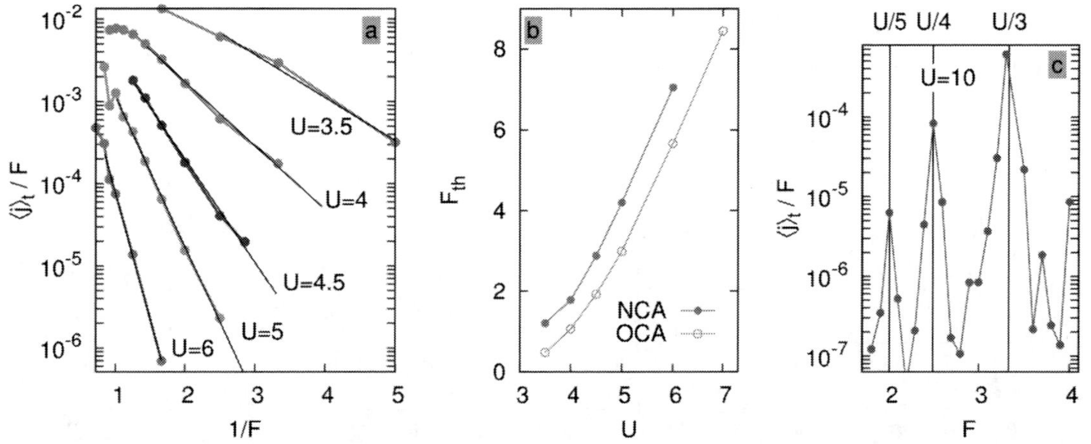

Figure 2. (a) Current $\langle j\rangle_t/F$ (averaged over times $40 < t < 50$) on a logarithmic scale as a function of $1/F$ for various values of U ($\beta = 5$). In this plot, black solid lines correspond to fits with Eq. (2), in the intervals covered by the lines. (b) The value $F_{\rm th}$, obtained from the linear fits (filled symbols). Open symbols show corresponding results from Ref. [1], which were obtained within the OCA. (c) Current $\langle j\rangle_t/F$ (averaged over times $20 < t < 30$) on a logarithmic scale, for large U and F (see discussion in Sec. 4). Vertical lines indicate resonances at integer U/F.

current $j(t)$ for various electric fields F as a function of time, for $U = 5$ and $\beta = 5$. The spectral function for these parameters, with a well-developed gap at the Fermi energy, identifies the initial state as a Mott insulator (inset of Fig. 1c). The temperature $1/\beta = 0.2$ is still too low to allow for a sizable number of thermally excited carriers, such that the linear dc conductivity is indistinguishable from zero on the scale of the plot. Nevertheless, the value of $j(t)$ at long times is nonzero and it strongly increases with F. (The large signal for $t \lesssim 3$, which is linear in F, is the current associated with the polarization of the insulator.) The highly nonlinear j-F characteristics of the Mott insulator is revealed by plotting the long-time average $\langle j\rangle_t$ of the current against F (Fig. 1c). The threshold-like increase of $\langle j\rangle_t/F$ around $F = 0.6$ is the hallmark of the dielectric breakdown of the Mott insulator.

Following Ref. [1], we fit the j-F curves with Eq. (2) in order to determine the threshold field $F_{\rm th}$ (Fig. 2). Note that Eq. (2) can only be expected to hold asymptotically for small F. For large F, the j-F curves can even behave non-monotonously (see, e.g., $F > 1$ for $U = 5$), which will be explained in Sec. 4 below. On the other hand, the exponentially small current for $F \ll 1$ cannot easily be resolved numerically. This restricts the fits with Eq. (2) to an intermediate range of F, as indicated by the black solid lines in Fig. 2a. The resulting threshold field $F_{\rm th}$, which is shown in Fig. 2b, decreases when U is decreased from the insulating regime towards the metal-insulator transition (which is only a crossover at $\beta = 5$).

In Fig. 2b, we have also included OCA data from Ref. [1] for the same parameters (open symbols). One can see that the higher-order corrections of the OCA are clearly important to get a quantitatively correct value of $F_{\rm th}$. The NCA threshold curve is shifted to smaller interactions compared to data from Ref. [1]. This behavior resembles the difference between OCA and NCA in equilibrium, e.g., for the location of the metal-insulator transition line. Apart from that, however, we find that the physics obtained with NCA throughout the insulating regime is qualitatively similar to the OCA solution (compare, e.g., Fig. 2 with Fig. 4c of Ref. [1]). Further discussions in this paper will thus only be based on NCA results. This allows us to study in detail several relaxation processes with and without dissipation, and to obtain high-resolution

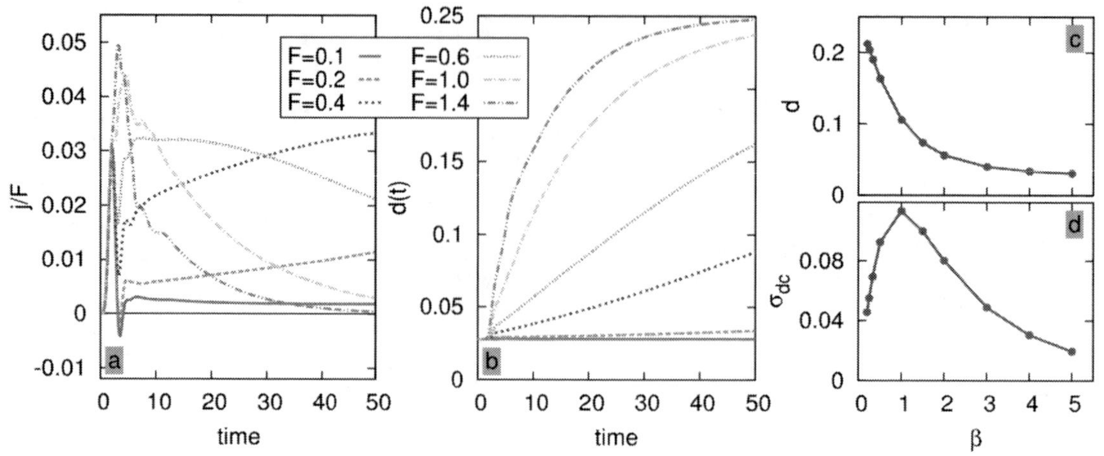

Figure 3. (a) Time-dependent current $j(t)$ for $U = 3$ and $\beta = 5$ and various electric fields F. (b) Double occupancy for the same parameters. (c) Double occupancy $d(T)$, and (d), linear response conductivity $\sigma_{\text{dc}}(T)$, in thermal equilibrium at $U = 3$ and temperature $T = 1/\beta$.

results for the spectral function. The calculation of the latter requires data at long times that would be accessible within the OCA only at large numerical expense.

3.2. Failure of the quasi-equilibrium description

Although the current in Fig. 1a becomes almost stationary at long times, the system is not in a true steady state. This is already clear from a very general energy consideration, because the total energy of any closed system in an external field F must increase at the rate given by Eq. (8). Because the energy per particle is bounded from above in a single-band model, a time-independent current cannot last forever. In general, the energy increase involves both kinetic and interaction energy. The double-occupancy $d(t)$, which is proportional to the interaction energy, is plotted in Fig. 1b. We find that $d(t)$ indeed increases almost linearly with time, while $j(t)$ does not change substantially (compare also Fig. 1b of Ref. [1]). This behavior is observed (in a quasi-steady manner) for rather long times, although in principle d cannot exceed the value 0.5. As mentioned in the introduction, it is this peculiar quasi-steady state from which properties of the dielectric breakdown are usually inferred. In this section we investigate to what extent a simple quasi-equilibrium description can account for the observed behavior.

In many situations, it is a valid assumption that a system rapidly evolves to a new equilibrium state at elevated temperature after the energy is increased. In accordance with this, the simplest possibility to describe an isolated system in an external field is to assume that all its properties can be obtained from an effective equilibrium state, with a time-dependent temperature $T_{\text{eff}}(t)$ which follows from the equation $C_V \dot{T}_{\text{eff}}(t) = j(t)F(t)$. In particular, the current at small fields would be given by

$$j_{\text{quasi-eq.}}(t) = F\sigma_{\text{dc}}(T_{\text{eff}}), \tag{12}$$

where σ_{dc} is the linear conductivity which becomes nonzero for $T_{\text{eff}} > 0$. This argument was found to apply to various correlated systems [31, 32, 33]. In the Hubbard model at weak-coupling, e.g., the quasi-equilibrium description works if U exceeds some critical interaction, while for smaller interaction scattering of the particles is too slow to establish the equilibrium state, and the system performs long-lived Bloch oscillations [33].

From the time-independence of $j(t)$ in Fig. 1a we can already infer that a simple quasi-equilibrium description is not valid for the dielectric breakdown at $U = 5$: Because the linear-response current in the insulator depends exponentially on temperature, a quasi-equilibrium state which is consistent with the increase of $d(t)$ would imply a current which is far too large. In the case of the Mott insulator at $U = 5$ and $F = 1$, e.g., the double occupancy increases by 0.006 from $t = 10$ to $t = 35$ (Fig. 1b), while the conductance remains at $j/F \approx 0.001$. On the other hand, in order to increase d by 0.006 *in equilibrium*, one would have to increase the temperature from $\beta = 5$ ($d \approx 0.010$) to $\beta = 2$ ($d \approx 0.016$), but at $\beta = 2$, the linear response conductivity is already more than one order of magnitude larger than the quasi-steady value at $F = 1$ [$\sigma_{\text{dc}}(\beta = 2) \approx 0.0155$].

The lack of thermalization in the Hubbard model at $U \gg t^*$ has been encountered previously [34, 35], and it implies the existence of various interesting metastable states in the Hubbard model [36, 37, 38]. Slow thermalization is ultimately related to the fact that the recombination time of doublons and holes is exponentially long for $U \gg t^*$ [39, 40], and hence kinetic energy and interaction energy cannot efficiently be redistributed. The long lifetime of doublons has been suggested earlier as a prerequisite for the existence of a quasi-steady current [1, 18]. Because the thermalization time τ_{th} of the double occupancy in the paramagnetic Mott insulator has recently been determined within DMFT (cf. Fig. 2b of Ref. [26], for a final temperature $\beta = 2$), we can make the argument more quantitative in the following. For $U = 5$, a value $\tau_{\text{th}} \approx 4000$ has been found (within NCA), which clearly exceeds the scale of Fig. 1. On the other hand, τ_{th} can become as short as a few inverse hoppings when U is in the metal-insulator crossover regime, and it will thus be interesting to see what a more rapid thermalization implies for the behavior of the current. Figure 3 shows $j(t)$ for $U = 3$ (where Ref. [26] gives $\tau_{\text{th}} \approx 40$ at $\beta = 2$.) In this case the system does indeed no longer establish a time-independent current. Instead, $j(t)$ increases with time as long as $d(t)$ is still small ($F = 0.1, 0.2, 0.4$), and it decreases when $d(t)$ approaches the value 1/4 ($F = 1.2, 1.4$). Qualitatively, this is consistent with the quasi-equilibrium behavior: At low temperatures, $U = 3$ shows an insulating character, such that both the double occupancy and the dc-conductivity increase with temperature (Fig. 3c and d). At high temperatures, σ_{dc} always *decreases* with T ($\sigma_{\text{dc}} \to 0$ for $T \to \infty$), while d increases to the uncorrelated value $d = 1/4$. Quantitatively, the current $j(t)$ in the driven system at $U = 3$ remains below the quasi-equilibrium value, in particular for large F, where the energy increase is faster than the thermalization rate.

In summary, the above discussion confirms that the existence of a quasi-steady current is tied to the lack of thermalization at large $U \gg t^*$. Since the thermalization time depends exponentially on U/t^*, the loss of the quasi-steady current occurs quite abruptly when U is decreased. (A different regime of fast thermalization to infinite temperature has been encountered at $U = F$, when doublon-hole production requires only a single hopping process [18].) That there is no steady current at smaller U also explains why we could so far not really determine F_{th} down to the metal-insulator transition. From now on, we will focus on regimes of large U where the steady current does exist.

4. Spectral function

Further insight into the quasi-steady current-carrying state can be gained from the spectral function, Eq. (9). In Fig. 4a and b, we plot $A(\omega, t)$ for $U = 10$. The values of F in Fig. 4a and b are well below F_{th} (which is large for $U = 10$), such that there is essentially no current after the initial polarization of the system, and no change of either energy or double occupancy within the numerical accuracy. The state is thus truly stationary for all practical purposes, and $A(\omega, t)$ has no measurable t-dependence for $t \gtrsim 10$. The spectrum consists of upper and lower Hubbard bands, separated by U. For large F, these Hubbard bands split into isolated peaks with a spacing $\Delta\omega = F$, which are narrowed with respect to the original Hubbard bands.

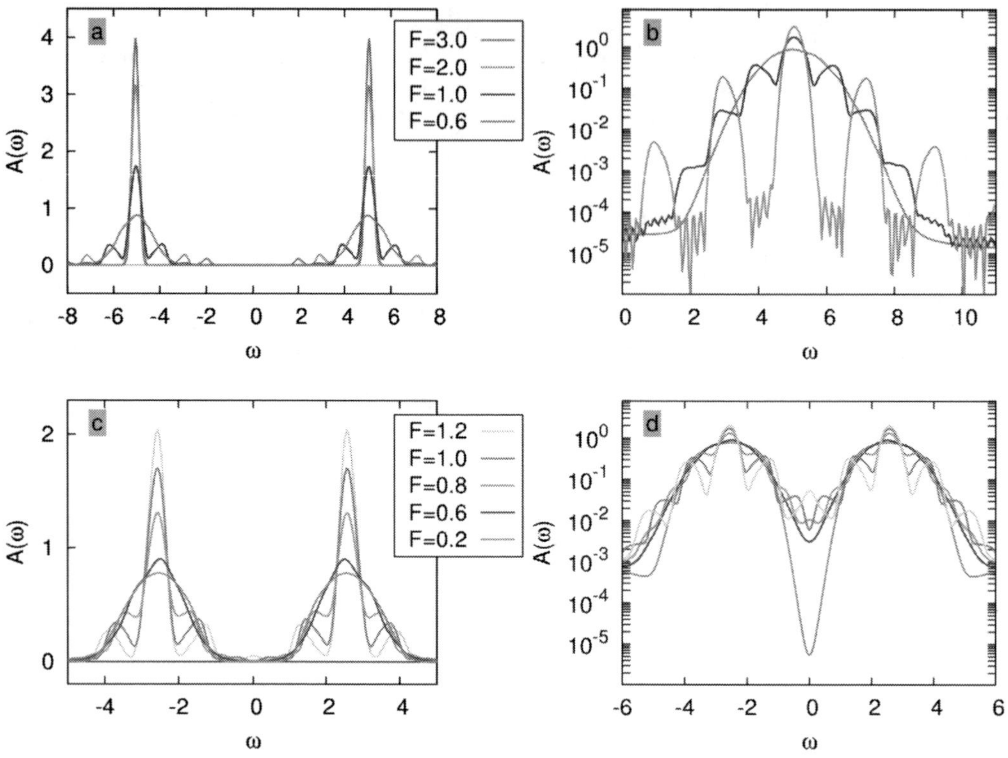

Figure 4. (a) Spectrum $A(\omega, t)$ for $t = 10$, $U = 10$, $\beta = 5$, and various fields. (b) Same data on a logarithmic scale. The low-amplitude noise is due to a finite cutoff $s_{\max} = 40$ of the Fourier integral Eq. (9). (c) and (d) Spectrum $A(\omega, t)$ for $t = 10$, $U = 5$, $\beta = 5$.

The side-bands resemble the Wannier-Stark ladder for an electron in a tight-binding band with applied homogeneous field [41, 22]. However, while all single-particle eigenstates in a tight-binding model with applied field are rigorously localized, we note that this is not true for the many-body situation of a doublon in the Mott insulator. A single doublon or hole which is added to the Mott insulator remains mobile even at large F, because its potential energy can be transferred to other particles or spin-excitations [42]. Nevertheless, the Hubbard side bands still reflect the localization which remains on short times, i.e., the side-peak at $\omega = U - nF$ may be interpreted as adding an electron at site i into a many-body Wannier-Stark "resonance" that is localized at a different site j, with $(\boldsymbol{R}_j - \boldsymbol{R}_i)\boldsymbol{F}/|\boldsymbol{F}| = (\boldsymbol{R}_j - \boldsymbol{R}_i)\boldsymbol{\eta} = n$.

At small fields, Wannier Stark side-bands are not clearly resolved, and we observe only a broadening of the spectrum ($F \lesssim 0.8$ in Fig. 4b and d). At smaller U, this field-induced broadening eventually leads to a filling-in of the gap (Fig. 4c and d), which indicates the possibility of the system to create a doublon-hole pair at no cost of energy. The Mott insulating ground state becomes unstable in the field. As soon as the Wannier-Stark side-bands split off, $A(\omega = 0)$ behaves non-monotonously as a function of F. This goes along with a non-monotonous j-F curve, and explains the failure of Eq. (2) at large F. At large U, where the system can tolerate strong enough fields such that the side-bands are well resolved, the non-monotonous behavior of the current even reveals clear resonances at integer U/F (Fig. 2c), although in this regime the steady current is superimposed with more and more irregular and longer lived transient oscillations. We note that the case $U/F = 1$ is particularly interesting because the

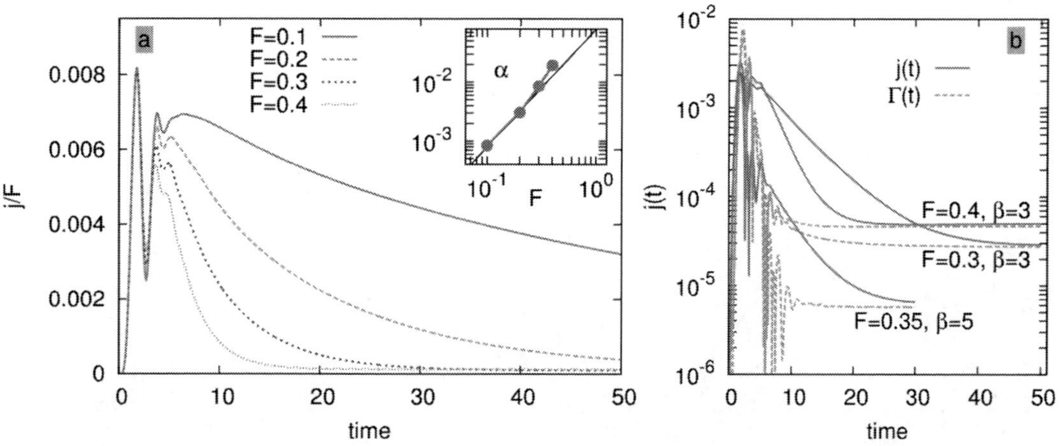

Figure 5. (a) Current for $U = 4.5$ and high temperature $\beta = 3$. *Inset:* The curves of the main plot are fit with an exponential $j(t) = A\exp(-\alpha t) + B$, and the rate α is plotted as a function of F. The solid line corresponds to the relation $\alpha \propto F^2$. (b) Current $j(t)$ (solid line) and doublon creation current $\Gamma_{\rm dh}(t)$ (dashed line) on a logarithmic scale, for $U = 4.5$ and indicated values of β and F.

degenerate manifold of states is then described by an effective spin model with a quantum phase transition [43], but this physics shall not be addressed here.

5. Influence of temperature

We now turn to one of the most important goals of this paper, i.e., to clarify why doublons and holes that are created by field-induced tunneling do not lead to an increase of the current with time, and to address the influence of the initial temperature on the dielectric breakdown. The two questions actually turn out to be closely related: In a hot initial state one has thermally excited carriers in addition to those which are induced by the field, and yet we will find that they do not influence the current at long times. In Fig. 5a we demonstrate this behavior for $U = 4.5$ and a rather high temperature, $\beta = 3$. For these parameters, the linear response current $F\sigma_{\rm dc}$ is actually larger than the field-induced current that is expected from Eq. (2). However, $j(t)$ is found to be close to $F\sigma_{\rm dc}$ only at early times, while it decreases almost to zero later on. This behavior was already described in Ref. [1], where we also found that the final value is close to the zero-temperature field-induced current, although the decay to that value is very slow for small F, which hindered a precise determination of the long-time limit for most parameters.

Within the NCA, it is possible to take a closer look at the long-time behavior. It is then useful to separately consider the component of the current that is related to the increase of the interaction energy, $(\frac{d}{dt}E_{\rm int})/F$. This turns out to be precisely the doublon-creation current that was defined ad-hoc in Eq. (1), $(\frac{d}{dt}E_{\rm int})/F = (1/F)\frac{d}{dt}Ud(t) = \Gamma_{\rm dh}(t)$. Remarkably, we find that the long time limit of $j(t)$ is very close to $\Gamma_{\rm dh}$, which by itself does not change much over the whole time interval (Fig. 5b). Hence the current at long times is essentially determined by the doublon production rate, while the produced doublons themselves seem not to contribute to the current. This observation is confirmed in Fig. 1c, where we plot the long-time average $\langle\Gamma_{\rm dh}\rangle_t$ as a function of F in addition to the time average $\langle j\rangle_t$.

To explain this behavior, one may wonder in the first place how charge carriers in correlated systems behave in the presence of large fields. This is well studied, e.g., in the tJ model, where

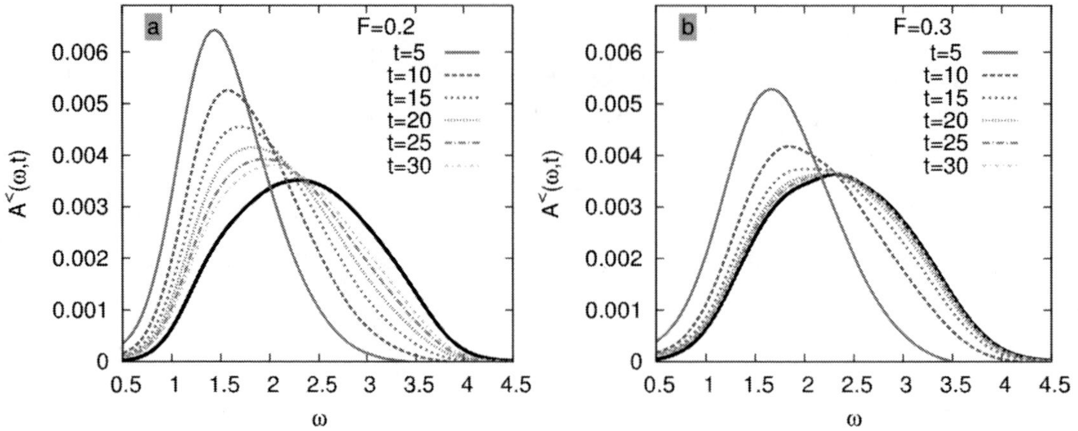

Figure 6. Occupation function $A^<(\omega,t)$ for $U = 4.5$, $\beta = 3$ in the upper Hubbard band, for various times. $F = 0.2$ for (a), and $F = 0.3$ for (b). The curves are compared to the spectral function $A(\omega, t)$ (rescaled by 0.0045), which is almost time-independent for those parameters (bold black line).

all terms changing the doublon or hole numbers have been projected out, and in a strong field one is left with the motion of quasiparticles. While noninteracting particles in a tight binding model are completely localized and perform Bloch oscillations (see, e.g., Ref. [44]), the motion of carriers in a many-body system is a competition between Wannier-Stark localization and transfer of potential energy to other degrees of freedom, such as spins in the tJ model [45, 42] or phonons in the Holstein model [47, 46]. Even in a large field, carriers remain mobile (the conductivity typically decreases with a power of F), and the decay of the linear response current in Fig. 5a cannot be related to Wannier-Stark localization.

On the other hand, the conductivity will always vanish at infinite temperature, because then potential energy can never be passed to other particles, spins, phonons etc. This motivates the explanation that for a low doublon production rate, the system can essentially establish a state of infinite temperature for the kinetic energy of doublons and holes, while their number is fixed (in contrast, infinite temperature for the interaction energy would imply $d = 0.25$ as in Sec. 3.2). More rigorously, to describe the short-time behavior one may project out all terms from the Hamiltonian that change the double occupancy (both field induced and interaction induced). The resulting model H_{dh} is a generalized tJ model, containing both doublons and holes in an external field F. Within this description, doublons and holes might then show quasi-equilibrium behavior with a conductivity σ_{dh}, and reach infinite doublon temperature T_{dh} much faster than the change of the doublon number happens via tunneling. For $T_{dh} \to \infty$ one again has $\sigma_{dh} = 0$, and the only remaining current is the dh-creation current.

Although we do not explicitly derive H_{dh} (we could not solve this model anyhow), two predictions from this simple argument can be verified directly: (i) Quite generally, from a high-temperature expansion we expect that when $T \to \infty$ one has both $E_{\text{tot}} \sim 1/T$ and $\sigma_{\text{dc}} \sim 1/T$. In a quasi-equilibrium description, $j(t)$ should thus decay exponentially with a rate $\propto F^{-2}$ for $F \to 0$, due to Eq. (8) [31, 33]. This behavior is indeed confirmed in the inset of Fig. 5a. (ii) To directly observe the "heating" of doublons and holes one can look at the occupation function, or photoemission spectrum Eq. (10), which is plotted in Fig. 6 for $U = 4.5$ and various times. The fields, $F = 0.2$ and 0.3, are chosen such that the total weight of $A^<(\omega, t)$ in the upper Hubbard band, which is roughly related to the total number of doublons, does not change much with

Journal of Physics: Conference Series **427** (2013) 012005 doi:10.1088/1742-6596/427/1/012005

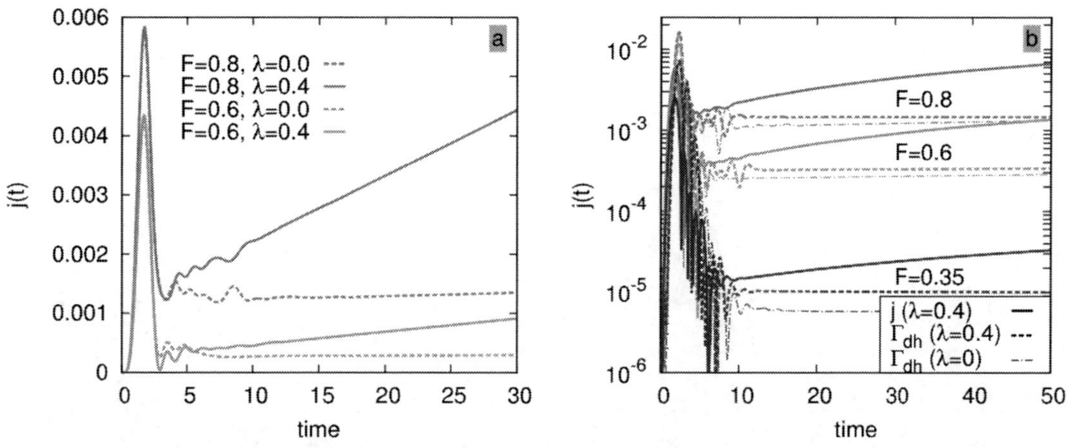

Figure 7. (a) Comparison of the current at $U = 4.5$, $\beta = 10$ with dissipation (solid lines) and without dissipation (dashed lines). (b) Comparison of the doublon-creation current (dashed lines) and the full electric current j (solid lines) for $U = 4.5$, $\beta = 10$, and various fields F ($\lambda = 0.4$).

time. The shape of $A^<(\omega, t)$, however, changes considerably. In the initial state and shortly after the switch-on of the field, the occupation is concentrated at low frequencies, as expected for a thermal state with $A^<(\omega) = A(\omega)f(\omega)$. As time increases, $A^<(\omega, t)$ approaches a curve that is proportional to the spectral function itself (bold lines in Fig. 6), corresponding to a flat distribution, $N(\omega) = const.$, characteristic of a state with zero kinetic energy (high-temperature state).

6. Influence of dissipation

In a condensed matter system, dissipation of energy to lattice-, spin-, and other degrees of freedom is unavoidable, and can happen on sub-picosecond timescales. It is thus essential to understand which of the findings of the previous section will be robust against such processes. The steady states of dissipative systems with dc driving is itself an interesting area of research [49, 48]. Within DMFT, dissipation terms can be included phenomenologically, either by coupling a local fermionic bath [50, 49, 51] or a bosonic ("phonon") bath [52] at fixed temperature. Here we use the latter approach, since it guarantees particle number conservation and dissipates only energy.

Following Ref. [52], the electronic self-energy in this case is supplemented by a bath contribution, which is the lowest order diagram for a Holstein-type electron-phonon coupling, $\Sigma_{\text{diss}}[G] = \lambda^2 G(t,t')D(t,t')$. here, λ measures the coupling strength, and $D(t,t')$ is the equilibrium propagator for a boson with energy ω_0; $D(t,t') = -i\text{Tr}[\text{T}_\mathcal{C} \exp(-i\int_\mathcal{C} dt\omega_0 b^\dagger b) b(t) b^\dagger(t')]/Z$. The temperature $1/\beta$ of the bath is fixed, such that the bath has no memory ($\omega_0 = t^*$ in the following). It is important to note that we choose $\omega_0 = t^* \ll U$ for the phonon frequency, to prevent opening a new channel for doublon-hole recombination via phonon emission. In general, the precise way of including the dissipation should not matter too much, and we chose parameters such that the equilibrium physics remains almost unchanged by the presence of dissipation. (In contrast, for a fermionic bath in the wide-band limit the equilibrium state would be modified due to mid-gap states that are created in the spectrum of the insulator with dissipation [51].)

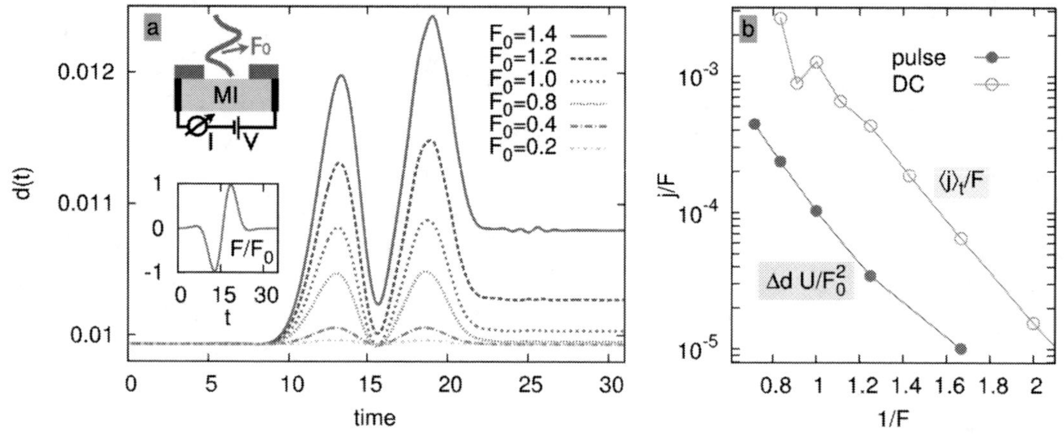

Figure 8. (a) $d(t)$ during a single-cycle THz-pulse with amplitude F_0, for $U = 5$ and $\beta = 5$. *Inset:* The electric field pulse, Eq. (13), with $\tau = 3.9269$. (b) Integrated doublon current, $\langle\Gamma_{\rm dh}\rangle_{\rm av}/F_0 = U\Delta d/F_0^2$ (filled symbols), where Δd is the change of the double occupancy over the field cycle in panel (a). The same parametrization as in Fig. 2a is used. For comparison, we include the dc curve $\langle j\rangle_t/F$ for $U = 5$, taken from Fig. 2a (open symbols). Both have the same slope, related to $F_{\rm th}$.

Figure 7a compares the current j at $U = 4.5$, with and without dissipation. In contrast to the isolated system ($\lambda = 0$), the current in a system with dissipation never becomes stationary, but it increases more or less linearly with time ($\lambda = 0.4$). Based on the results of the previous sections, the explanation is straightforward: If the system can dissipate energy, doublons and holes will never reach $T_{dh} = \infty$, and the distribution $N(\omega, t)$ never becomes flat. Hence doublons and holes contribute to the current at all times. Since their number is increasing due to the field-induced doublon-hole production, the current increases with time. The doublon-creation current $\Gamma_{\rm dh}$, on the other hand, is not expected to depend strongly on the coupling to phonons with an energy $\omega_0 \ll U$. This is in fact true, as apparent from Fig. 7b, where we compare the currents j and $\Gamma_{\rm dh}$ for dissipative systems with electron-phonon coupling $\lambda = 0.4$. The qualitative behavior of $\Gamma_{\rm dh}$ is the same with and without dissipation, i.e., the value is time-independent and increases exponentially with F.

7. Implication for experiments

To design an experiment that can potentially measure the threshold behavior, one should account for the coupling of the sample to leads [53, 10]. Doublons and holes can then escape through the boundary, such that their density in the sample becomes stationary and a true steady state is reached. (Otherwise, the quasi-steady state would be quite short-lived: extrapolating the linear increase of d in Fig. 1b, e.g., gives a time of a few 1000 until the uncorrelated value $d = 0.25$ would be reached, corresponding to a few picoseconds for $t^* = 1eV$.) Nevertheless, in the dc setup heating of the system is quite substantial, and the current will be a balance between non-perturbative doublon-hole production and dissipative effects. To avoid these complications, one could probe the strong field behavior by using short pulses, as already proposed for dielectrics (band insulators) [54]. If the pulse frequency $\Omega \ll U$ is below the Mott gap, e.g., in the THz range, the doublon production might be described by the dc results to a good approximation. If a weak external bias V is applied to the sample in addition to the strong THz pulse, the induced charge is collected at the leads. The current, averaged over many THz pulses, would then be proportional to the total number Δd of doublons created per pulse. In Fig. 8, we have simulated the outcome of such an experiment (as sketched in the inset). Figure 8a shows the time-evolution of the double occupancy during a single-cycle pulse

$$F(t) = F_0 \sin(\Omega(t - t_0))\, e^{-\frac{1}{2}\left(\frac{t-t_0}{\tau}\right)^2}, \tag{13}$$

centered around $t_0 = 2\pi/\Omega$, with frequency $\Omega = 0.4$, and a Gaussian envelope (see inset in Fig. 8a). Motivated by our results, we define an averaged doublon creation current $\langle\Gamma_{\rm dh}\rangle_{\rm av} = U/F_0\,\Delta d$. As seen in Fig. 8b, the threshold behavior of $\langle\Gamma_{\rm dh}\rangle_{\rm av}$ turns out to be the same as in the dc case (Fig. 2). This seems reasonable if the doublon production is determined by the largest field during the cycle.

Using time-resolved photoemission spectroscopy, one might be able to observe the heating effect demonstrated in Fig. 6: In a system with strong external bias, but not strong enough to lead to a dielectric breakdown, one may suddenly excite carriers created by photo-doping, monitor the evolution of the distribution function, and compare it to the result without applied field.

8. Summary and Conclusion

In conclusion, we have used DMFT with the NCA-based impurity solver to study the behavior of a Mott insulator in strong dc electric fields F. The electric current j is given by a contribution $\Gamma_{\rm dh}$ which can be associated with the doublon-hole production due to field-induced tunneling [Eq. (1)], and a contribution related to the conductance of doublons and holes which are either thermally excited or induced by the field. For an isolated system, however, these carriers accumulate energy from the field and rapidly reach an "infinite temperature" state with zero conductivity, as apparent from a flat occupied density of states. This explains the peculiar steady state that has been described earlier, in which the current is time-independent (and given by $\Gamma_{\rm dh}$) although the number of carriers constantly increases. In an isolated system, the current itself is thus a good measure for the field-induced tunneling. In contrast, when dissipation of energy is taken into account, carriers cannot reach infinite temperature. The current now *does* depend on the number of doublons, temperature, and the coupling to the environment, whereas the doublon-hole creation current $\Gamma_{\rm dh}$ still provides an intrinsic measure of the tunneling. For fields $F \lesssim t^*$, $\Gamma_{\rm dh}$ increases with F with a threshold behavior, Eq. (2). For larger fields, we find that the j-F characteristics reveals peaks at integer values of U/F, which are related to signatures of Wannier-Stark localization of carriers in the spectrum.

The threshold field might be determined experimentally by measuring the current induced by a field pulse under weak bias, while the heating of induced doublons should be observable with time-resolved photoemission.

Acknowledgments
We thank T. Oka for useful discussions. PW acknowledges support from FP7/ERC starting grant No. 278023.

References

[1] M. Eckstein, T. Oka, and Ph. Werner, Phys. Rev. Lett. **105**, 146404 (2010).
[2] M. Imada, A. Fujimori, and Y. Tokura, Rev. Mod. Phys. **70**, 1039 (1998).
[3] Y. Tokura, H. Okamoto, T. Koda, T. Mitani, and G. Saito, Phys. Rev. B **38**, 2215 (1988).
[4] S. Guénon, S. Scharinger, S. Wang, J. G Ramírez, D. Koelle, R. Kleiner, and I. K. Schuller, arXiv:1210.6648 (unpublished).
[5] K. Inagaki, I. Terasaki, H. Mori, T. Mori, J. Phys. Soc. Jpn. **73**, 3364 (2004).

[6] F. Sawano, I. Terasaki, H. Mori, T. Mori, M. Watanabe, N. Ikeda, Y. Nogami, Y. Noda, Nature **437**, 522 (2005).
[7] M. Greiner, O. Mandel, T. Esslinger, Th. W. Hänsch, and Immanuel Bloch, Nature **415**, 39 (2002).
[8] T. Oka, R. Arita, and H. Aoki, Phys. Rev. Lett. **91**, 066406 (2003).
[9] T. Oka, and H. Aoki, Phys. Rev. Lett. **95**, 137601 (2005).
[10] F. Heidrich-Meisner, I. González, K. A. Al-Hassanieh, A. E. Feiguin, M. J. Rozenberg, and E. Dagotto, Phys. Rev. B **82**, 205110 (2010).
[11] S. Kirino and K. Ueda J. Phys. Soc. Jpn. **79**, 093710 (2010).
[12] T. Oka and H. Aoki, Phys. Rev. B **81**, 033103 (2010).
[13] T. Oka, Phys. Rev. B **86**, 075148 (2012).
[14] Z. Lenarčič and P. Prelovšek, Phys. Rev. Lett. **108**, 196401 (2012).
[15] F. Queisser, P. Navez, and R. Schützhold, Phys. Rev. A **85**, 033625 (2012).
[16] L. D. Landau, Phys. Z. Sowjetunion **2**, 46 (1932); C. Zener, Proc. R. Soc. London A **137**, 696 (1932).
[17] T. Oka, N. Konno, R. Arita, and H. Aoki, Phys. Rev. Lett. **94**, 100602 (2005).
[18] K. Mikelsons, J. K. Freericks, and H. R. Krishnamurthy, arXiv:1210.5797 (unpublished).
[19] W. Metzner and D. Vollhardt, Phys. Rev. Lett. **62**, 324 (1989).
[20] A. Georges, G. Kotliar, W. Krauth, and M. J. Rozenberg, Rev. Mod. Phys. **68**, 13 (1996).
[21] V. Turkowski and J. K. Freericks, Phys. Rev. B **71**, 085104 (2005).
[22] J. H. Davies and J. W. Wilkins, Phys. Rev. B **38**, 1667 (1988).
[23] P. Schmidt and H. Monien, arXiv:cond-mat/0202046 (unpublished).
[24] J. K. Freericks, V. M. Turkowski, and V. Zlatić, Phys. Rev. Lett. **97**, 266408 (2006); J. K. Freericks, Phys. Rev. B **77**, 075109 (2008).
[25] M. Eckstein and Ph. Werner, Phys. Rev. B **82**, 115115 (2010).
[26] M. Eckstein and Ph. Werner, Phys. Rev. B **84**, 035122 (2011).
[27] M. Eckstein, M. Kollar, and P. Werner, Phys. Rev. B **81**, 115131 (2010).
[28] H. Keiter and J. C. Kimball, Intern. J. Magnetism **1**, 233 (1971); H. Keiter and J. C. Kimball, J. Appl. Phys. **42**, 1460 (1971); N. Grewe and H. Keiter, Phys. Rev. B **24**, 4420 (1981); Y. Kuramoto, Z. Phys. B **53**, 37 (1983).
[29] T. Pruschke and N. Grewe, Z. Phys. B **74**, 439 (1989).
[30] J. K. Freericks, H. R. Krishnamurthy, and T. Pruschke, Phys. Rev. Lett. **102**, 136401 (2009); M. Eckstein and M. Kollar, Phys. Rev. B **78**, 245113 (2008).
[31] M. Mierzejewski and P. Prelovšek, Phys. Rev. Lett. **105**, 186405 (2010).
[32] M. Mierzejewski, J. Bonča, and P. Prelovšek, Phys. Rev. Lett. **107**, 126601 (2011).
[33] M. Eckstein and P. Werner, Phys. Rev. Lett. **107**, 186406 (2011).
[34] C. Kollath, A. M. Läuchli, and E. Altman, Phys. Rev. Lett. **98**, 180601 (2007).
[35] M. Eckstein, M. Kollar, and P. Werner, Phys. Rev. Lett. **103**, 056403 (2009);
[36] D. Petrosyan, B. Schmidt, J. R. Anglin, and M. Fleischhauer, Phys. Rev. A **76**, 033606 (2007).
[37] A. Rosch, D. Rasch, B. Binz, and M. Vojta, Phys. Rev. Lett. **101**, 265301 (2008).
[38] Ph. Werner, N. Tsuji, and M. Eckstein, Phys. Rev. B **86**, 205101 (2012).
[39] N. Strohmaier, D. Greif, R. Jördens, L. Tarruell, H. Moritz, T. Esslinger, R. Sensarma, D. Pekker, E. Altman, and E. Demler, Phys. Rev. Lett. **104**, 080401 (2010).
[40] R. Sensarma, D. Pekker, E. Altman, E. Demler, N. Strohmaier, D. Greif, R. Jördens, L. Tarruell, H. Moritz, and T. Esslinger, Phys. Rev. B **82**, 224302 (2010).
[41] G. H. Wannier, Phys. Rev. **117**, 432 (1960).
[42] M. Mierzejewski, L. Vidmar, J. Bonča, and P. Prelovšek, Phys. Rev. Lett. **106**, 196401 (2011).
[43] S. Sachdev, K. Sengupta, and S. M. Girvin, Phys. Rev. B **66**, 075128 (2002).
[44] M. Glück, A. R. Kolovsky, and H. J. Korsch, Phys. Rep. **366**, 103 (2002).
[45] L. Vidmar, J. Bonča, and S. Maekawa, Phys. Rev. B **79**, 125120 (2009).
[46] L. Vidmar, J. Bonča, T. Tohyama, and S. Maekawa, Phys. Rev. Lett. **107**, 246404 (2011).
[47] L. Vidmar, J. Bonça, M. Mierzejewsk, P. Prelovšek, and S. A. Trugman, Phys. Rev. B **83**, 134301 (2011).
[48] J. E. Han, arXiv:1210.6297 (unpublished).
[49] A. Amaricci, C. Weber, M. Capone, and G. Kotliar, Phys. Rev. B **86**, 085110 (2012).
[50] N. Tsuji, T. Oka, and H. Aoki, Phys. Rev. Lett. **103**, 047403 (2009).
[51] C. Aron, Phys. Rev. B **86**, 085127 (2012).
[52] M. Eckstein and Ph. Werner, arXiv:1207.0402 (unpublished).
[53] S. Okamoto, Phys. Rev. Lett. **101**, 116807 (2008).
[54] V. Apalkov and M. I. Stockman Phys. Rev. B **86**, 165118 (2012).

The generalized Kadanoff-Baym ansatz. Computing nonlinear response properties of finite systems

K. Balzer
University of Hamburg, Max Planck Research Department for Structural Dynamics, Building 99 (CFEL), Luruper Chaussee 149, 22761 Hamburg, Germany
E-mail: karsten.balzer@mpsd.cfel.de

S. Hermanns and M. Bonitz
University of Kiel, Institut für Theoretische Physik und Astrophysik, Leibnizstrasse 15, 24098 Kiel, Germany

Abstract. For a minimal Hubbard-type system at different interaction strengths U, we investigate the density-response for an excitation beyond the linear regime using the generalized Kadanoff-Baym ansatz (GKBA) and the second Born (2B) approximation. We find strong correlation features in the response spectra and establish the connection to an involved double excitation process. By comparing approximate and exact Green's function results, we also observe an anomalous U-dependence of the energy of this double excitation in 2B+GKBA. This is in accordance with earlier findings [K. Balzer *et al.*, EPL **98**, 67002 (2012)] on double excitations in quantum wells.

1. Introduction
The study of strongly correlated quantum systems and materials, e.g., [1] and references therein, is a very rapidly evolving field in both experimental and theoretical physics. Especially the out-of-equilibrium dynamics is of great current interest in solid-state, atomic and molecular physics, in nanoelectronics, quantum transport etc. In all these fields, the availability of intense and coherent radiation, combined with ultra-short laser pulses, has triggered many key experiments that allow one to investigate matter under extreme nonequilibrium conditions where correlation and nonlinear effects are important. As examples, consider the photoionization of multi-electron atoms and molecules [2], the many-body dynamics of particles in optical lattices [3] or quantum interference effects in Mott insulators [4].

In the recent decade, the nonequilibrium Green's function (NEGF) approach has developed into a powerful numerical tool and is able to address the dynamics of very different quantum systems. Examples include the excitation of non-ideal (semiconductor) plasmas [5, 6], nuclear collisions [7], the carrier dynamics and carrier-phonon interaction in quantum dots, wells and lattice networks contacted to leads [8, 9, 10, 11], and the problem of baryogenesis in cosmology [12]. In all these cases, the central quantity is the one-particle nonequilibrium Green's function defined on the complex Keldysh contour $\mathcal{C}$,

$$\begin{aligned} g^{\sigma}_{ss'}(t,t') &= -\mathrm{i}\langle T_{\mathcal{C}}\, \hat{c}_{s,\sigma}(t)\, \hat{c}^{\dagger}_{s',\sigma}(t')\rangle \\ &= \theta_{\mathcal{C}}(t-t')\, g^{\sigma,>}_{ss'}(t,t') + \theta_{\mathcal{C}}(t'-t)\, g^{\sigma,<}_{ss'}(t,t')\,, \end{aligned} \tag{1}$$

with non-temporal degrees of freedom s, s' and σ. For our purpose, s and s' will refer to spatial degrees of freedom, and $\sigma =\uparrow,\downarrow$ will indicate the spin, assuming diagonality in the spin subspace. Further, the operator $T_{\mathcal{C}}$ accounts for contour ordering of the times t and t', and $\langle\ldots\rangle$ means averaging in the grand canonical ensemble.

In principle, the two-time dependence of the Green's function enables the systematic and exact treatment of dynamic correlation effects independently of the strength of the interaction and regardless of the presence of weak or strong external fields. Even interaction quenches can be studied, e.g. [13]. However, the application to large system sizes and the investigation of the system's long time behavior usually demand additional approximations and simplifications to the equations of motions of the NEGF [14].

The most straightforward approach is to perform a many-body approximation (MBA) on the self-energy. Due to diagram expansions, this can be done in a systematic and conserving way and leads, e.g., to the Hartree-Fock (HF), second(-order) Born (2B), GW or T-matrix approximations. Correlation effects are then partially treated beyond the level of Hartree-Fock. Additional simplifications typically concern the two-time structure which renders NEGF calculations demanding [15, 16]. Within the generalized Kadanoff Baym ansatz (GKBA) [17, 18], it is possible to study the equations of motion for the Green's function in the single-time limit, which drastically reduces the numerical effort.

In general, the application of MBAs and other simplifications such as the GKBA require thorough justification as they can lead to unphysical effects such as self-interaction errors and spurious dynamical excitations [19], bistability [11] or artificial steady states [20]. For this reason, we, in this contribution, extend previous work on weak perturbations [21], which included the discussion of double excitations (see also [22]), to the nonlinear regime and analyze the performance of the GKBA for a finite system with the two-body interactions being treated in the second Born approximation.

2. Equations of motion and GKBA

Introducing the one-particle self-energy $\Sigma^{\sigma}_{ss'}(t,t')$, the Green's function (1) obeys the Kadanoff-Baym equation (KBE) [23],

$$\left(-\mathrm{i}\frac{\partial}{\partial t}\delta_{s\bar{s}} - h^{\sigma}_{s\bar{s}}(t)\right) g^{\sigma}_{\bar{s}s'}(t,t') = \delta_{\mathcal{C}}(t-t')\delta_{ss'} + \int_{\mathcal{C}} \mathrm{d}\bar{t}\; \Sigma^{\sigma}_{s\bar{s}}(t,\bar{t}) g^{\sigma}_{\bar{s}s'}(\bar{t},t')\,, \tag{2}$$

and its adjoint with $t \leftrightarrow t'$. Here, $h^{\sigma}_{ss'}$ denotes the single-particle (kinetic plus potential) energy, and summation over $\bar{s}$ is implied.

Using the generalized Kadanoff-Baym ansatz, we transform the set of KBEs into a single equation for the reduced density matrix $\rho^{\sigma,<}_{ss'}(t) = -\mathrm{i}g^{\sigma,<}_{ss'}(t,t)$ or equivalently $\rho^{\sigma,>}_{ss'}(t) = -\mathrm{i}g^{\sigma,>}_{ss'}(t,t)$ retaining time causality and important conservation laws such as the conservation of total energy, momentum and density. In the course of this, the time-off-diagonal components of the NEGF are reconstructed as (we again sum over $\bar{s}$),

$$g^{\sigma,\gtrless}_{ss'}(t,t') = -g^{\sigma,\mathrm{ret}}_{s\bar{s}}(t,t')\,\rho^{\sigma,\gtrless}_{\bar{s}s'}(t') + \rho^{\sigma,\gtrless}_{s\bar{s}}(t)\, g^{\sigma,\mathrm{adv}}_{\bar{s}s'}(t,t')\,, \tag{3}$$

and inserted into the time-diagonal collision integral (the r.h.s. of the KBEs for $t = t'$ in the second Born approximation). For an explicit expression of $\Sigma^{\sigma}_{ss'}(t,t')$ as functional of the

Journal of Physics: Conference Series **427** (2013) 012006 doi:10.1088/1742-6596/427/1/012006

NEGF and for the corresponding collision term, see, e.g., the contribution of S. Hermanns *et al.* in the present conference proceedings. Finally, the treatment of the two-time retarded and advanced propagators $g^{\sigma,\mathrm{ret}}_{ss'}(t,t')$ and $g^{\sigma,\mathrm{adv}}_{ss'}(t,t')$ in Hartree-Fock approximation renders the ansatz practical:

$$g^{\sigma,\mathrm{ret/adv}}_{ss'}(t,t') = \mp \mathrm{i}\theta_{\mathcal{C}}(\pm[t-t'])T\exp\left(-\mathrm{i}\int_{t'}^{t}\mathrm{d}\bar{t}\; h^{\sigma}_{\mathrm{HF}}(\bar{t})\right)\Bigg|_{ss'} . \tag{4}$$

Here, $h^{\sigma}_{\mathrm{HF}}(t)$ is the single-particle time-dependent Hartree-Fock Hamiltonian, and T denotes the ordinary time-ordering operator.

We emphasize that the reconstruction of the greater and lesser components of the Green's function with real t and t' is sufficient as long as the method of adiabatic switching is applied to generate the correlated initial (ground) state by time propagation, for details see [18].

3. Nonlinear density-response

As outlined above, we are interested in the dynamical behavior of a finite quantum system beyond the regime of linear response. To this end, we consider a one-dimensional Hubbard model at half-filling with hopping T and on-site interaction U and strongly perturb it by changing the site energies in time [24].

The initial Hamiltonian for times $t < 0$ reads,

$$\hat{H}_0 = -T\sum_{<s,s'>}\sum_{\sigma=\uparrow,\downarrow}\hat{c}^{\dagger}_{s,\sigma}\hat{c}_{s',\sigma} + U\sum_{s}\hat{n}_{s,\uparrow}\hat{n}_{s,\downarrow} , \tag{5}$$

where s and s' range from 0 to $L-1$ for a chain of length L, and $< s, s' >$ indicates nearest-neighbor sites. Further, $\hat{n}_{s,\sigma} = \hat{c}^{\dagger}_{s,\sigma}\hat{c}_{s,\sigma}$ denotes the density operator, and the energy (time) is measured in units of T (the inverse hopping T^{-1}). Generally, we study the chain for open boundary conditions but this is irrelevant for the special case of two sites.

The system is excited by an instantaneous change of the site energies from zero to a positive finite value ϵ_s. If two or more sites differ in energy after the switching process, a correlated electron motion is initiated in the chain. For simplicity, we detune only the energy of the first site which leads to the perturbation,

$$\hat{H}_1 = \epsilon_0\,\theta(t)\sum_{\sigma=\uparrow,\downarrow}\hat{n}_{0,\sigma} . \tag{6}$$

For $t > 0$ and arbitrary ϵ_0, the perturbed system $\hat{H}_0 + \hat{H}_1$ will initially show a depopulation of the first site followed by an accumulation of density on the second. Subsequently, also the density on the remaining sites will change with time, and, finally, all occupations will start to oscillate. In the case of $\epsilon_0 \ll 1$, the population change will be small, such that the dynamics should be well characterized by the linear response properties of the chain. For $\epsilon \gtrsim 1$, however, we expect nonlinear effects to be of importance. In the following, we consider the case $\epsilon_0 = 5.0$ and resort to the zero-temperature limit $\beta \to \infty$. Initially, at $t = 0$, the system is prepared in the ground state of $\hat{H}_0$.

We measure the response of the Hubbard chain with respect to the time-dependent electron density on the first site ($s = 0$),

$$\gamma^{\sigma}(t) = \langle\hat{n}^{\sigma}_0\rangle(t) - \frac{1}{t^*}\int_0^{t^*}\mathrm{d}\bar{t}\;\langle\hat{n}^{\sigma}_0(\bar{t})\rangle , \tag{7}$$

$$\gamma^{\sigma}(\omega) = \int_0^{t^*}\mathrm{d}t\;\exp(-\mathrm{i}\omega t)\,\gamma^{\sigma}(t) , \tag{8}$$

where $\langle \hat{n}_0^\sigma \rangle(t) = -\mathrm{i}g_{00}^{\sigma;<}(t,t)$. Due to the spin symmetry obeyed by Eq. (6), it is $\gamma(\omega) = \gamma^\uparrow(\omega) = \gamma^\downarrow(\omega)$. Furthermore, the time t^* indicates a finite propagation time used in the numerics and is chosen sufficiently long such that it only affects the basic width of the peaks in $\gamma(\omega)$ but not their position.

3.1. Exact dynamics

In order to benchmark the approximate solution of the KBEs under the use of the GKBA and the second Born approximation, we compare to exact reference data computable for small L. The exact electron dynamics are obtained from propagating the one-particle nonequilibrium Green's function independently of the KBEs. To this end, we rewrite the average in the definition (1) of the NEGF as,

$$g_{ss'}^\sigma(t,t') = -\frac{\mathrm{i}}{Z_0} \sum_{N_\uparrow,N_\downarrow=0}^{L} \sum_n \mathrm{e}^{-\beta(E_{N_\uparrow,N_\downarrow;n}-\mu N)} \langle N_\uparrow, N_\downarrow; n | T_{\mathcal{C}}\, \hat{c}_{s,\sigma}(t)\, \hat{c}_{s',\sigma}^\dagger(t') \, | N_\uparrow, N_\downarrow; n \rangle \,, \tag{9}$$

where $Z_0 = \sum_{N_\uparrow,N_\downarrow} \sum_n \mathrm{e}^{-\beta(E_{N_\uparrow,N_\downarrow;n}-\mu N)}$ is the partition function, $N = N_\uparrow + N_\downarrow$, and n labels the states in the N-particle subspace. For small system sizes, the action of the time-dependent creation and annihilation operators,

$$\hat{c}_{s,\sigma}^{(\dagger)}(t) = \hat{U}(0,t)\, \hat{c}_{s,\sigma}^{(\dagger)}\, \hat{U}(t,0) \,, \tag{10}$$

with,

$$\hat{U}(t',t) = T \exp\left(-\mathrm{i} \int_t^{t'} \mathrm{d}\bar{t}\, (\hat{H}(\bar{t}) - \mu \hat{N}) \right) \,, \tag{11}$$

on the energy eigenstates $|N_\uparrow, N_\downarrow; n\rangle$ (with energy $E_{N_\uparrow,N_\downarrow;n}$) can be computed numerically exactly. For each particle sector involved in Eq. (9), we work in the trivial site basis which involves the states $|0\rangle$, $|\uparrow\rangle$, $|\downarrow\rangle$ and $|\uparrow\downarrow\rangle$ and apply the Krylov method [25] together with a high-order commutator-free exponential time propagation, see [26].

In Figure 1 a) and b), we show exact results based on Eqs. (9), (10) and (11), where the consideration of the thermodynamic ground state limits the trace in Eq. (9) to $N_\uparrow = N_\downarrow = 1$. The thick curves show the response $\gamma(\omega)$ for a two-site chain ($L = 2$) at different repulsive interaction strengths U according to Eq. (8). For $U = 0$, see the black curve in the upper panel, there exist two peaks—one at $\omega_0 = 5.385$ and one at $2\omega_0 = 10.770$. Interestingly, for $U > 0$, the energetically lowest peak at ω_0 develops into a double-peak structure of which the left part steadily has a larger spectral weight. For larger U there is even a weight difference of several orders of magnitude, cf. Figure 1 b). On the contrary, the peak at $2\omega_0$ changes only slightly with U and almost maintains its spectral weight.

To understand the spectrum, we recall the switch between two time-independent Hamiltonians given by Eqs. (5) and (6) without a ramp function. In this context, the initial state (given by the ground state of $\hat{H}_0$) can be expressed as a superposition of eigenstates of $\hat{H}_2 = \hat{H}_0 + \hat{H}_1$ for $t > 0$. Consequently, the transition frequencies between the eigenstates of $\hat{H}_2$ determine the system's nonequilibrium dynamics, and we expect them therefore to appear in $\gamma(\omega)$. The inset in Figure 1 a) shows the eigenenergies in the asymmetric chain with $\epsilon_0 = 5.0$ and $\epsilon_1 = 0$ as a function of U, and the thin vertical lines in Figure 1 a) and b) indicate the associated excitation frequencies for the selected interaction strengths. Some of the vertical lines indeed coincide with the peaks in $\gamma(\omega)$. More precisely, we find three out of six transitions: ω_{AC}, ω_{CD} and ω_{AD}, see the arrows in the inset. In the limit $U \to 0$, ω_{AC} and ω_{CD} become degenerate which explains the double-peak structure for small, finite U.

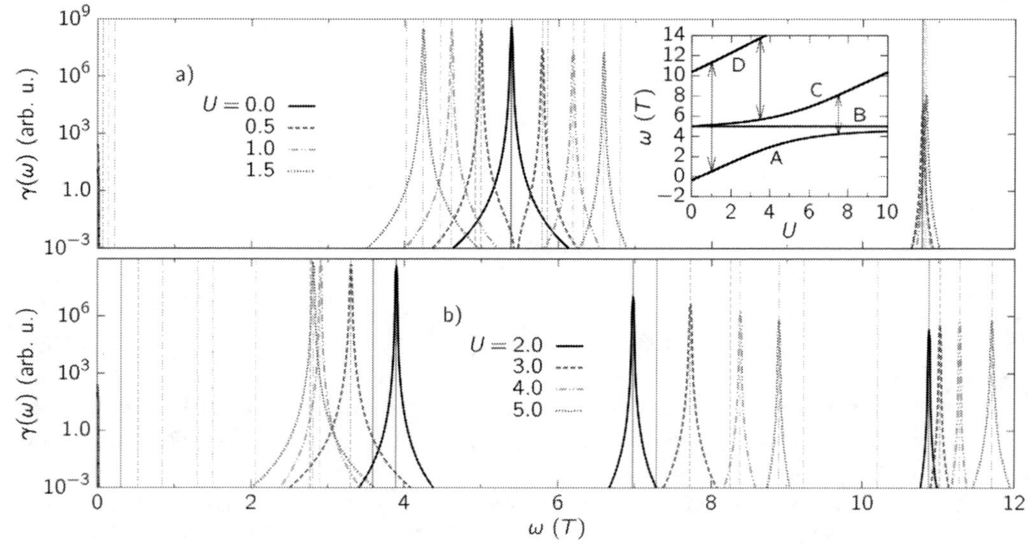

Figure 1. Exact nonlinear density-response spectra $\gamma(\omega) = \gamma^{\uparrow}(\omega) = \gamma^{\downarrow}(\omega)$ for a two-site chain at $\epsilon_0 = 5.0$ and half-filling. a) The noninteracting case $U = 0$ and low to moderate interaction strengths $U = 0.5$, 1.0 and 1.5. b) Moderate towards strong interaction $U = 2.0$, 3.0, 4.0 and 5.0. Inset in a): Static energy spectrum of the asymmetric two-site chain with $\epsilon_0 = 5.0$ and $\epsilon_1 = 0$. From left to right, the arrows indicate the transition frequencies ω_{AD}, ω_{CD} and ω_{AC}.

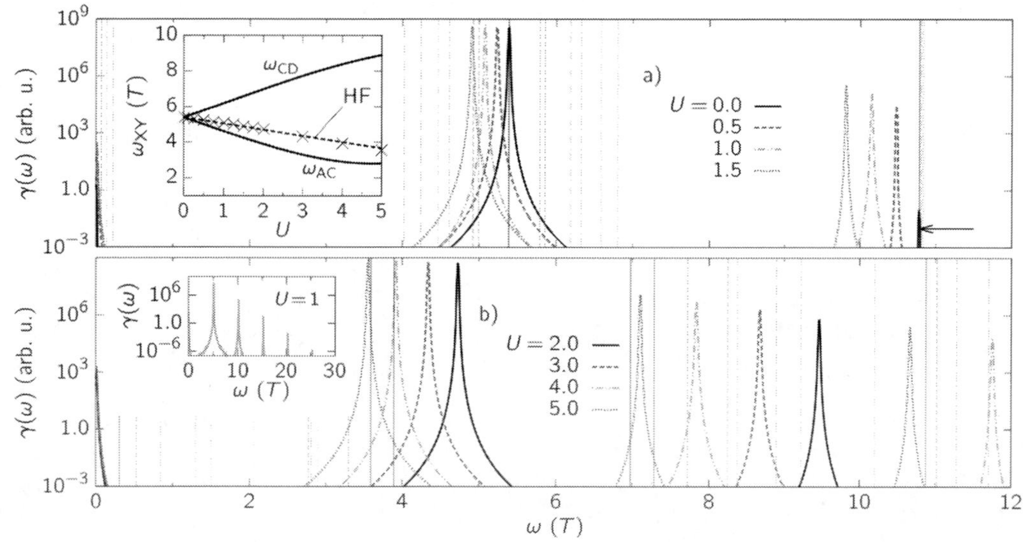

Figure 2. a) and b): Nonlinear density-response spectra $\gamma(\omega)$ in Hartree-Fock (HF) approximation. The parameters and the thin vertical lines indicating the exact transition frequencies are the same as in Figure 1. The inset in a) compares the U-dependence of the exact transition frequencies ω_{AC} and ω_{CD} (solid lines) to the Hartree-Fock result (crosses). The dashed line is a linear fit, cf. also Figure 5 and Eq. (14). For $U = 1.0$, the inset in b) shows the emergence of "higher harmonics" of the main peak in the HF solution. Note that, by accident, the first harmonic for $U = 0$ coincides with the excitation $\omega_{\mathrm{AD}} = 10.770$, see the arrow in a).

3.2. Approximate dynamics

In this section, we investigate the density-response $\gamma(\omega)$ in Hartree-Fock as well as in second Born approximation propagating the KBEs (2). In the latter case, we in addition apply the GKBA of Eq. (3).

The results of Hartree-Fock (HF) calculations for the two-site chain are summarized in Figure 2 a) and b). What immediately becomes evident is the fact that there is no double-peak structure developing for finite $U > 0$ as in Figure 1. Instead, we observe only a single peak below $\omega = 6$ which moves towards smaller energies as U is increased. The inset in the upper panel indicates that the corresponding peak position changes linearly with the on-site interaction U. Moreover, its proximity to the lower black solid line referring to ω_{AC} as a function of U suggests that the HF approximation covers only the transition A $\leftrightarrow$ C, compare with Figure 1 a) (inset). On top of that, the HF approximation seems not to describe the structure of $\gamma(\omega)$ for frequencies around $\omega = 11$. Here, the observed peaks are obviously "higher harmonics" of the main peak, cf. the inset in Figure 2 b).

Next, we perform GKBA calculations including correlations on the second Born (2B) level. Figure 3 shows the result (dots) for $U = 0.5$ and compares to the HF solution (dashed line) and the exact data (solid line). Interestingly, the 2B+GKBA calculation drastically improves the HF result in the sense that, instead of a single peak at $\omega_{\mathrm{HF}} = 5.234$, there appears a double-peak structure with practically vanishing spectral weight around ω_{HF} and almost the correct energy difference. However, the individual peaks are broadened [27], and in addition there emerge side peaks at distances comparable to the double-peak energy spacing $\Delta\omega$. Furthermore, we obtain a similar structure for $\gamma(\omega)$ between $\omega = 9$ and 12 but with one central peak. Importantly, the position of this peak is not a multiple of the frequency of one of the peaks belonging to the double-peak at lower frequencies. Hence, it is not a higher harmonics artifact as observed in the Hartree-Fock solution (see also Figure 5).

In Figure 4, we show the nonlinear density-response $\gamma(\omega)$ in 2B+GKBA for different interaction strengths and investigate the noninteracting limit $U \to 0$. For practically all values of U shown, we are able to identify the double-peak structure which is absent in Hartree-Fock. This is the case, though the broadening of the peaks strongly increases with U. In the opposite case, for U approaching zero, the double-peak structure correctly merges into a single peak indicating the degenerate transition frequencies ω_{AC} and ω_{CD}. In the course of this, also the additional side peaks (which throughout have frequency offsets of multiples of $\Delta\omega$) vanish. In general, the resolution of the side peaks depends on the window function used in the Fourier transform of

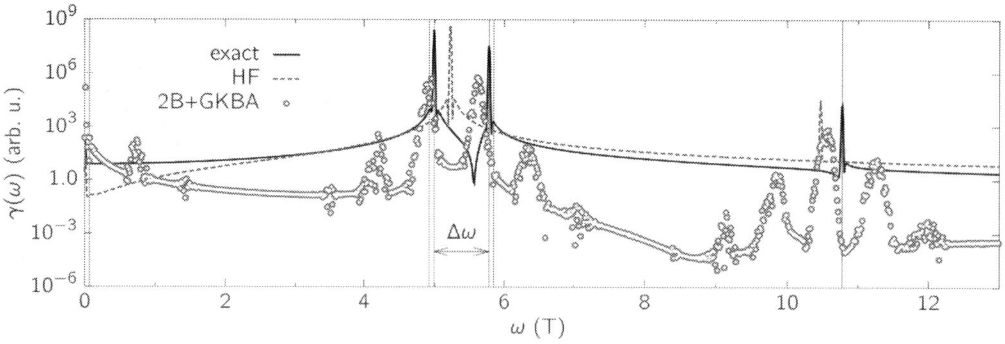

Figure 3. Dynamics of the two-site chain as in Figures 1 and 2 for the case of $U = 0.5$. Comparison of the exact density response $\gamma(\omega)$ to the result in Hartree-Fock (HF) and second Born (2B+GKBA) approximation. The thin black vertical lines are the same as the red dashed lines in Figures 1 a) or 2 a).

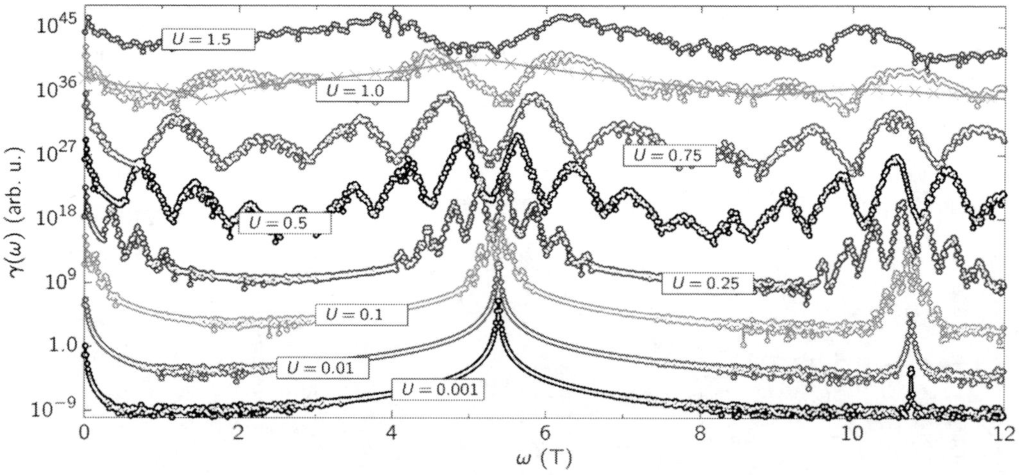

Figure 4. Nonlinear density-response spectrum $\gamma(\omega)$ for a two-site chain at $\epsilon_0 = 5.0$ and half-filling for selected interaction strengths U in second Born approximation and using the GKBA. For the case of $U = 1.0$, the crosses (and the green line) indicate the result for full, self-consistent two-time second Born calculation without applying the GKBA. In the time domain, the corresponding function $\gamma(t)$ is known to be strongly damped, compare with [24].

Eq. (8)—while in Figure 3 we use a Hamming window, the data shown in Figure 4 are obtained by applying a Hann window. However, we note that the characteristic form and broadening of the peaks in 2B+GKBA is independent of the window function and also independent of t^*, cf. Eq. (7).

Finally, in Figure 4, we compare the second Born result for $U = 1.0$ also to full, self-consistent two-time NEGF calculations without the GKBA for $\beta = 100$ (see the green crosses). Here, $\gamma(\omega)$ is extremely broad such that we do not obtain precise information about the dynamical properties and barely can extract frequencies for the involved transitions. The reason for this extreme broadening is a strong damping of the time-dependent occupations $\langle \hat{n}_i^\sigma \rangle(t)$ in 2B, for details see [24].

4. Discussion

The comparison of Figure 1 with Figures 2 to 4, allows for the statement that 2B+GKBA seems to capture the correct physics (in contrast to the Hartree-Fock results and full 2B calculations) although there appear additional artifacts in $\gamma(\omega)$ such as broadening and side peaks. The main result is summarized in the left panel of Figure 5. It shows the exact and approximate transition frequencies ω_{AC}, ω_{CD} and ω_{AD} as function of the interaction strength U.

In order to explain the failure of Hartree-Fock in more detail, we investigate the character of the excitation processes involved, recall the energy spectrum in the inset of Figure 1 a) and see the right panel in Figure 5. For the asymmetric chain with $\epsilon_0 = 5.0$ and $\epsilon_1 = 0$ described by Hamiltonian $\hat{H}_2 = \hat{H}_0 + \hat{H}_1$ for $t > 0$, it is easily shown [28] that the excited state D is a doubly-excited state relative to the ground state A, whereas the states B and C are singly-excited states. With the same argument, state D is furthermore also a doubly-excited state relative to the ground state (gs) of $\hat{H}_0$. This has the following consequences:

(i) As transition A $\leftrightarrow$ D is a double excitation process, it will not appear in the HF solution. This is because a Hartree-Fock response function involves solely single-particle excitations in the case of linear response [21] and includes only linear combinations of these for

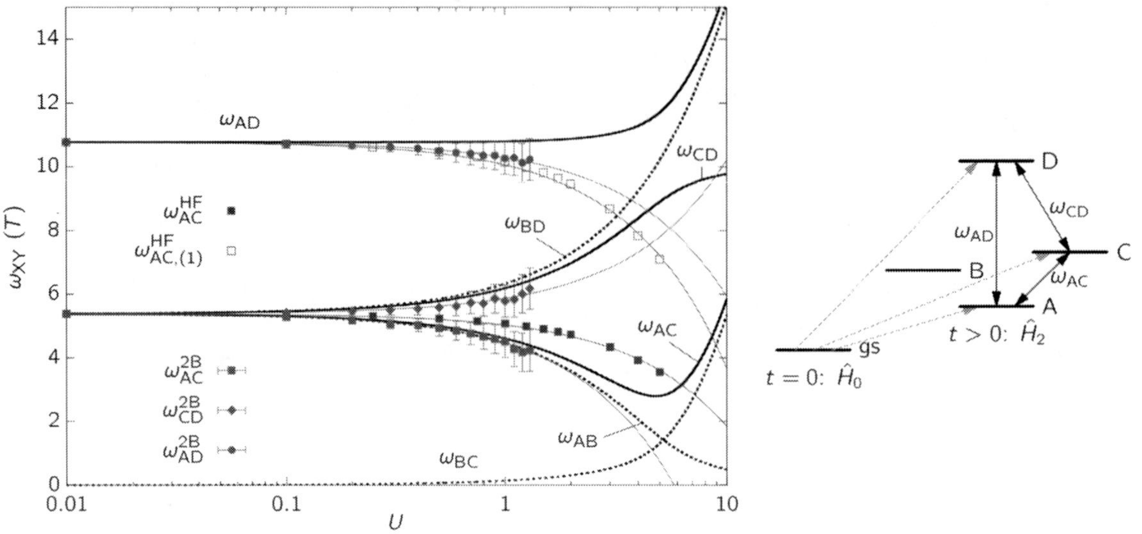

Figure 5. Left: Comparison of the exact transition frequencies ω_{AC}, ω_{CD} and ω_{AD} (black solid lines) to the results obtained in HF (blue) and 2B+GKBA (red). The thin blue and red lines are linear fits to the data points: $\omega_{XY}(U) = \alpha U + \omega_{XY,U=0}$, cf. Eqs. (14) and (13). Right: Schematic view of the excitation dynamics starting from the ground state (gs) of Hamiltonian $\hat{H}_0$. For $t > 0$, we switch to the new Hamiltonian $\hat{H}_2 = \hat{H}_0 + \hat{H}_1$, cf. Eqs. (5) and (6).

contributions of higher order, see, e.g., [29] and references therein.

(ii) As transitions A $\leftrightarrow$ C and C $\leftrightarrow$ D are both of one-electron character, they are in principle describable by Hartree-Fock. The reason why, nonetheless, one observes only the transition A $\leftrightarrow$ C is that the function $\gamma(\omega)$ combined with the excitation (6) only probes energy differences between states that are populated already at time $t = 0$. According to (i), however, the state D is never populated in HF.

Points (i) and (ii) well elucidate the simplicity and shortcomings of the nonlinear response spectra $\gamma(\omega)$ in Hartree-Fock. Moreover, point (i) explains the emergence of higher harmonic effects in the nonlinear regime: Indeed, the frequency $\omega^{\mathrm{HF}}_{\mathrm{AC},(1)}$ of the first harmonic in the HF spectra refers to two single-particle excitations A $\leftrightarrow$ C, see also Eq. (14) below. We emphasize that only by accident the frequencies $\omega^{\mathrm{HF}}_{\mathrm{AC},(1)}$ (of the harmonic) and ω_{AD} (of the double excitation) coincide in the limit $U \to 0$, compare Figures 2 a) and 5.

Furthermore, an analysis of the exact eigenstates of the two Hamiltonians $\hat{H}_0$ and $\hat{H}_2$ reveals that there is finite overlap only between the ground state of $\hat{H}_0$ and the states A, C and D of $\hat{H}_2$ (all being singlets). The vanishing overlap with the eigenstate B (the triplet with $S_z = 0$ [30]) is responsible for the fact that we do not observe transitions involving state B in the density-response. In terms of the wave function, the dynamics of the system for $t > 0$ is therefore,

$$|\Psi(t)\rangle = c_{\mathrm{A}}\, \mathrm{e}^{-\mathrm{i}E_{\mathrm{A}}t}|\mathrm{A}\rangle + c_{\mathrm{C}}\, \mathrm{e}^{-\mathrm{i}E_{\mathrm{C}}t}|\mathrm{C}\rangle + c_{D}\, \mathrm{e}^{-\mathrm{i}E_{\mathrm{D}}t}|\mathrm{D}\rangle\,, \tag{12}$$

with $|\mathrm{X}\rangle$ being the eigenstates of $\hat{H}_2$ (having energy E_{X}) and $c_{\mathrm{X}} = \langle \mathrm{X}|\mathrm{gs}\rangle$ denoting the expansion coefficients with respect to the ground state $|\mathrm{gs}\rangle$ of $\hat{H}_0$.

In contrast to the HF approximation, a second-order self-energy (as 2B) has the ability to describe double excitations, see [22]. For this reason, the 2B+GKBA calculations capture the transition $\omega_{\rm AD}$ and, accordingly, also $\omega_{\rm CD}$. In this context, we again emphasize that the obtained frequency for the transition A $\leftrightarrow$ D is not a multiple of any other frequency observed in 2B+GKBA, compare the red dots to the red squares as well as the empty to the full blue squares in the left panel of Figure 5. The linear fits to the data points are,

$$\omega_{\rm AC}^{\rm 2B} = -0.928\,U + 5.385\,, \quad \omega_{\rm CD}^{\rm 2B} = 0.483\,U + 5.385\,, \quad \omega_{\rm AD}^{\rm 2B} = -0.385\,U + 10.770\,. \tag{13}$$

In Hartree-Fock, we obtain,

$$\omega_{\rm AC}^{\rm HF} = -0.355\,U + 5.385\,, \quad \omega_{\rm AC,(1)}^{\rm HF} = -0.710\,U + 10.770 = 2\,\omega_{\rm AC}^{\rm HF}\,. \tag{14}$$

With the result of Eq. (13), the second Born approximation plus GKBA gives a complete picture of the nonequilibrium dynamics in the considered Hubbard chain. However, there is one additional effect in 2B+GKBA we cannot ignore. This concerns the approximate value for the double excitation frequency $\omega_{\rm AD}$ which behaves differently as function of U than the values for the single excitations $\omega_{\rm AC}$ and $\omega_{\rm CD}$: While the approximate frequencies $\omega_{\rm AC}$ and $\omega_{\rm CD}$ follow the exact results as U is increased, we observe a negative slope (-0.385 in the linear fit) as function of U for $\omega_{\rm AD}$. This is opposite to the exact solution which reveals a clearly positive slope and minor absolute value of $\mathrm{d}\omega_{\rm AD}/\mathrm{d}U$ for small $U \lesssim 2$, see the upper solid black line in Figure 5 (left panel).

5. Summary

In conclusion, we have investigated a small Hubbard-type system with respect to its excitation properties beyond the linear response regime by propagating the Kadanoff-Baym equations in HF and 2B+GKBA. In the course of this, we found clear signals of electron-electron correlations, and the absence of damped solutions for the density-response $\gamma(t)$ in 2B+GKBA allowed us to compute the approximate transition frequencies as function of the interaction strength. Comparisons with the exact Green's function furthermore showed that 2B+GKBA produces reasonable results with only minor deficiencies regarding broadening and side peak effects. Finally, our analysis underlined the importance of double excitations in finite systems.

Acknowledgments

We thank C. Verdozzi for a fruitful discussion and acknowledge the North-German Supercomputing Alliance (HLRN) for providing computing time.

References

[1] Pavarini E, Koch E, Vollhardt D and Lichtenstein A (Eds.) 2011 The LDA+DMFT approach to strongly correlated materials (Forschungszentrum Jülich GmbH: Zentralbibliothek Verlag)
[2] Becker W, Liu X J, Ho P J and Eberly H J 2012 *Rev. Mod. Phys.* **84** 1011
[3] Bloch I, Dalibard J and Zwerger W 2008 *Rev. Mod. Phys.* **80** 885
[4] Wall S, Brida D, Clark S R, Ehrke H P, Jaksch D, Ardavan A, Bonora S, Uemura H, Takahashi Y, Hasegawa T, Okamoto H, Cerullo G and Cavalleri A 2011 *Nature Physics* **7** 114
[5] Kwong N H, Bonitz M, Binder R and Köhler H S 1998 *phys. stat. sol.* (b) **206** 197
[6] Kwong N H and Bonitz M 2000 *Phys. Rev. Lett.* **84** 1768
[7] Rios A, Barker B, Buchler M and Danielewicz P 2011 *Annals of Physics* **326** 1274
[8] Gartner P, Seebeck J and Jahnke F 2006 *Phys. Rev.* B **73** 115307
[9] Lorke M, Nielsen T R, Seebeck J, Gartner P and Jahnke F 2006 *Phys. Rev.* B **73** 085324
[10] Uimonen A-M, Khosravi E, Stan A, Stefanucci G, Kurth S and van Leeuwen R and Gross E K U 2011 *Phys. Rev.* B **84** 115103
[11] Khosravi E, Uimonen A-M, Stan A, Stefanucci G, Kurth S, van Leeuwen R and Gross E K U 2012 *Phys. Rev.* B **85** 075103
[12] Garny M, Kartavtsev A and Hohenegger A 2011 *Preprint* arXiv:1112.6428
[13] Eckstein M, Kollar M and Werner P 2010 *Phys. Rev.* B **81** 115131

[14] Bonitz M 1998 Quantum Kinetic Theory (Stuttgart/Leipzig: Teubner)
[15] Balzer K, Bauch S and Bonitz M 2010 *Phys. Rev.* A **81** 022510
[16] Balzer K, Bauch S and Bonitz M 2010 *Phys. Rev.* A **82** 033427
[17] Lipavský P, Špička V and Velický B 1986 *Phys. Rev.* B **34** 6933
[18] Hermanns S, Balzer K and Bonitz M 2012 *Phys. Scr.* (*Preprint* arXiv:1205.4427)
[19] Hermanns S, Balzer K and Bonitz M 2013 in preparation
[20] Puig von Friesen M, Verdozzi C and Almbladh C-O 2010 *Phys. Rev.* B **82** 155108
[21] Balzer K, Hermanns S and Bonitz M 2012 *EPL* **98** 67002
[22] Säkkinen N, Manninen M and van Leeuwen R 2012 *New J. Phys.* **14** 013032
[23] Kadanoff L P and Baym G 1962 Quantum Statistical Mechanics (New York: Benjamin)
[24] Puig von Friesen M, Verdozzi C and Almbladh C-O 2009 *Phys. Rev. Lett.* **103** 176404
[25] Hochbruck M and Lubich C 1997 *SIAM J. Numer. Anal.* **34** 1911
[26] Alvermann A and Fehske H 2011 *J. Comput. Phys.* **230** 5930
[27] We note that the peak broadening is not determined by the length of the $\gamma(t)$-time series but is a property of the second Born approximation in combination with the GKBA, cf. also Figure 4.
[28] Consider for example the most simple case $U = 0$, which leads to the single particle energies $E_1 = -0.193$ and $E_2 = 5.193$. According to the inset in Figure 1, the transition frequency $\omega_{\rm AD}$ measures $2E_2$, which means that both electrons are excited.
[29] Elliott P, Goldson S, Canahui C and Maitra N T 2011 *Chem. Phys.* **391** 110
[30] Jafari S A 2008 *Iranian J. Phys. Res.* **8** 113

Progress in Nonequilibrium Green's Functions V (PNGF V) IOP Publishing
Journal of Physics: Conference Series **427** (2013) 012007 doi:10.1088/1742-6596/427/1/012007

Time-dependent restricted active space Configuration Interaction theory applied to the photoionization of neon

David Hochstuhl and Michael Bonitz
Institut für Theoretische Physik und Astrophysik, D-24098 Kiel, Germany
E-mail: hochstuhl@theo-physik.uni-kiel.de

Abstract. We give a presentation on the time-dependent restricted active space Configuration Interaction method, which has been recently introduced [D. Hochstuhl, M. Bonitz, PRA, in press] to efficiently tackle the few-particle nonequilibrium many-body problem. Particularly, we apply it to the calculation of photoionization cross sections and angular resolved ionization yields of groundstate neon.

1. Introduction

The rapidly increasing quality of experimental techniques in the generation of sub-femtosecond laser pulses has led to a lot of new insights and surprises. For example, in a recently performed experiment using the attosecond streak camera, i.e. a combination of attosecond extreme ultraviolet (XUV) pump and infrared (IR) probe spectroscopy, a time-delay in the photoemission of electrons from the $2s$ and $2p$-shells of neon has been observed, with the $2s$-electron leaving the neon atom around 21 attoseconds before the $2p$-electron [1]. Similar results have been also obtained using different attosecond techniques for the argon atom [2]. Another recent combined experimental and theoretical study considering as well XUV pump pulses but terahertz pulses for the probe revealed the existence of a chirp in the emission of Auger electrons [3, 4].

A general framework for the theoretical description of such explicitly time-dependent photoionization processes is given by the time-dependent Schrödinger equation (TDSE) which, due to the lack of analytical solutions, must be tackled through computer simulations. For one- and two-particle systems, the numerical approach is nowadays routinely applied, as can be observed by a plethora of investigations on helium and atomic and molecular hydrogen. For larger electron numbers, however, direct numerical solutions are unfeasible for all but very small grid sizes, which is a direct consequence of the exponentially growing effort required to store and process the discretized wavefunction. This fact is well known in computer science under the name *exponential wall* or the *course of dimensionality*, and beside the Schrödinger equation does affect several other disciplines such as combinatorics, optimal control or graph theory.

The basic problem can be readily illustrated: consider the N-particle fermionic wavefunction $\Psi(\mathbf{x}_1, \cdots, \mathbf{x}_N, t)$, where $\mathbf{x}_k$ is a complete observable of electron k at time t, and assume that the state $\mathbf{x}_k$ is described by a number of N_b attributes, the basis functions. Then, a general wavefunction would have to assign a complex number to each of the $\binom{N_b}{N}$ possible ordered combinations of $(\mathbf{x}_1, \cdots, \mathbf{x}_N)$. The accurate simulation of photoionization applications now can easily require basis sizes of $N_b \sim 10000$, as the electrons may travel to large distances from the ionic nucleus, so that the corresponding wavefunction becomes too difficult to manage in numerical calculations. The same is true for the case of a moderate basis size but a large number of particles, as it is encountered, for instance, in quantum chemistry.

Several workarounds have been designed to avoid this fundamental issue. Multiconfigurational time-dependent Hartree-Fock theory, for instance, uses a smaller single-particle basis which are time-dependent, and in this way tries to reduce the determinant size [5–7]. Time-dependent density-functional theory (TDDFT) replaces the wavefunction by the single-particle density $\rho(\mathbf{x}, t)$, i.e. by a much reduced quantity. Density-matrix functional theories instead employ the generalized one-particle density $\rho(\mathbf{x}, \mathbf{x}', t)$, whereas nonequilibrium Greens functions (NEGFs) $G(\mathbf{x}t, \mathbf{x}'t')$ even keep the information of a second time variable, which is basically related to the system's ionization energies and electron affinities. The exact equations of motion for these kind of approximations usually involve a coupling to higher order density matrices or Greens functions, which, in order to obtain a closed scheme, must be approximated in terms of the respective reduced quantity. Depending on the contained information, this reconstruction can be performed more (NEGFs) or less (TDDFT) systematically. In any case, the error of the result can hardly be controlled.

The time-dependent restricted active space Configuration Interaction (TD-RASCI) method presented in this work follows a different approach, which is set completely on the N-particle level. In order to reduce the exponential problem, the domain of the wavefunction is restricted. In the above mentioned example, one could for instance allow only one electron to travel to large distances from the core, whereas the remaining $(N-1)$ particles are restricted to a narrow region in the vicinity of the nucleus. Given that the $(N-1)$ particles can be described by a reduced number of basisfunctions of say $N_b \sim 100$, the discretized domain $(\mathbf{x}_1, \cdots, \mathbf{x}_N)$ becomes smaller by roughly a factor of 100^{N-1} as compared to the previous example using 10000 basisfunctions for all particles. This enables the treatment of systems having a much larger number of particles.

In the following, we summarize the TD-RASCI method, which has been proposed in our former work [8]. Thereafter, in section 3, we apply the method to the calculation of angular resolved photoionization cross sections and asymmetry parameters of the Neon atom.

2. TD-RASCI method

The time-dependent restricted active space Configuration Interaction method begins with an ansatz to the N-particle wavefunction in terms of Slater determinants,

$$|\,\Psi(t)\,\rangle \;=\; \sum_{I\in\Omega} C_I(t)\,|\,\psi_{i_1}\psi_{i_2}\cdots\psi_{i_N}\,\rangle\,. \tag{1}$$

where the multi-index $I = (i_1, \cdots, i_N) \in \Omega$ specifies the occupied single-particle spin-orbitals $|\,\psi_k\,\rangle$. The included Slater determinants are determined by the index set $\Omega \subset \mathbb{N}^N$. By choosing all anti-symmetrized indices which are possible for N-particles and N_b orbitals,

$$\Omega \;=\; \{(i_1, \cdots, i_N)\,|\,1 \le i_1 < \cdots < i_N \le N_b\}\,, \tag{2}$$

one obtains the Full Configuration Interaction (FCI) ansatz: the wavefunction is completely correlated and exact up to the finite size of the spin-orbital basis. In many cases, however, the FCI expansion is unnecessary complex as a large number of Slater determinants never get occupied during the propagation. The idea of the TD-RASCI scheme is therefore to drop those determinants from the expansion which are considered to be of minor importance. The selection procedure must thereby be performed in advance, i.e. one should at best be aware of the physical processes that are expected to occur and then try to incorporate these processes in a way that the calculation becomes manageable.

In practice, the selection can hardly be performed manually, but rather a systematic procedure should be applied. Such one is presented in our former work [8]. Thereby, the single-particle basis space is first divided into a chosen number of partitions, and then restrictions are imposed onto the occupation of these partitions. To take up the example of the introduction, for the calculation of single-electron photoionization it is convenient to divide the coordinate space into an inner bound region and outer continuum region and allow for at most a single electron in the outer region. The corresponding wavefunction is then given by

$$| \Psi(t) \rangle = \sum_{I \in \Omega_{N-1}} \sum_{j=1}^{N_b} C_{I,j}(t) \hat{\mathcal{A}} | \phi_{i_1} \cdots \phi_{i_{N-1}} \rangle | \phi_j \rangle , \tag{3}$$

which is set up in terms of a localized basis set $\{| \phi_k \rangle\}$, of which some basis functions are located only in the inner region and others only in the outer region. The index set Ω_{N-1} then is taken to include only indices corresponding to inner basis functions, while the second sum runs over the whole orbital basis. The total size of the wavefunction expansion (3) is given by $N_{\rm det} = N_b |\Omega_{N-1}|$ and requires therefore, at least for small cardinality of Ω_{N-1}, a similar effort than a single-particle problem. In case Ω_{N-1} contains only a single $(N-1)$-particle determinant, one recovers the single-active electron approximation. Note, however, that in contrast to the commonly employed SAE approaches the inter-shell interaction is treated in a consistent way and no pseudo-potentials are required whatsoever. It is further straightforward to consider more sophisticated approximations, e.g., allowing for two electrons outside, or even for more elaborate choices. Using a reasonable computer implementation, such as the one presented in Ref. [8], this is accomplished simply by a change of the input parameters.

For the present work, we consider only single ionization and therefore apply the wavefunction expansion given in Eq. (3). The single-particle basis set is given by

$$\phi_{klm\sigma}(\mathbf{r}, m_s) = \frac{\chi_k(r)}{r} Y_{lm}(\theta, \phi) \, \sigma(m_s) , \tag{4}$$

where Y_{lm} denote spherical harmonics, σ the spin function ($\sigma \in \alpha, \beta$), and the radial functions χ_k are taken to be finite-element discrete variable representation (FEDVR) functions [9]. In short, a FEDVR constitutes a basis of Lagrange interpolating polynomials constructed over a Gauss-Lobatto grid, which on the one hand is flexible in describing arbitrary functions, and on the other leads to a sparse representation of the electron integrals. If we now applied the basis (4), we would observe that a large number of determinants were required in the inner region to obtain an appropriate initial state, which is caused by the fact that Lagrange polynomials do not resemble atomic orbitals. Hartree-Fock orbitals, on the other hand, provide a reasonable initial state with a few determinants, but have the disadvantage that the electron integrals are not sparse at all, which poses a severe restriction on the applicable size of the basis. The solution is to use a mixed basis of Hartree-Fock and FEDVR functions, where the Hartree-Fock problem is solved in the first finite element using the FEDVR basis, and the thus obtained eigenfunctions replace the FEDVR functions in the first element. The arising basis set is illustrated in Fig. 1. The mixed basis has the advantage that it provides an appropriate initial state in the region near the nucleus, while in the continuum it exploits all the advantages of a grid-like treatment. Note that, due to the unitary transformation in the first element, no reflections or similar problems arise at the element boundary.

Having constructed an appropriate TD-RASCI ansatz, the further steps are straightforwardly similar to any direct approach such as Full Configuration Interaction. Insertion of the ansatz into the time-dependent Schrödinger equation leads to the equation of motion for the expansion

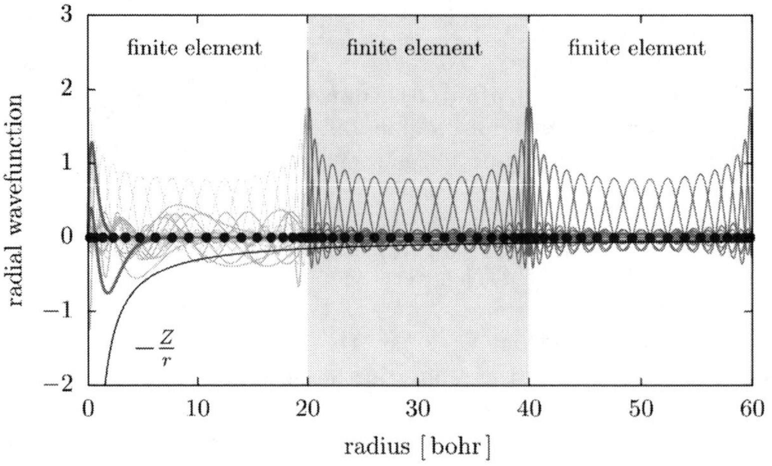

Figure 1. Mixed basis applied in the TD-RASCI method for the efficient treatment of photoionization. It consists of a Hartree-Fock basis in the region near the atomic core and a finite-element discrete variable representation in the region outside $r = 20$ bohr, the continuum. The mixed basis, on the one hand, provides an appropriate initial state which is set up by a small number of determinants and, on the other, exploits the advantages of the grid-like treatment in the continuum.

coefficients,

$$i\,\dot{C}_I(t) \;=\; \sum_{J\in\Omega} \left\langle\, \phi_{i_1}\cdots\phi_{i_N} \,\middle|\, \hat{H}(t) \,\middle|\, \phi_{j_1}\cdots\phi_{j_N} \,\right\rangle C_J(t)\,, \tag{5}$$

where we apply a Hamiltonian of the form

$$\hat{H}(t) \;=\; \sum_{k=1}^{N}\left\{\frac{\hat{\mathbf{p}}_k^2}{2} - \frac{Z}{r_k} + \mathbf{E}(t)\cdot\mathbf{r}_k\right\} + \frac{1}{2}\sum_{k\neq l}\frac{1}{|\mathbf{r}_k-\mathbf{r}_l|}\,, \tag{6}$$

which consists of kinetic and potential energy, the Coulomb interaction and the action of an electromagnetic field coupled in dipole approximation. The arising Hamiltonian matrix elements between Slater determinants are evaluated using Slater-Condon rules [10] and stored in sparse matrices. The initial state is found using the Lanczos method which iteratively constructs the groundstate using only matrix-vector products, and the thus obtained groundstate is then propagated with the time-dependent version of the Lanczos method [11]. At the end of the propagation, we use the ionized part of the wavefunction [the factor $|\,\phi_j\,\rangle$ in Eq. (3)] to extract the relevant information of the ionized particle.

3. Single-photoionization of Neon

Neon is a common target in photoionization studies, which has been investigated several times in the previous half century. Theoretically, its closed-shell structure facilitates the application of theoretical methods, and several methods such as perturbation theory [12], Hartree-Fock [13] or random-phase methods [14], R-matrix theory [15] or the Stieltjes-Chebyshev moment technique [16] have been used to calculate cross sections. Furthermore, experimentally measured cross sections can be found in Refs. [17–19]. In this work, we calculate total and

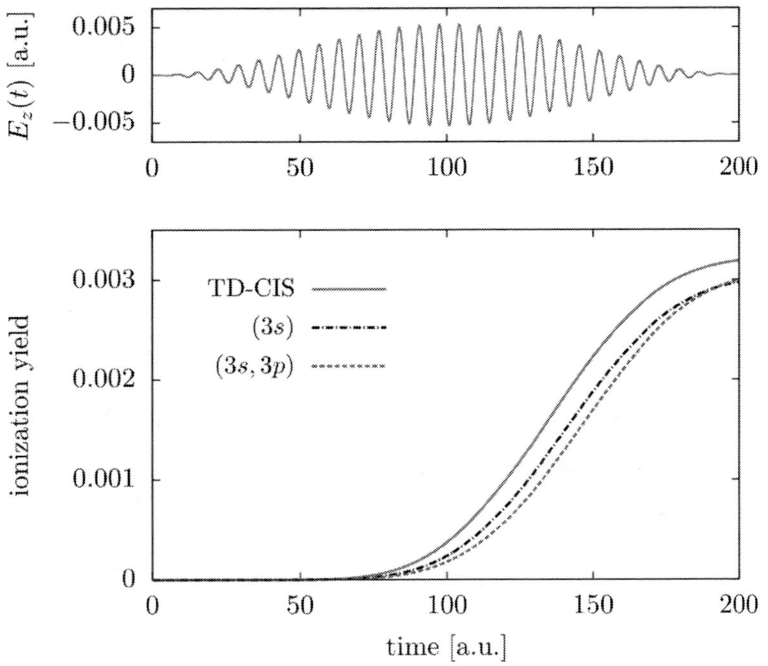

Figure 2. Ionization yield plotted versus time for photoionization from the groundstate of neon subjected to the laser field shown in the top panel. The photon energy is 25 eV and the intensity is 10^{12} W/cm^2. The TD-RASCI method predicts a slightly larger ionization yield than the $(3s)$ and $(3s, 3p)$ approximations (see text).

angular-resolved photoionization yields with the TD-RASCI method. The basic form of the wavefunction is given in Eq. (3), i.e. we consider single-ionization, where only one particle is allowed in the continuum (outside $r > 20$ bohr). For the wavefunction in the inner region, which is determined by the set Ω_{N-1}, we consider three approximations: (i) the TD-CIS approximation where only the Hartree-Fock groundstate plus all single-excitations is included, (ii) the approximation where double-excitations up to the $3s$ orbital are allowed which is denoted $(3s)$ in the following, and (iii) the approximation where double-excitations to the $3s$ orbital and further to the $3p$ orbitals are allowed [denoted $(3s, 3p)$]. The single-particle basis we use consists of partial waves up to $l = 3$ and a radial grid that extends up to 400 bohr and consists of finite elements of length 4 bohr which include 10 basisfunctions each (except for the first one which includes 20 basisfunctions). The total number of Slater determinants used to describe the wavefunction is 32447 for the TD-CIS approximation, 237922 for the $(3s)$ approximation, and 454151 for the $(3s, 3p)$ approximation. The neon atom is propagated for $T = 200$ a.u. under the action of an laser pulse with a sine-squared envelope (also with duration T). The calculations take in between 90 minutes on a single processor for the TD-CIS approximation and 48 hours for the $(3s, 3p)$ approximation.

Figure 2 depicts the ionization dynamics for photoionization from the neon groundstate with the 25 eV pulse shown in the top panel. A particle is thereby considered as being ionized if it is located outside a radius of 20 bohr from the nucleus. All three TD-RASCI approximations show the same qualitative behavior: a steep rise around $t = 100$ a.u., which is delayed with respect to the field due to the time needed to travel to $r = 20$ bohr, followed by a saturation around the end of the pulse. The TD-CIS approximation produces the largest yield and a slightly earlier

Journal of Physics: Conference Series **427** (2013) 012007 doi:10.1088/1742-6596/427/1/012007

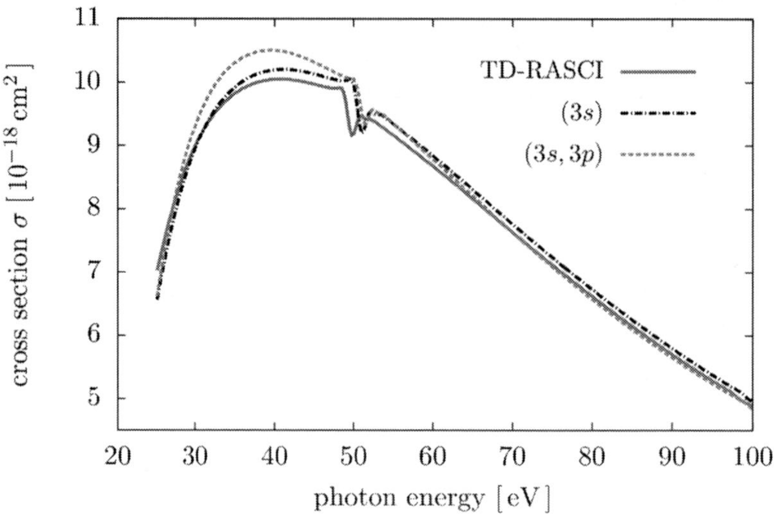

Figure 3. Total single-photoionization cross section of neon plotted against the photon energy for photoionization from the groundstate. The duration of the laser field's squared-sine envelope is $T = 200$ a.u. and the intensity is 10^{12} W/cm^2.

rise than the more sophisticated TD-RASCI approximations.

We now turn to the photoionization cross sections. The cross section σ is calculated from the ionization yield Γ with the formula derived in Ref. [20],

$$\sigma[\text{cm}^2] = 1.032 \cdot 10^{-4} \frac{\omega^2 \Gamma}{n I_0}, \tag{7}$$

where n represents the number of cycles in the pulse, ω is the frequency in eV and I_0 is the intensity in W/cm^2. The formula determines the cross section through the ionization rate (the yield divided by time) by assuming an adiabatic approximation for the laser pulse, i.e. the laser pulse is treated as if it had a constant intensity which acts for a certain effective time. The ionization yield Γ is determined as the norm of the wavefunction outside $r = 20$ bohr at the end of the pulse. In Fig. 3, we plot the single-photoionization total cross section against the photon energy, for a laser pulse with duration 200 a.u. and an intensity of 10^{12} W/cm^2. All three approximations produce cross sections which rise around the Hartree-Fock ionization energy which is 23.15 eV for Neon. One further obtains a monotonic decrease towards larger photon energies. Around 50 eV, we observe the $2s\,2p^6\,3s$ resonance, the exact location of which is 45.547 eV according to the NIST database. The better approximations shift the position of the resonance towards the wrong direction. When comparing the calculated cross section to experimental results, we further observe that the calculations are 10% larger than the measured values in Ref. [19]. This reason for this discrepancy remains a subject of further investigation, since the TD-RASCI scheme usually performs well in estimating the cross sections [8].

The momentum distribution of the ionized particle for a photon energy of 70 eV is depicted in Fig. 4. The momentum is calculated through a Fourier transform of the ionized part of the wavefunction ($r > 20$ bohr). The left picture shows the two-dimensional momentum distribution which arises from the $(3s, 3p)$ approximation. Due to the linear polarization of the laser it is symmetric with respect to rotations around the z-axis. One clearly notices two rings which correspond to ionization from the $2s$ and $2p$ shells. The inner ring around 30 eV

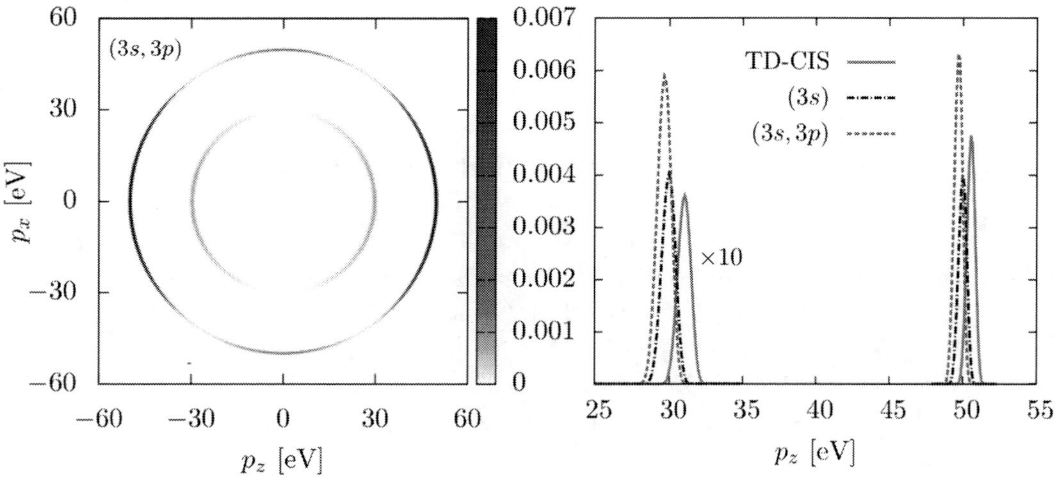

Figure 4. Momentum distribution of the ionized particle for a photon energy of 70 eV. *Left:* Momentum distribution for the $(3s, 3p)$ approximation. One clearly notices two rings which correspond to ionization from the $2s$-orbital (inner ring) and the $2p$-orbital (outer ring). *Right:* Momentum distribution in forward direction for the three TD-RASCI approximations. The peaks which correspond to ionization from the $2s$ shell are multiplied by a factor of 10.

exhibits p-symmetry and stems from the ionization from the $2s$-orbital. The outer ring at 50 eV corresponds to the ionization from the $2p$ orbitals and is dominantly of d-symmetry, but also has a contribution of s-symmetry, as can be seen since the distribution does not go down to zero, but remains at a finite value. The right panel shows the momenta from the three approximations in forward direction along the z-axis, i.e. $\theta = 0$. It is directly obvious that the accuracy of the wavefunction has a distinct impact on the spectra. The TD-CIS results yield the highest photoelectron momentum. With increasing accuracy, both peaks shift towards a lower momentum. Interestingly, the $2s$-peak shows a larger relative shift than the peak corresponding to the $2p$ ionization. This is likely to have a major impact in the calculation of time-delays using the streak camera principle, and indicates the importance of applying a correlated description.

4. Conclusion

In this proceeding paper, we applied the recently introduced time-dependent restricted active space Configuration Interaction method to the photoionization of the neon atom. After a recapitulation of the method, we have calculated the total cross sections for three different TD-RASCI approximations, and further considered the time-resolved photoionization at 25 eV as well as the momentum distribution at 70 eV. Particularly the momentum distribution is significantly affected by the increased amount of correlation present in the more accurate TD-RASCI wavefunctions. We therefore expect correlations to be of similar importance for the topic we plan to consider in the future, namely the calculation of time-delays in the photoionization of neon.

Acknowledgments
We gratefully acknowledge support by the Bundesministerium für Bildung und Forschung via the project FLASH and by a CPU grant at the North German Supercomputer Center (HLRN, grant shp0006).

References
[1] Schultze M, Fieß M, Karpowicz N, Gagnon J, Korbman M, Hofstetter M, Neppl S, Cavalieri A, Komninos Y, Mercouris T *et al.* 2010 *Science* **328** 1658–1662
[2] Klünder K, Dahlström J, Gisselbrecht M, Fordell T, Swoboda M, Guénot D, Johnsson P, Caillat J, Mauritsson J, Maquet A *et al.* 2011 *Phys. Rev. Lett.* **106** 143002
[3] Schütte B, Bauch S, Frühling U, Wieland M, Gensch M, Plönjes E, Gaumnitz T, Azima A, Bonitz M and Drescher M 2012 *Phys. Rev. Lett.* **108**(25) 253003
[4] Bauch S and Bonitz M 2012 *Phys. Rev. A* **85** 053416
[5] Hochstuhl D, Bauch S and Bonitz M 2010 *J. Phys.: Conf. Ser.* **220** 012019
[6] Hochstuhl D and Bonitz M 2011 *J. Chem. Phys.* **134** 084106
[7] Bonitz M, Hochstuhl D, Bauch S and Balzer K 2010 *Contrib. Plasm. Phys.* **50** 54–59
[8] Hochstuhl D and Bonitz M 2012 *Phys. Rev. A* **86** 053424
[9] Rescigno T N and McCurdy C W 2000 *Phys. Rev. A* **62** 032706
[10] Helgaker T, Joergensen P and Olsen J 2000 *Molecular electronic structure theory* (New York: Wiley)
[11] Park T and Light J 1986 *J. Chem. Phys.* **85** 5870
[12] Chang T N 1977 *Phys. Rev. A* **16**(3) 1171–1178
[13] Sewell K G 1965 *Phys. Rev.* **138**(2A) A418–A421
[14] Johnson W and Cheng K 1979 *Phys. Rev. A* **20** 978
[15] Burke P G and Taylor K T 1975 *J. Phys. B* **8** 2620
[16] Gokhberg K, Vysotskiy V, Cederbaum L S, Storchi L, Tarantelli F and Averbukh V 2009 *J. Chem. Phys.* **130** 064104
[17] Marr G and West J 1976 *Atomic Data and Nuclear Data Tables* **18** 497 – 508 ISSN 0092-640X
[18] Chan W F, Cooper G, Guo X and Brion C E 1992 *Phys. Rev. A* **45**(3) 1420–1433
[19] Samson J and Stolte W 2002 *J. Electron. Spectrosc. Relat. Phenom.* **123** 265–276
[20] Foumouo E, Kamta G L, Edah G and Piraux B 2006 *Phys. Rev. A* **74** 063409

Progress in Nonequilibrium Green's Functions V (PNGF V) IOP Publishing
Journal of Physics: Conference Series **427** (2013) 012008 doi:10.1088/1742-6596/427/1/012008

Few-particle quantum dynamics—comparing nonequilibrium Green functions with the generalized Kadanoff–Baym ansatz to density operator theory

S Hermanns[1], **K Balzer**[2] **and M Bonitz**[1]

[1] Institut für Theoretische Physik und Astrophysik, Universität Kiel, D-24098 Kiel, Germany
[2] University of Hamburg, Max Planck Research Department for Structural Dynamics Building 99 (CFEL), Luruper Chaussee 149, D-22761 Hamburg, Germany

E-mail: hermanns@theo-physik.uni-kiel.de

Abstract. The nonequilibrium description of quantum systems requires, for more than two or three particles, the use of a reduced description to be numerically tractable. Two possible approaches are based on either reduced density matrices or nonequilibrium Green functions (NEGF). Both concepts are formulated in terms of hierarchies of coupled equations—the Bogoliubov-Born-Green-Kirkwood-Yvon (BBGKY) hierarchy for the reduced density operators and the Martin-Schwinger-hierarchy (MS) for the Green functions, respectively. In both cases, similar approximations are introduced to decouple the hierarchy, yet still many questions regarding the correspondence of both approaches remain open.

Here we analyze this correspondence by studying the generalized Kadanoff–Baym ansatz (GKBA) that reduces the NEGF to a single-time theory. Starting from the BBGKY-hierarchy we present the approximations that are necessary to recover the GKBA result both, with Hartree-Fock propagators (HF-GKBA) and propagators in second Born approximation. To test the quality of the HF-GKBA, we study the dynamics of a 4-electron Hubbard nanocluster starting from a strong nonequilibrium initial state and compare to exact results and the Wang-Cassing approximation to the BBGKY hierarchy presented recently by *Akbari et al.* [1].

1. Introduction

The *ab initio* time-dependent description of quantum many-body systems has been a major focus in the physics and chemistry communities since the invention of quantum mechanics. Despite many efforts, up to the present time, the exact (analytic or numerical) solution of the underlying equation of motion for the wavefunction $|\Psi\rangle_S(t)$ of the system—the Schrödinger equation—is, in general, restricted to only very few particles, due to the exponentially growing complexity with the particle number, e.g. [2]. To overcome this limitation a great variety of methodologies have been developed to approximately describe systems of larger particle numbers. This includes time-dependent density functional theory [3], time-dependent Hartree-Fock [4, 5, 6, 7, 8], multi-configuration time-dependent Hartree-Fock [9], time-dependent coupled cluster theory [10], Møller-Plesset many-body perturbation theory [11], density matrix renormalization group based approaches [12] or dynamical mean field theory, e.g. [13].

In this article, we focus on two different closely related powerful methods: nonequilibrium Green functions (NEGF) [14] and reduced density operators (DO), e.g. [15, 16, 17, 18], respectively. Both methods involve the solution of a coupled hierarchy—the MS and BBGKY hierarchy, respectively—of reduced quantities, where the solution of the full hierarchy is equivalent to the solution of the N-particle Schrödinger equation or von Neumann equation, in the case of pure or mixed states, respectively. To reduce the complexity, both methods aim at simplifications via a physically motivated truncation of the hierarchy. Since both approaches are selfcontained and independent, comparisons of the two are of great interest for applications. NEGF are commonly regarded as more accurate, however, nonequilibrium solutions of the two-time equations of motion—the Keldysh-Kadanoff–Baym equations (KBE)—are computationally very expensive. While there has been remarkable progress during the last decade for homogeneous [19, 20] and inhomogeneous systems, e.g. [21, 22, 23], the two-time structure puts strong limits on the achievable propagation time. In contrast, the single-time density operator approach does not suffer this problem. Furthermore, recent solutions of the KBE for finite systems [24, 25] indicated unphysical long-time behavior (damping). Here, again solutions using single-time equations [26] are, apparantly, closer to the exact result. It is, therefore, important to understand how the two-time and single-time approximations to the hierarchy are related to each other and whether and when single-time solutions are justified.

The recipe how to derive the single-time approximation from the NEGF for an arbitrary selfenergy is the so-called generalized Kadanoff–Baym ansatz (GKBA) derived by Lipavsky, Spicka and Velicky [27, 28]. The GKBA has, so far been used for spatially homogeneous systems, for electron-phonon scattering and for Coulomb scattering in the second Born approximation. Numerical comparisons for the case of Coulomb scattering have shown satisfactory agreement [29, 30, 31]. Similar observations have been recently made for finite systems [26]. Nevertheless, a systematic analysis is still missing. The aim of this paper is two-fold. First, we discuss the relation between NEGF and DO (on the level of the second Born approximation). We take the opposite route, compared to Lipavsky et al. [27, 28] and derive the GKBA result, starting from the BBGKY-hierarchy and identifying the approximations necessary to obtain the GKBA. Second, we consider, as an example of the quantum dynamics of a finite system, the evolution of a 4-site Hubbard nano-cluster—computed with Green functions and the GKBA—and compare them to BBGKY-based results by *Akbari et al.* [1] using the Wang/Cassing approximation.

2. Theory

In this section, we give a brief overview of the theoretical foundations of the two aforementioned methods for the statistical description of quantum many-body systems. One of the most basic differences between these two approaches is the incorporation of the spin statistics of the particles, which, e.g., for Fermions leads to the effect of Pauli blocking [34]. While the Green function method, which will be described in Sec. 2.2, has the spin statistics intrinsically build in by use of bosonic/fermionic creation- and annhilation-operators, the reduced density operator theory is formulated for spinless particles, which requires an explicit (anti-)symmetrization of the equations. This advantage of the former approach is at least partly counterfeited by the more complicated structure of the equations for the Green function $G(t, t')$, which depends—apart from the physical time t—on an additional time argument t', e.g., providing access to the corresponding ionized system. Numerically, this leads to comparatively more involved calculations with the Green function approach. To make up for that, the introduction of the GKBA for the single-particle Green function allows for a simple approximate reconstruction of the two-time Green function from its time-diagonal value, by which a scaling of computation time comparable to density operator theory is achieved.

Journal of Physics: Conference Series **427** (2013) 012008 doi:10.1088/1742-6596/427/1/012008

2.1. Density operator theory

For an N-particle quantum system in a mixed state (in contact with a stationary environment), the proper fundamental quantity is the N-particle density operator

$$\rho_{1\ldots N} = \sum_k W_k \left|\Psi^{(k)}_{1\ldots N}\right\rangle \left\langle\Psi^{(k)}_{1\ldots N}\right| \qquad \sum_k W_k = 1, \tag{1}$$

where $\left|\Psi^{(k)}_{1\ldots N}\right\rangle$ is a possible micro-state compatible with a given macro-state of the whole system including the bath, and W_k denotes the probability of its realization. This description is valid if the interaction of the system with the bath is weak. The density operator obeys the von Neumann equation

$$\mathrm{i}\hbar\partial_t\rho_{1\ldots N} - \left[H_{1\ldots N}\,,\,\rho_{1\ldots N}\right](t) = 0\,, \tag{2}$$

where $H_{1\ldots N}$ is the Hamiltonian of the N-particle system and $\left[\cdot\,,\,\cdot\cdot\right]$ denotes the standard commutator. For a given initial state $\rho_{1\ldots N}(t_0) = \rho^{(0)}_{1\ldots N}$, the solution of this equation completely determines the time evolution of the system and is equivalent to the solution of the Schrödinger equation for all possible micro-states.

2.1.1. Nonequilibrium quantum BBGKY-Hierarchy. Instead of the full N-particle density operator, it is usually useful to consider reduced density operators, the s-particle operator $F_{1\ldots s}$ being defined as a partial trace over the N-body density operator $\rho_{1\ldots N}$ [17],

$$F_{1\ldots s} = \mathcal{V}^s \mathrm{Tr}_{s+1\ldots N}\rho_{1\ldots N}\,, \qquad \frac{1}{\mathcal{V}^s}\mathrm{Tr}_{1\ldots s}F_{1\ldots s} = 1\,, \tag{3}$$

where $\mathcal{V}$ denotes the volume, and the partial trace $\mathrm{Tr}_{1\ldots s}A$ of an operator A in an arbitrary basis of states $|x\rangle = |x_1\rangle|x_2\rangle\ldots|x_N\rangle$ is defined as

$$\mathrm{Tr}_{1\ldots s}A_{1\ldots N} = \sum_{x_1\ldots x_s} A(x_1\ldots x_s, x_{s+1}\ldots x_N; x_1\ldots x_s, x'_{s+1}\ldots x'_N)\,. \tag{4}$$

The density operators obey a system of equations of motion—the BBGKY hierarchy that follows from taking the partial trace over Eq. (2)

$$\mathrm{i}\hbar\partial_t F_1 - \left[H_1\,,F_1\right] = n\mathrm{Tr}_2\left[V_{12}\,,F_{12}\right], \tag{5}$$

$$\mathrm{i}\hbar\partial_t F_{12} - \left[H_{12}\,,F_{12}\right] = n\mathrm{Tr}_3\left[V_{13}+V_{23}\,,F_{123}\right], \tag{6}$$

$$\ldots \quad \ldots \quad \ldots$$

$$F_1(t_0) = F_1^0\,, \qquad F_{12}(t_0) = F_{12}^0\,, \qquad \ldots$$

and so on. It is obvious that the whole hierarchy is equivalent to the von Neumann equation. Note that this system of equations is local in time, all functions depend on a single physical time t, and it has to be complemented by initial conditions $F_1(t_0), F_{12}(t_0)$ and so on. Here H_1 is the single-particle Hamiltonian whereas the two-particle Hamiltonian is given by $H_{12} = H_1 + H_2 + V_{12}$. As mentioned before, Eqs. (5) and (6) are written for spinless particles to make the mathematical structure more transparent. We will add the appropriate exchange contributions in the next section where we consider the (anti-)symmetrization and approximations to the BBGKY hierarchy via a cluster expansion.

2.1.2. (Anti-)Symmetrization and Cluster expansion of the BBGKY hierarchy. For the practical analytical or numerical solution of the BBGKY hierarchy, it is obvious that it has to be truncated to become tractable. A suitable approach consists in performing the so-called cluster expansion to separate the two-particle, three-particle and higher correlations from the ideal part of the density operator which is given by a product of single-particle operators,

$$\begin{aligned} F_{12}(t) &= F_1(t)F_2(t) + c_{12}(t)\,, \\ F_{123}(t) &= F_1(t)F_2(t)F_3(t) + F_1(t)c_{23}(t) + F_2(t)c_{13}(t) + F_3(t)c_{12}(t) + c_{123}(t)\,, \end{aligned} \tag{7}$$

and analogously for the higher-order density operators.

To correctly account for the spin statistics of bosons (fermions) we now (anti-)symmetrize all expressions. This can be done by introducing matrix representations with respect to an (anti-)symmetrized system of basis states defined in Fock space. Alternatively, (anti-)symmetric expectation values of observables can be computed with standard Hilbert space states when the respective operators are (anti-)symmetrized [35]. Here we follow the latter idea as it leads to more compact expressions. The (anti-)symmetrization of the density operators is achieved by replacing

$$\begin{aligned} F_{12} &\longrightarrow F_{12}\Lambda^{\pm}_{12}\,, \\ c_{12} &\longrightarrow c_{12}\Lambda^{\pm}_{12}\,, \\ F_{123} &\longrightarrow F_{12}\Lambda^{\pm}_{123}\,, \\ c_{123} &\longrightarrow c_{123}\Lambda^{\pm}_{123}\,, \end{aligned} \tag{8}$$

where the binary/ternary (anti-)symmetrization operator is defined by its action on an arbitrary two-particle/three-particle state $|12\rangle$ and $|123\rangle$, respectively,

$$\begin{aligned} \Lambda^{\pm}_{12}\,|12\rangle &= (1 \pm P_{12})\,|12\rangle = |12\rangle \pm |21\rangle\,, \\ \Lambda^{\pm}_{123}\,|123\rangle &= \Lambda^{\pm}_{12}(1 \pm P_{13} \pm P_{23})\,|123\rangle \end{aligned} \tag{9}$$

and the upper/lower sign applies to bosons/fermions.

We now introduce these cluster expansions (7) with the (anti-)symmetrized density operators into the BBGKY-hierarchy (5,6). We limit ourselves to the first three equations where the decoupling is achieved by neglecting four-particle correlations, $c_{1234} = 0$, (for details, see Ref. [17]),

$$\begin{aligned} \mathrm{i}\hbar\partial_t F_1 - \left[\bar{H}^0_1\,, F_1\right] &= n\mathrm{Tr}_2\left[V_{12}\,, c_{12}\right]\Lambda^{\pm}_{12}\,, && (10)\\ \mathrm{i}\hbar\partial_t c_{12} - \left[\bar{H}^0_{12}\,, c_{12}\right] &= \hat{V}_{12}F_1F_2 - F_1F_2\hat{V}^{\dagger}_{12} + n\mathrm{Tr}_3\left[V_{13} + V_{23}\,, c_{123}\right]P_{13;23} && (11)\\ &\quad + L_{12} + \Pi_{12}\,, && (12)\\ \mathrm{i}\hbar\partial_t c_{123} - \left[\bar{H}^0_{123}\,, c_{123}\right] &= \hat{V}^{\dagger}_{12}F_1F_2F_3 + \left(\hat{V}^{\dagger}_{13} + \hat{V}^{\dagger}_{23}\right)F_3c_{12} \\ &\quad \mp nF_3\left(F_1V_{13} + F_2V_{23}\right)c_{12} \mp n\left(c_{13}V_{13} + c_{23}V_{23}\right)c_{12} && (13)\\ &\quad + \Pi_{123} + L_{123} + \mathcal{P}_{123}(\mathrm{rhs.}) - \mathrm{h.c.}(\mathrm{rhs.})\,, \\ F_1(t_0) &= F^0_1\,, \quad c_{12}(t_0) = c^0_{12}\,, \quad c_{123}(t_0) = c^0_{123}\,, \qquad P_{13;23} = (1 \pm P_{13} \pm P_{23})\,, \\ L_{12}, L_{123} &= \text{ladder terms}\,, \qquad \Pi_{12}, \Pi_{123} = \text{polarization terms}\,, \\ \mathcal{P}_{123}(\mathrm{rhs.}) &= \text{cyclic permutation of 1, 2, 3 in all terms on the rhs.}\,, \\ \mathrm{h.c.}(\mathrm{rhs.}) &= \text{hermitean conjugate of all terms on rhs. including } \mathcal{P}_{123}(\mathrm{rhs.})\,. \end{aligned}$$

Since below we will focus on the second Born approximation we do not explicitly write out the ladder and polarization terms [17] since they will be neglected. On the left-hand sides we introduced mean field Hamiltonians which are renormalized by a Hartree-Fock potential $U^{\rm HF}$,

$$\bar{H}_1^0 = H_1 + U_1^{\rm HF}\,, \qquad U_1^{\rm HF} = n{\rm Tr}_2 V_{12} F_2 \Lambda_{12}^{\pm}\,, \tag{14}$$

$$\bar{H}_{12}^0 = \bar{H}_1^0 + \bar{H}_2^0\,, \qquad \bar{H}_{123}^0 = \bar{H}_1^0 + \bar{H}_2^0 + \bar{H}_3^0 \tag{15}$$

and the non-hermitian operator $\hat{V}_{12} = (1 \pm nF_1 \pm nF_2)V_{12}$ which takes into account the exchange renormalization of the pair interaction giving rise, e.g., to Pauli blocking.

Below we will consider two approximations: first, $c_{123} \equiv 0$, which leads to the second Born (2B) approximation of NEGF theory together with the Hartree-Fock GKBA (HF-GKBA, i.e. with HF propagators), see Sec. 2.2. Second, we include in c_{123} all relevant terms that give rise to 2B plus GKBA with full propagators (HF plus 2B). We start with the second approximation since the first follows from it as a special case. To this end, we solve the third hierarchy equation by retaining on the r.h.s. only terms that are proportional to c_{12}:

$$\left\{\left(\hat{V}_{13}^\dagger + \hat{V}_{23}^\dagger\right) F_3 \mp nF_3\left(F_1 V_{13} + F_2 V_{23}\right) \mp n\left(c_{13}V_{13} + c_{23}V_{23}\right)\right\} c_{12} - \text{h.c.} \tag{16}$$

These terms describe the coupling of the pair 1–2 to third particles including medium effects which will give rise to energ renormalization (selfenergy). This expression can be rewritten using the definition of $\hat{V}$, and neglecting the ladder-type corrections involving products $c_{13}V_{13}$ and $c_{23}V_{23}$, yielding

$$\begin{aligned}&\left\{(1 \pm nF_1)(1 \pm nF_3) V_{13} F_3 \mp F_1 F_3 V_{13}(1 \pm nF_3)\right\} c_{12} - \text{h.c.} + (1 \Leftrightarrow 2)\\ &=: \left(S_{13}^> \mp S_{13}^< + S_{23}^> \mp S_{23}^<\right) c_{12} - \text{h.c.}\,,\end{aligned} \tag{17}$$

where we defined

$$S_{ab}^{\gtrless} := F_a^{\gtrless} F_b^{\lessgtr} V_{ab} F_b^{\gtrless}\,. \tag{18}$$

We now rewrite Eq. (11) in static second Born approximation, neglecting all polarization and ladder terms and retaining, on the r.h.s., only the selfenergy contributions 17,

$$\mathrm{i}\hbar\partial_t c_{123} - \left\{\bar{H}_{123}^{0,\rm eff} c_{123} - c_{123}\bar{H}_{123}^{0,\rm eff\dagger}\right\} = \left(S_{13}^> \mp S_{13}^< + S_{23}^> \mp S_{23}^<\right) c_{12} - \text{h.c.}\,, \tag{19}$$

where we have introduced an effective (non-hermitian) three-particle Hamiltonian $\bar{H}_{123}^{0,\rm eff} = \bar{H}_1 + \bar{H}_2 + \bar{H}_3$. We can now formally solve this equation for c_{123} in terms of c_{12},

$$\begin{aligned}c_{123}(t) &= U_{123}^{0+}(tt_0)c_{123}^0 U_{123}^{0-}(t_0 t)\\ &+ \frac{1}{\mathrm{i}\hbar}\int_{t_0}^t \mathrm{d}\bar{t}\, U_{123}^{0+}(t\bar{t})\left\{\left(S_{13}^> \mp S_{13}^< + S_{23}^> \mp S_{23}^<\right) c_{12} - \text{h.c.}\right\}\Big|_{\bar{t}} U_{123}^{0-}(\bar{t}t)\,,\end{aligned} \tag{20}$$

where we introduced the propagators $U_{123}^{0\pm}$ with the properties

$$U_{123}^{0\pm}(tt') = \left[U_{123}^{0\mp}(t't)\right]^\dagger, \qquad U_{123}^{0\pm}(tt') = U_1^{\pm}(tt')U_2^{\pm}(tt')U_3^{\pm}(tt')\,. \tag{21}$$

The single-particle propagators obey effective one-particle Schrödinger-type equations

$$\left\{\mathrm{i}\hbar\partial_t - \bar{H}_1(t)\right\} U_1^+(tt') = 0\,, \qquad U_1^+(tt) = 0\,, \tag{22}$$

$$U_1^-(t't)\left\{\mathrm{i}\hbar\partial_{t'} + \bar{H}_1^\dagger(t')\right\} = 0\,, \qquad U_1^-(tt) = 0\,, \tag{23}$$

where, in the second equation, the time derivative acts onto operators placed left of it. The definition of the effective one-particle Hamiltonian $\bar{H}_1$ will be derived in the following.

We now turn to the second hierarchy equation (11), inserting the formal result for $c_{123}(t)$ on the r.h.s. and again neglect the polarization and ladder terms,

$$i\hbar\partial_t c_{12} - \left[\bar{H}_1^0 + \bar{H}_2^0\,,\, c_{12}\right] = I_{12}^{>} - I_{12}^{<} + n\mathrm{Tr}_3\left\{\left[U_{123}^{0+}(tt_0)c_{123}^0 U_{123}^{0-}(t_0t)\right]\right\}\Lambda_{123}^{\pm}$$
$$+\frac{n}{i\hbar}\int_{t_0}^{t} d\bar{t}\,\mathrm{Tr}_3\left\{\left[V_{13}+V_{23}\,,\,U_{123}^{0+}(t\bar{t})\left\{(S_{13}^{>}\mp S_{13}^{<}+S_{23}^{>}\mp S_{23}^{<})\,c_{12}-\mathrm{h.c.}\right\}\Big|_{\bar{t}}U_{123}^{0-}(\bar{t}t)\right]\right\}\Lambda_{123}^{\pm}\,. \tag{24}$$

Here, the first term on the r.s.h. of Eq. (11) has been transformed according to

$$\hat{V}_{12}F_1F_2 - F_1F_2\hat{V}_{12}^{\dagger} = (1\pm nF_1)(1\pm nF_2)V_{12}F_1F_2 - F_1F_2V_{12}(1\pm nF_1)(1\pm nF_2) =: I_{12}^{>} - I_{12}^{<}\,, \tag{25}$$

where we introduced the greater- and less-collision integral operators, $I_{12}^{>}$ and $I_{12}^{<}$

$$\begin{aligned} I_{ab}^{\gtrless} &= F_a^{\gtrless}F_b^{\gtrless}V_{ab}F_a^{\lessgtr}F_b^{\lessgtr}\,,\\ F_a^{<} &= F_a\,,\\ F_a^{>} &= 1\pm nF_a\,. \end{aligned} \tag{26}$$

Inspection of the integral term in Eq. (24) reveals that it has the structure of a selfenergy operator, $\tilde{\Sigma}_{12}$, acting on c_{12}:

$$\tilde{\Sigma}_{12}(t)c_{12}(t) = \int_{t_0}^{t} d\bar{t}\left\{\Sigma_{12}^{+}(t\bar{t})c_{12}(\bar{t})U_{12}^{-}(\bar{t}t) - \mathrm{h.c.}\right\}\,, \tag{27}$$

which we can decompose into one- and two-particle contributions,

$$\tilde{\Sigma}_{12} = \tilde{\Sigma}_1 + \tilde{\Sigma}_2 + \tilde{\Sigma}_{12}^{\mathrm{cor}}\,, \qquad \Sigma_{12}^{\pm} = \Sigma_1^{\pm}U_2^{\pm} + \Sigma_2^{\pm}U_1^{\pm} + \Sigma_{12}^{\pm,\mathrm{corr}} \tag{28}$$

which are, in turn, given by (cf. Eq. (24)),

$$\begin{aligned} \Sigma_1^{+}(t\bar{t}) &= \frac{n}{i\hbar}\mathrm{Tr}_3\left\{V_{13}U_{13}^{0+}(t\bar{t})\,(S_{13}^{>}\mp S_{13}^{<})\Big|_{\bar{t}}U_3^{-}(\bar{t}t)\right\}\Lambda_{13}^{\pm}\,,\\ \Sigma_2^{+}(t\bar{t}) &= \frac{n}{i\hbar}\mathrm{Tr}_3\left\{V_{23}U_{23}^{0+}(t\bar{t})\,(S_{23}^{>}\mp S_{23}^{<})\Big|_{\bar{t}}U_3^{-}(\bar{t}t)\right\}\Lambda_{23}^{\pm}\,,\\ \Sigma_{12}^{\mathrm{cor}+}(t\bar{t}) &= \frac{n}{i\hbar}\mathrm{Tr}_3\left\{V_{23}U_{123}^{0+}(t\bar{t})\,(S_{13}^{>}\mp S_{13}^{<})\Big|_{\bar{t}}U_3^{-}(\bar{t}t)\right\}\Lambda_{123}^{\pm} + 1\Leftrightarrow 2\,. \end{aligned} \tag{29}$$

With this, we can rewrite Eq. (24) by collecting all terms acting on c_{12} into an effective two-particle hamiltonian,

$$i\hbar\partial_t c_{12}(t) - \left\{H_{12}^{0,\mathrm{eff}}(t)c_{12}(t) - c_{12}(t)H_{12}^{0,\mathrm{eff}}(t)\right\} = I_{12}^{>}(t) - I_{12}^{<}(t)\,, \tag{30}$$

with the definition

$$H_{12}^{0,\mathrm{eff}}(t)c_{12}(t) = \bar{H}_{12}^{0}(t)c_{12}(t) + \int_{t_0}^{t} d\bar{t}\,\Sigma_{12}^{+}(t\bar{t})c_{12}(\bar{t})U_{12}^{0-}(\bar{t}t). \tag{31}$$

Journal of Physics: Conference Series **427** (2013) 012008 doi:10.1088/1742-6596/427/1/012008

This hamiltonian consists of three parts, $H_{12}^{0,\text{eff}} = \bar{H}_1 + \bar{H}_2 + \bar{H}_{12}^{0\text{cor}}$,

$$\begin{aligned}
\bar{H}_1(t)c_{12}(t) &= \bar{H}_1^0(t)c_{12}(t) + \int_{t_0}^{t} d\bar{t}\,\Sigma_1^+(t\bar{t})U_2^+(t\bar{t})c_{12}(\bar{t})U_{12}^{0-}(\bar{t}t)\,,\\
\bar{H}_2(t)c_{12}(t) &= \bar{H}_2^0(t)c_{12}(t) + \int_{t_0}^{t} d\bar{t}\,U_1^+(t\bar{t})\Sigma_2^+(t\bar{t})c_{12}(\bar{t})U_{12}^{0-}(\bar{t}t)\,,\\
\bar{H}_{12}^{0\text{cor}}(t)c_{12}(t) &= \int_{t_0}^{t} d\bar{t}\,\Sigma_{12}^{\text{cor}+}(t\bar{t})c_{12}(\bar{t})U_{12}^{0-}(\bar{t}t)\,.
\end{aligned} \tag{32}$$

To preserve the additivity of $H_{12}^{0,\text{eff}}$ resulting from the Born approximation, it is necessary to neglect the term $\bar{H}_{12}^{0\text{cor}}c_{12}$, which yields a result that is consistent with our previous definition of the effective three-particle hamiltonian,

$$\begin{aligned}
H_{12}^{0,\text{eff}}c_{12} &= \left(\bar{H}_1 + \bar{H}_2\right)c_{12}\,,\\
H_{123}^{0,\text{eff}}c_{123} &= \left(\bar{H}_1 + \bar{H}_2 + \bar{H}_3\right)c_{123}\,.
\end{aligned} \tag{33}$$

This result can now be used to solve the equations of motion, Eq. (22), for the renormalized one-particle propagator $U_1^+(tt')$ which transforms into

$$\left\{i\hbar\partial_t - \bar{H}_1^0\right\}U_1^+(tt') - \int_{t_0}^{t} d\bar{t}\,\Sigma_1^+(t\bar{t})U_1^+(\bar{t}t') = 0\,. \tag{34}$$

Using the definition of $S^{\gtrless}$, we can write out the selfenergy Σ_1^+ in second Born approximation explicitly (neglecting initial value terms),

$$\Sigma_1^+(tt') = \frac{n}{i\hbar}\text{Tr}_3\left\{V_{13}U_1^+(tt')U_3^+(tt')\left[F_1^> F_3^> V_{13}F_3^< \mp F_1^< F_3^< V_{13}F_3^>\right]U_3^-(t't)\Big|_{t'}\right\}\,. \tag{35}$$

The structure of Σ_1^+ suggests to define new quantities

$$\begin{aligned}
g_a^>(tt') &= U_a^+(tt')F_a^>(t') - F_a^>(t)U_a^-(tt')\,,\\
g_a^<(t't) &= U_a^+(tt')F_a^<(t') - F_a^<(t)U_a^-(tt')\,,
\end{aligned} \tag{36}$$

giving rise to a compact and symmetric expresssion

$$\begin{aligned}
\Sigma_1^+(tt') &= \frac{n}{i\hbar}\text{Tr}_3V_{13}\left\{g_1^>(tt')g_3^>(tt')V_{13}g_3^<(t't) \mp g_1^<(tt')g_3^<(tt')V_{13}g_3^>(t't)\right\}\\
&=: \Sigma_1^>(tt') \mp \Sigma_1^<(tt')\,,
\end{aligned} \tag{37}$$

where we defined the greater- and less-selfenergy $\Sigma_1^>(tt')$ and $\Sigma_1^<(tt')$. With these definitions one can finally write down the equation of motion for the single-particle density operator F_1, which reads according to Eq. (10),

$$\begin{aligned}
i\hbar\partial_t F_1 - \left[\bar{H}_1^0, F_1\right] &= n\text{Tr}_2\left[V_{12}, c_{12}\right]\Lambda_{12}^{\pm}\\
&= \left(n\text{Tr}_2\left\{V_{12}U_{12}^{0+}(t,t_0)c_{12}^0U_{12}^{0-}(t_0,t)\right\} - n\text{Tr}_2\left\{U_{12}^{0+}(t,t_0)c_{12}^0U_{12}^{0-}(t_0,t)V_{12}\right\}\right)\Lambda_{12}^{\pm}\\
&\quad + \frac{in}{\hbar}\int_{t_0}^{t} d\bar{t}\text{Tr}_2\Big(\left\{V_{12}U_{12}^{0+}(t\bar{t})\left[F_1^> F_2^> V_{12}F_1^< F_2^< - F_1^< F_2^< V_{12}F_1^> F_2^>\right]|_{\bar{t}}U_{12}^{0-}(\bar{t}t)\Lambda_{12}^{\pm}\right\}\\
&\qquad - \left\{U_{12}^{0+}(t\bar{t})\left[F_1^> F_2^> V_{12}F_1^< F_2^< - F_1^< F_2^< V_{12}F_1^> F_2^>\right]|_{\bar{t}}U_{12}^{0-}(\bar{t}t)V_{12}\Lambda_{12}^{\pm}\right\}\Big)\\
&= \int_{t_0}^{t} d\bar{t}\,\left\{\Sigma_1^>(t,\bar{t})g^<(\bar{t},t) - \Sigma_1^<(t,\bar{t})g^>(\bar{t},t)\right\}\,.
\end{aligned} \tag{38}$$

Journal of Physics: Conference Series **427** (2013) 012008 doi:10.1088/1742-6596/427/1/012008

Here we have used, after the first line, the formal solution $c_{12}(t)$ of Eq. (30) that includes initial correlations, c^0_{12} and scattering contributions (the time integral).

In the next section, we will see, that this equation of motion for the reduced single-particle density operator has the same form as the result obtained within the NEGF formalism for the time-diagonal limit of the function $G^<$ in second Born approximation, after applying the GKBA. Thereby it turns out that the full result, as derived above, corresponds to the GKBA with full second-Born propagators. In contrast, the HF-GKBA follows if the propagator equation (34) is solved after neglecting the selfenergy correction, which is equivalent to decoupling the BBGKY-hierarchy by requiring $c_{123} = 0$, cf. Sec. 2.2.

2.2. Nonequilibrium Green functions

In contrast to density operator theory, presented in the preceding section, the basic quantities in the nonequilibrium Green function formalism are the creation/annihilation operators a_i and $a_i^\dagger$, which create/annihilate a particle in the i-th one-particle orbital and obey the canonical commutator/anticommutator relations for bosonic or fermionic particles,

$$\left[\hat{c}_i^{(\dagger)}, \hat{c}_j^{(\dagger)}\right]_\mp = 0, \quad \left[c_i, \hat{c}_j^\dagger\right]_\mp = \delta_{i,j}\,. \tag{39}$$

With this, one can define the ensemble average of the combination N creation and annihilation operators at $2N$ different points in time,

$$\begin{aligned}&\left\langle c_{j_1}(t'_1)\ldots c_{j_N}(t'_N)c^\dagger_{i_N}(t_N)\ldots c^\dagger_{i_1}(t_1)\right\rangle =: G^{N,<}_{i_1\ldots i_N;j_1\ldots j_N}(t_1\ldots t_N;t'_1\ldots t'_N)\\ &= \mathrm{Tr}\,\left\{\rho c_{j_1}(t'_1)\ldots c_{j_N}(t'_N)c^\dagger_{i_N}(t_N)\ldots c^\dagger_{i_1}(t_1)\right\},\end{aligned} \tag{40}$$

as the less-part of the real-time N-particle thermal Green function $G^{N,<}$. Here ρ is the density operator of the system and the notation "$<$" refers to the particular ordering of the operators. For equal times ($t_1 = \ldots = t_N = t'_1 = \ldots = t'_N$) the quantity $\mathrm{i}^N G^{N,<}$ is just the N-particle density operator $\rho_{1\ldots N}$ in the one-particle orbital basis. Analogous to the density operator theory, one can define reduced Green functions involving fewer operators, so that the most basic quantity, the single-particle two-time less Green function $G^<_{ij}(t_1, t_2)$ can be defined as

$$G^<_{ij}(t_1, t_2) = \left\langle c_j^\dagger(t_2)c_i(t_1)\right\rangle\,. \tag{41}$$

The Green functions also obey a hierarchy of equations of motion, the Martin-Schwinger hierarchy (MSH) [36], which for all N connects the N-particle Green function to the $(N-1)$- and the $(N+1)$-particle Green functions. Since the complexity of the whole hierarchy—due to the additional time-arguments— is computationally even more demanding, one again resorts to the closure of the hierarchy equation on the single-particle level by a cluster expansion, stating the two-particle Green function in terms of the one-particle Green function by introduction of a suitable selfenergy functional $\Sigma(G)$. With this, the equation of motion for the less Green function attains the well known form of the Keldysh/Kadanoff–Baym equation (KBE) (initial correlations can be included either via the density operator ρ, cf. Eq. (40), or by the technique of adiabatic switching, for details see, e.g., [26, 32]),

$$\mathrm{i}\partial_{t_1} G^<_{ij}(t_1, t_2) = \sum_k h_{ik}(t_1)G^<_{kj}(t_1, t_2) + \int \mathrm{d}t_3\, \Sigma^{\mathrm{R}}_{ik}(t_1, t_3)G^<_{kj}(t_3, t_2) + \int \mathrm{d}t_3\, \Sigma^<_{ik}(t_1, t_3)G^{\mathrm{A}}_{kj}(t_3, t_2)\,. \tag{42}$$

The quantities G^R and G^A—the retarded and advanced Green functions—are defined as

$$G_{ij}^{\mathrm{R/A}}(t_1,t_2) = \pm\theta\left(\pm(t_1-t_2)\right)\left[G_{ij}^{>}(t_1,t_2) - G_{ij}^{<}(t_1,t_2)\right], \tag{43}$$

$$G_{ij}^{\gtrless}(t_1,t_2) = -\mathrm{i}\left\langle \hat{c}_i(t_1)\hat{c}_j^{\dagger}(t_2)\right\rangle, \tag{44}$$

where h_{ij} is the ij-th matrix element of the Hartree-Fock part of the Hamiltonian. The corresponding components of the selfenergy read in the second order Born approximation:

$$\begin{aligned}\Sigma_{ij}^{\mathrm{2B},<}(t_1,t_2) &= \sum_{klmnrs} w_{ikms}(\sigma\, w_{rnlj} - w_{rlnj}) G_{kl}^{<}(t_1,t_2) G_{mn}^{<}(t_1,t_2) G_{rs}^{>}(t_2,t_1)\,,\\ \Sigma_{ij}^{\mathrm{2B,R}}(t_1,t_2) &= \sum_{klmnrs} w_{ikms}(\sigma\, w_{rnlj} - w_{rlnj}) G_{kl}^{\mathrm{R}}(t_1,t_2) G_{mn}^{\mathrm{R}}(t_1,t_2) G_{rs}^{\mathrm{A}}(t_2,t_1)\,,\end{aligned} \tag{45}$$

where $\sigma = \{1,2\}$ for spin-polarized/spin-restricted systems. As an alternative formulation of the KBE *Lipavskii et al.* proposed [27]

$$\begin{aligned}G^{<}(t_1,t_2) &= \int_{t_2}^{t_1} \mathrm{d}t_3 \int_{t_0}^{t_2} \mathrm{d}t_4\, G^{\mathrm{R}}(t_1,t_3)\Sigma^{<}(t_3,t_4)G^{\mathrm{A}}(t_4,t_2)\\ &+ \Theta(t_1-t_2)\left[-G^{\mathrm{R}}(t_1,t_2)\rho(t_2) + \int_{t_2}^{t_1} \mathrm{d}t_3 \int_{t_0}^{t_2} \mathrm{d}t_4\, G^{\mathrm{R}}(t_1,t_3)\Sigma^{\mathrm{R}}(t_3,t_4)G^{<}(t_4,t_2)\right]\\ &+ \Theta(t_2-t_1)\left[\rho(t_1)G^{\mathrm{A}}(t_1,t_2) + \int_{t_2}^{t_1} \mathrm{d}t_3 \int_{t_0}^{t_2} \mathrm{d}t_3\, G^{<}(t,t_3)\Sigma^{\mathrm{A}}(t_3,t_4)G^{\mathrm{A}}(t_4,t_2)\right].\end{aligned} \tag{46}$$

To further simplify this equation, they introduced the generalized Kadanoff–Baym ansatz (GKBA), which is equivalent to solving Eq. (46) in first order,

$$\begin{aligned}G^{(1),<}(t_1,t_2) &= \int_{t_2}^{t_1} \mathrm{d}t_3 \int_{t_0}^{t_2} \mathrm{d}t_4\, G^{\mathrm{R}}(t_1,t_3)\Sigma^{(0),<}(t_3,t_4)G^{\mathrm{A}}(t_4,t_2)\\ &+ \Theta(t_1-t_2)\left[-G^{\mathrm{R}}(t_1,t_2)\rho^{(0)}(t_2) + \int_{t_2}^{t_1} \mathrm{d}t_3 \int_{t_0}^{t_2} \mathrm{d}t_4\, G^{\mathrm{R}}(t_1,t_3)\Sigma^{\mathrm{R}}(t_3,t_4)G^{(0),<}(t_4,t_2)\right],\\ &+ \Theta(t_2-t_1)\left[\rho^{(0)}(t_1)G^{\mathrm{A}}(t_1,t_2) + \int_{t_2}^{t_1} \mathrm{d}t_3 \int_{t_0}^{t_2} \mathrm{d}t_3\, G^{(0),<}(t,t_3)\Sigma^{\mathrm{A}}(t_3,t_4)G^{\mathrm{A}}(t_4,t_2)\right],\end{aligned} \tag{47}$$

with the result

$$G^{(0),<}(t_1,t_2) = \Theta(t_1-t_2)\left[-G^{\mathrm{R}}(t_1,t_2)\rho^{(0)}(t_2)\right] + \Theta(t_2-t_1)\left[\rho^{(0)}(t_1)G^{\mathrm{A}}(t_1,t_2)\right], \tag{48}$$

$$\rho^{(0)}(t_1) = -\mathrm{i}\, G^{(0),<}(t_1,t_1)\,. \tag{49}$$

One notices that Eq. (47) is only formally closed in terms of the single-particle density matrix ρ, since the propagators $G^{\mathrm{R/A}}$ still obey a two-time equation of a similar degree of complexity as the original KBE. To circumvent this we introduce a further approximation, replacing the full propagators by the HF propagators $G_{\mathrm{HF}}^{\mathrm{R/A}}$, which are defined as

$$G_{\mathrm{HF}}^{\mathrm{R/A}}(t_1,t_2) = \mp\mathrm{i}\theta[\pm(t_1-t_2)]\mathcal{T}\exp\left(-\mathrm{i}\int_{t_2}^{t_1}\mathrm{d}t_3\, h(t_3)\right), \tag{50}$$

Correspondingly, this approximation will be called HF-GKBA.

Then the numerical solution of Eq. (47) can be obtained, using a finite time step Δ, in the following way [21],

$$\begin{aligned} G^<(t+\Delta,t+\Delta) &= U(t)G^<(t,t)U^\dagger(t) \\ &\quad - \mathrm{i}\Delta\, U(t)I^<(t,t)U^\dagger(t) - \mathrm{i}\Delta\, U(t)\left[I^<(t,t)\right]^\dagger U^\dagger(t)\,, \end{aligned} \tag{51}$$

where

$$X(t+\Delta,t) = \exp\left(-\mathrm{i}\int_t^{t+\Delta} \mathrm{d}\bar t\, h(\bar t)\right) = e^{-\mathrm{i}h(t)\Delta} =: U(t)\,, \tag{52}$$

for a small time step $\Delta \ll 1$, and the HF-Hamiltonian h is assumed not to change between t and $t+\Delta$. The collision integral $I(t,t)$ is given by

$$I(t,t) = \int_{t_0}^t \mathrm{d}\bar t\,\left\{\Sigma^{>,0}(t,\bar t)G^{<,0}(\bar t,t) - \Sigma^{<,0}(t,\bar t)G^{>,0}(\bar t,t)\right\}\,, \tag{53}$$

where in all the two-time quantities under the integral the GKBA reconstruction according to Eq. (48) is used.

When comparing Eq. (53) with the right-hand side of Eq. (38), one immediately recognizes that it is of exactly the same structure, with the identifications

$$\begin{aligned} \Sigma^{\gtrless,0}(t,\bar t) &= \Sigma_{\bar 1}^{\gtrless}(t,\bar t)\,, \\ G^{\mathrm{R}}(t,\bar t) &= -\Theta(t-\bar t)U^+(t,\bar t)\,, \\ G^{\gtrless,0}(t,\bar t) &= g^{\gtrless}(t,\bar t)\,. \end{aligned} \tag{54}$$

Moreover, the GKBA, Eq. (48) appeared naturally in our density operator theory, cf. Eq. (36). While this equivalence holds for general propagators U^+ defined by Eq. (34), neglect of the renormalization (selfenergy term)—which is equivalent to decoupling the BBGKY-hierarchy by using $c_{123} = 0$—directly leads to HF-propagators and to the HF-GKBA of nonequilibrium Green functions theory.

3. Numerical example: 4-site Hubbard model

To illustrate the findings obtained in sections 2.1 and 2.2, in this section we show some results for a 4-site Hubbard nano-cluster obtained from the Green function method with the GKBA applied and compare to density operator results by *Akbari et al.* [1]. The Hubbard model [37, 38] is a commonly used simplified description of a narrow-band solid-state system, where the motion of the electrons in the solid is mapped onto a hopping process between adjacent atomic sites with just one orbital for each spin projection. The Coulomb interaction between the electrons is assumed to be shielded so that it is mainly restricted to on-site interaction. Despite of these simplifications, this description is able to cover much of the rich behavior of these systems, for example the phase transition between a conductor and Mott insulator through the interplay of hopping and on-site interaction in two and three dimensions. Also in 1D, the system is—depending on the interaction strength—strongly coupled and, therefore, the theoretical description needs to treat correlations between the electrons.

The Hamiltonian of a one-dimensional Hubbard cluster comprised of N sites at electronic half-filling is given by

$$\hat H = -t\sum_{ij}^{N}\sum_{\alpha} h_{ij}\,\hat c^\dagger_{i\alpha}\hat c_{j\alpha} + U\sum_i^N \hat c^\dagger_{i\uparrow}\hat c_{i\uparrow}\hat c^\dagger_{i\downarrow}\hat c_{i\downarrow}\,, \tag{55}$$

Journal of Physics: Conference Series **427** (2013) 012008 doi:10.1088/1742-6596/427/1/012008

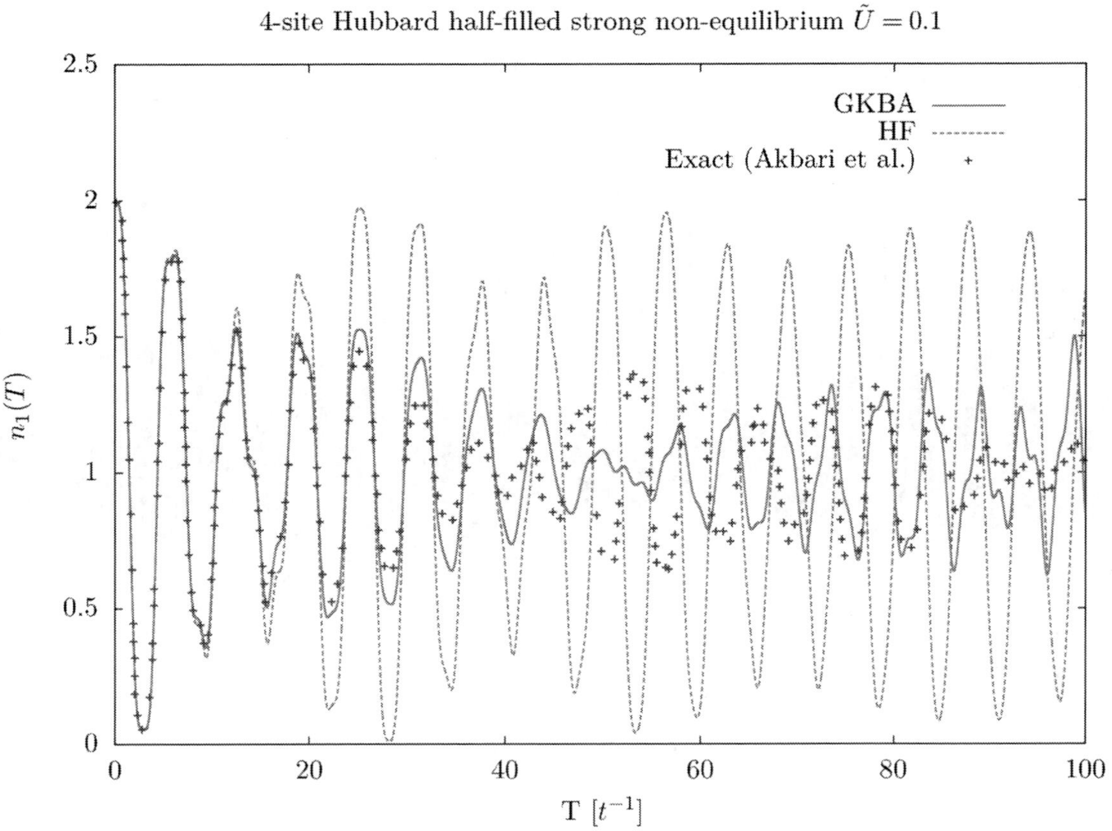

Figure 1. Density response on the first site n_1 from strong nonequilibrium initial state for the 4-site Hubbard chain with a coupling strength $\tilde{U} = 0.1$. The GKBA and the HF results are depicted by the solid red lines and green dashed lines, respectively. In blue the exact results from *Akbari et al.* [1] is shown.

where $\hat{c}_{i\alpha}^{(\dagger)}$ denotes the annihilation (creation) operator in the single-particle orbital on site i with spin α and $h_{ij} = \delta_{\langle i,j \rangle}$ is the hopping matrix between nearest neighbor sites with the convention, $\delta_{\langle i,j \rangle} = 1$, if (i, j) are nearest neighbors, and $\delta_{\langle i,j \rangle} = 0$, otherwise. The first term originates from the single-particle energies in the periodic lattice structure and incorporates the hopping amplitude $-t$. The second term describes the on-site interaction of the electrons, which is given in terms of the interaction strength U.

In the following, we consider a $(N = 4)$-site cluster with weak interaction, $U = 0.1$, at zero temperature with periodic boundary conditions. For this system we are interested in the study of strong nonequilibrium situations. To this end, we prepare the system in an initial state, where all the 4 electrons are forced to the left-most two sites and afterwards examine the free evolution of the system. For this setup, Green function solutions within the HF–GKBA, as well as HF results are easily achieved. In Figs. 1 and 2, the time-evolution of the density $n_1(T)$ on the first site—summed over the two orbitals for different spin orientations—measured in terms of the inverse hopping amplitude t^{-1} are shown. In Fig. 1 we present HF-GKBA results together with the exact solutions and time-dependent HF simulations. In Fig. 2 the HF-GKBA data are compared to exact results as well as to density operator results [1] within the Wang-Cassing (WC) decoupling of the BBGKY-hierarchy [18].

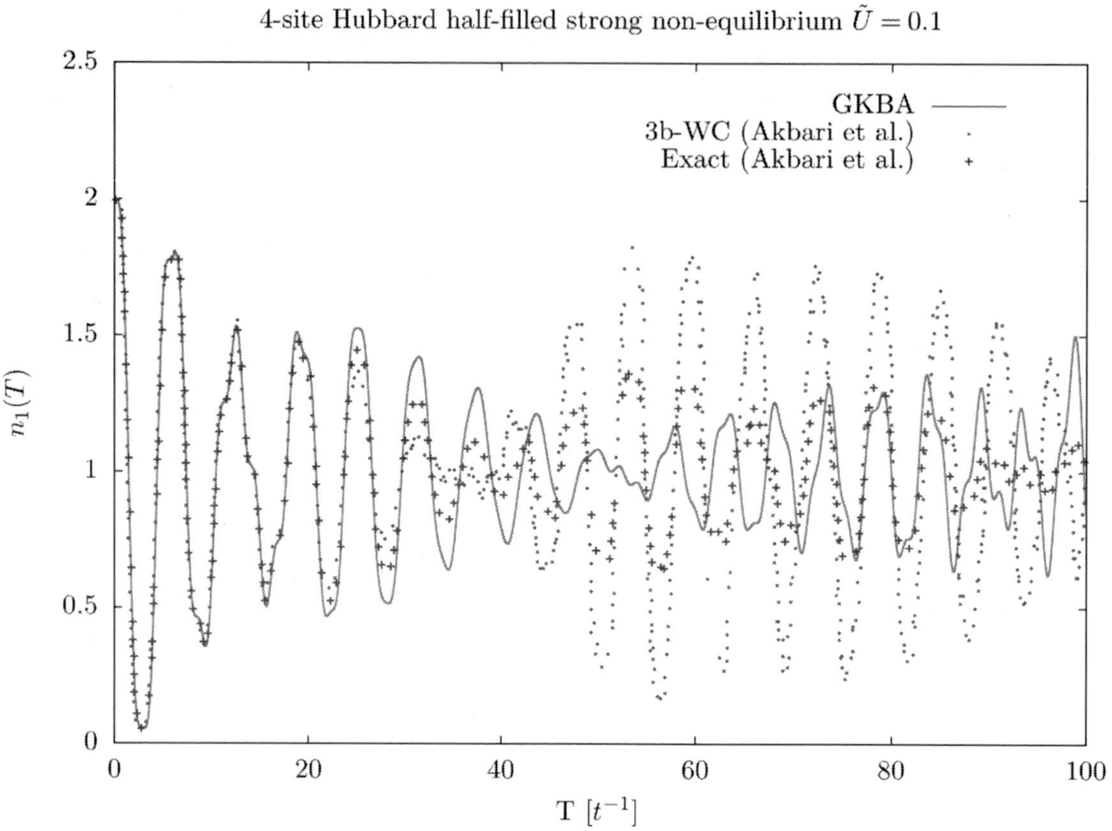

Figure 2. Density response on the first site n_1 from strong nonequilibrium initial state for the 4-site Hubbard chain with a coupling strength $\tilde{U} = 0.1$. The GKBA results are depicted by the solid red lines. The blue pluses and brown dots represent the exact results and the ones obtained from the WC approximation to the BBGKY hierarchy by *Akbari et al.* [1].

4. Discussion

The results show that the HF-GKBA performs very well for the propagation of a strong nonequilibrium initial state, especially in comparison with time-dependent Hartree-Fock, that does not show a decrease of the oscillation amplitude at all, present in the exact solution. Comparing the amplitudes with the exact ones, the GKBA overestimates them, though. In the part from $T = 40$ on, the GKBA cannot describe the oscillations sufficiently well, but always keeps a good phase agreement with the exact solution. Comparing with the WC approximation in Fig. 2, it is obvious that the WC solution has a very good agreement of the oscillation frequency up to $T = 80$, but the height of the peaks is much overestimated for time $T > 40$, both compared to the exact ones as well as to those computed with the GKBA.

A main issue of the BBGKY based methods like the WC approximation, pointed out by *Akbari et al.* [1], is the occurrence of negative or above-maximum site occupations and, for sufficiently long time propagation, eventual divergence. The divergence occurred for times around $500\,t^{-1}$ (for $U = 0.1$) and even earlier, for large values of U. In contrast, we have tested the GKBA–HF algorithm for a propagation time of $1000\,t^{-1}$ and find that it is stable and the total energy is conserved with no measurable deviation. The site occupancies stay within the allowed boundaries (0 to 2) for the whole time propagation with only one exception, where they are violated by no more than 2 percent. This seems to have no negative effect on the subsequent propagation, though. A more detailed discussion, including a broader range of values of U, will be given in a forthcoming paper [39], in which we analyze the abilities of the GKBA–HF method for the time-dependent description of Hubbard nano-clusters far from equilibrium.

5. Conclusion and Outlook

In this paper we have shown that, within the second order Born approximation, we can establish a one-to-one correspondence between nonequilibrium Green functions within the GKBA and reduced density operator theory, yielding the same formulas for the equation of motion of the single-particle density matrix. On the example of the free evolution of a four-electron quantum dot from a strong nonequilibrium initial state we have shown the overall satisfactory agreement of the HF-GKBA to the exact dynamics of the system far away from equilibrium, for weak interaction strength. This is encouraging since the HF-GKBA allows for long propagation times that are impossible to achieve with full two-time calculations and it retains most of the attractive properties of NEGF, such as time reversibility, total energy conservation and memory effects, e.g. [40, 17]. In the future, it will be very interesting to examine the potential of the GKBA for higher order selfenergy schemes such as GW- or T-matrix (ladder) approximation. For the T-matrix approximation for homogeneous macroscopic systems, a similar correspondence has been found previously [41, 17]. It will be interesting to extend this correspondence to inhomogeneous finite systems in order to achieve long propagations also for strongly coupled systems and to verify whether the GKBA is here able, as well, to remove the unphysical damping observed in two-time calculations [24].

Acknowledgments

This work is supported in part by the Deutsche Forschungsgemeinschaft via project BO1366/9 and by a grant for CPU time at the HLRN.

References

[1] Akbari A, Hashemi M J, Nieminen R M, van Leeuwen R and Rubio A 2012 `arXiv:1204.4395`
[2] Bauch S and Bonitz M 2008 *Phys. Rev. A* **78** 043403
[3] Runge E and Gross E K U 1984 *Phys. Rev. Lett.* **52** 997
[4] Kulander K C 1987 *Phys. Rev. A* **36**(6) 2726
[5] Bunge C F, Barrientos J A, Bunge A V and Cogordan J A 1992 *Phys. Rev. A* **46**(7) 3691
[6] Doyle P A and Turner P S 1968 *Acta Crystallographica Section A* **24** 390–397
[7] Thouless D and Valatin J 1962 *Nucl. Phys.* **31** 211 – 230
[8] Bonche P, Koonin S and Negele J W 1976 *Phys. Rev. C* **13**(3) 1226–1258
[9] Hochstuhl D, Bauch S and Bonitz M 2010 *J. Phys. Conf. Ser.* **220** 012019 and Hochstuhl D, and Bonitz M 2011, J. Chem. Phys. **134**, 084106
[10] Bartlett R J 1981 *Annual Review of Physical Chemistry* **32** 359
[11] Møller C and Plesset M S 1934 *Phys. Rev.* **46**(7) 618–622
[12] Schollwöck U 2005 *Rev. Mod. Phys.* **77**(1) 259
[13] Georges A, Kotliar G, Krauth W and Rozenberg M J 1996 *Rev. Mod. Phys.* **68**(1) 13–125
[14] Kadanoff L P and Baym G *Quantum Statistical Mechanics*, (Benjamin, New York, 1962)
[15] Born M and Green H S 1946 *Royal Society of London Proceedings Series A* **188** 10–18
[16] Kirkwood J G 1947 *J. Chem. Phys.* **15** 72–76
[17] Bonitz M 1998 *Quantum Kinetic Theory* (Teubner)
[18] Cassing W and Pfitzner A 1992 *Z. Physik A: Hadrons and Nuclei* **342** 161 ISSN 0939-7922
[19] N.-H. Kwong and M. Bonitz, Phys. Rev. Lett. **84**, 1768 (2000).
[20] M. Lorke, T.R. Nielsen, J. Seebeck, P. Gartner and F. Jahnke, Phys. Rev. B **73**, 085324 (2006).
[21] Stan A, Dahlen N E and van Leeuwen R 2009 *J. Chem. Phys.* **130** 224101
[22] Balzer K, Bauch S and Bonitz M 2010 *Phys. Rev. A* **81** 022510
[23] Balzer K, Bauch S and Bonitz M 2010 *Phys. Rev. A* **82** 033427
[24] M. Puig von Friesen, C. Verdozzi and C.-O. Almbladh, Phys. Rev. Lett. **103**, 176404 (2009).
[25] M. Puig von Friesen, C. Verdozzi and C.-O. Almbladh, Phys. Rev. B **82**, 155108 (2010).

[26] Hermanns S, Balzer K and Bonitz M 2012 *Physica Scripta* (*Preprint* `arXiv.1205.4427`)
[27] Lipavsky, Spicka V and Velicky 1986 *Phys. Rev. B* **34** 6933
[28] Spicka V, Velicky B and Kalvova A 2005 *Physica E: Low-dimensional Systems and Nanostructures* **29** 154–174
[29] Bonitz M, Kremp D, Scott D C, Binder R, Kraeft W D and Köhler S 1996 *J. Phys. Cond. Matt.* **8** 6057
[30] Binder R, Köhler S, Bonitz M and Kwong N H 1997 *Phys. Rev. B* **55** 5110
[31] Kwong N H, Bonitz M, Binder R and Köhler S 1998 *Phys. Status Solidi B* **206** 197
[32] Balzer K, Hermanns S and Bonitz M 2012 *J. Phys.: Conf. Ser., same issue* (*Preprint* `arXiv.1211.3036`)
[33] Balzer K, Hermanns S and Bonitz M 2012 *EPL* **98** 67002
[34] Fetter A and Walecka J 1971 *Quantum Theory of Many–Particle Systems* (McGraw–Hill Book Company, New York)
[35] Boercker D B and Dufty J W 1979 *Ann. Phys. (N.Y.)* **119** 43
[36] Martin P C and Schwinger J 1959 *Phys. Rev.* **115** 1342–1373
[37] Hubbard J 1963 *Royal Society of London Proceedings Series A* **276** 238
[38] Lieb E H and Wu F 2003 *Phys. A* **321** 1
[39] Hermanns S and Bonitz M 2013, to be published
[40] Bonitz M, and Kremp D 1996 *Phys. Lett. A* **212**, 83
[41] Kremp D, Bonitz M, Kraeft W-D, and Schlanges M, 1997 *Ann. Phys.* **258**, 320

Progress in Nonequilibrium Green's Functions V (PNGF V) IOP Publishing
Journal of Physics: Conference Series **427** (2013) 012009 doi:10.1088/1742-6596/427/1/012009

Self-consistent Green's functions calculation of the nucleon mean-free path

A. Rios[1] **and V. Somà**[2,3]

[1] Department of Physics, Faculty of Engineering and Physical Sciences, University of Surrey, Guildford, Surrey GU2 7XH, United Kingdom
[2] Institut für Kernphysik, Technische Universität Darmstadt, 64289 Darmstadt, Germany
[3] ExtreMe Matter Institute EMMI, GSI Helmholtzzentrum fur Schwerionenforschung GmbH, 64291 Darmstadt, Germany

E-mail: a.rios@surrey.ac.uk

Abstract. Transport coefficients provide a unique insight into the near-equilibrium behavior of quantum many-body systems. The mean-free path, λ, of a particle within a dense medium is a basic transport coefficient, at the basis of several theoretical concepts and closely related to experimentally measured quantities. Green's functions techniques are particularly well suited to study such transport properties, since they are naturally formulated in the time domain. We present a calculation of the mean-free path of a nucleon in symmetric nuclear matter using self-consistent ladder self-energies extended to the complex energy plane. Our results indicate that, for energies above 50 MeV at densities close to saturation, a nucleon has a mean-free path of 4 to 5 femtometers.

1. Introduction

Over the years, the nuclear many-body community has primarily concentrated on the study of the equilibrium properties of extended hadronic systems [1]. A focal point of these efforts has been the calculation of the Equation of State (EoS) of neutron matter [2]. This determines to a large extent the mass-radius relation of neutron stars, which is expected to be experimentally constrained by near-future astronomical observations. The hope is that precise simultaneous measurements of both the radius and the mass of pulsars will indirectly provide precious information on nucleon-nucleon interactions and more in general on strongly interacting nuclear systems [3, 4].

Transport properties characterize the behaviour of systems when driven out of equilibrium [5]. Viscosities and mean-free paths determine the dynamics of damping (or potential growth) of various instability modes and give insight on the microscopic structure of the medium. Unlike those of equilibrium properties, calculations of transport coefficients of homogeneous hadronic matter are relatively scarce. These coefficients, however, are extremely important to describe the dynamics and time evolution of astrophysical compact objects and affect the analysis of present generation astronomical observations. The rapid cooling observed in Cassiopeia A, for instance, is thought to be related to a modification in the specific heat associated to pairing phenomena [6] and it is sensitive to transport in dense matter [7, 6]. Shear and bulk viscosities are often needed in neutron star modeling [8, 9, 10]. In particular, the potential instability to general relativistic modes is very sensitive to the bulk viscosity [11].

In some cases, the theoretical calculations of transport coefficients involve an artificial coupling between the many-body treatment (or the equilibrium problem) and the transport (or non-equilibrium) approach. A typical example is provided by recent calculations of the shear viscosity [10, 12]. The equilibrium problem is treated using many-body techniques, like the Correlated Basis Functions or the Brueckner-Hartree-Fock approaches. These provide access to scattering properties in equilibrated dense matter, in particular in-medium cross sections. The latter can then be used as input in an Abrikosov-Khalatnikov-type calculation for the viscosity [13]. Yet, the scattering processes described by equilibrium many-body calculations need not be the same as those originally considered within Abrikosov theory. One can therefore argue that, by coupling both approaches, uncontrolled errors are being introduced in the calculation.

In contrast, Green's functions are well suited to compute transport coefficients from first principles. Time dependence is treated explicitly in the theory [14], to the extent that one can even consider non-equilibrium dynamics in a meaningful way [15]. Pauli principle and beyond mean-field correlations can be consistently taken into account. Kubo relations give access to several non-equilibrium properties from correlators computed near equilibrium [16] and have been used to compute a handful of transport coefficients in a wide variety of systems [17]. In principle, two-time Kadanoff-Baym calculations could also provide reliable calculations of transport properties beyond the near-equilibrium assumption. In the following, we concentrate on the application of equilibrium Green's functions techniques to compute the mean-free path of particles in the nuclear medium.

2. Nucleon mean-free path

The mean-free path, λ, is perhaps the best studied transport coefficient in nuclear physics, as it is directly linked to the medium's absorptive properties [18, 19, 20, 21, 22]. This absorption is closely related to nuclear optical potentials, which can be fitted or measured in a variety of experiments. In the past, the lack of microscopic propagators in nuclear systems has hampered the calculation of transport coefficients using Green's function techniques. Consequently, most previous attempts to compute the mean-free path have been derived from extensions of Brueckner's theory of nuclear matter [18, 19, 21, 23, 24, 25]. These calculations have been performed at energies arbitrarily close to the real axis and, as a consequence, ad-hoc non-locality corrections have had to be introduced, generally through the $k-$mass approximation [18].

Our philosophy has been that of a fully *ab initio* calculations and we have sought to introduce as few approximations as possible. Our starting point are state-of-the-art self-energies, computed within the ladder approximation both at zero and finite temperature [26, 27]. We take into account the underlying model dependence associated to different nucleon-nucleon (NN) interactions. We shall also briefly comment on the dependence on temperature and density of our results. As we shall see in the following, the extension of Green's functions techniques to the complex energy domain provides a consistent approach to compute quasi-particle (qp) properties [28, 18, 14]. Direct calculations on the complex energy plane have been performed in electronic systems since the early 1960's [29] and have been recently employed to describe microscopic excitations in solid state applications [30, 31]. To our knowledge, this represents the first application of these techniques within nuclear physics.

Our many-body approximation of choice is the *ladder approximation* implemented self-consistently [15]. Its formal and computational implementation is a demanding task and the interested reader can find more details in Refs. [32, 27]. Here, we summarise briefly the different stages of the approach. First, an energy- and momentum-dependent in-medium interaction, or $T-$matrix, is obtained from a Lippman-Schwinger equation, which succinctly reads:

$$T = V + VG_{II}T\,. \tag{1}$$

The in-medium two-body propagator, G_{II}, is a convolution of fragmented one-body propagators,

$\mathcal{G}(k,\omega)$, which naturally takes into account intermediate particle-particle and hole-hole states [33]. Further, the self-energy, $\Sigma(k,\omega)$, is obtained from the $T-$matrix by closing a fermionic line. The Dyson equation can then be used to find the in-medium propagator, $\mathcal{G}$. This gives rise to a self-consistent problem, as G_{II} depends on $\mathcal{G}$. For a given density and temperature, these equations are solved in an iterative fashion, taking into account the full energy and momentum dependence of all quantities. Three-body forces (3BF), which are necessary in nuclear systems, are included effectively via an average over a third, correlated nucleon [27].

2.1. Complex energy self-energy

The propagation of an excitation in nuclear matter is described by the retarded propagator, $\mathcal{G}_R(k,t) \equiv \Theta(t)\left\langle\left\{a(k,t),a^\dagger(k,0)\right\}\right\rangle$. In a uniform system in thermal equilibrium, $\mathcal{G}_R$ only depends on the time difference, t, and the momentum modulus, k. For a system with well-defined quasi-particles, the retarded one-body Green's functions has the following long-time asymptotic form [14, 34]:

$$\mathcal{G}_R(k,t) \xrightarrow[t>>\Gamma^{-1}]{} -i\,\eta(k)e^{-i\varepsilon(k)t}e^{-|\Gamma(k)|t}\,. \tag{2}$$

While $\eta(k)$ represents the strength of the excitation, the qp spectrum $\varepsilon(k)$ determines the oscillation frequency of the propagator. The spectrum is closely related to the group velocity,

$$v(k) = \frac{\partial\varepsilon(k)}{\partial k} = \frac{k}{m^*(k)}\,, \tag{3}$$

which is often studied in terms of an effective mass, $m^*(k)$. The inverse lifetime, $\Gamma(k)$, determines the timescale of the quasi-particle decay. Within Fermi liquid theory, one expects $\Gamma(k)$ to vanish near the Fermi surface. The knowledge of both the group velocity and the inverse lifetime is enough to compute the mean-free path through the fundamental relation:

$$\lambda(k) = \frac{v(k)}{|\Gamma(k)|}\,. \tag{4}$$

The question is how to extract both $\varepsilon(k)$ and $\Gamma(k)$ from the retarded propagator, $\mathcal{G}_R(k,\omega)$. Note that, in nuclear physics, the latter is traditionally computed in real time rather than in, say, Matsubara space [33]. From a Fourier transform point of view, one is tempted to associate Eq. (2) to a pole of order 1 in the lower half-plane [14]. The argument goes as follows. To Fourier transform to time-space, one integrates the propagator slightly above the real axis. If this contour is swapped to a contour that includes the (so-far hypothetical) pole in the lower complex plane, Cauchy's theorem guarantees the final result:

$$\mathcal{G}_R(k,t) = \int_{-\infty}^{\infty}\frac{d\omega}{2\pi}e^{-i\omega t}\mathcal{G}(k,\omega_+) \sim \int_{C'}\frac{dz}{2\pi}e^{-izt}\frac{\eta(z)}{z-(\varepsilon(k)-i|\Gamma(k)|)} = -i\eta(k)e^{-i\varepsilon(k)t}e^{-|\Gamma(k)|t}\,, \tag{5}$$

where $\omega_+ \equiv \omega + i\eta$ with $\eta \to 0$. The single pole description is particularly attractive due to its simplicity and its relation to Fermi liquid theory [35].

This description, however, is somewhat at odds with the complex extension derived from the Lehmann representation,

$$\mathcal{G}_R(k,\omega) = \int\frac{d\omega'}{2\pi}\frac{\mathcal{A}(k,\omega')}{\omega_+-\omega'}\,. \tag{6}$$

Journal of Physics: Conference Series **427** (2013) 012009 doi:10.1088/1742-6596/427/1/012009

The extension to the complex energy plane is obtained by replacing ω_+ with a complex energy, z. Such extension exists and is unique, but gives rise to a complex variable function that is analytic off the real axis [36]. In other words, the extension of $\mathcal{G}$ to complex energies implies that no poles are present in the complex plane. An illustration of the complex energy dependence of the propagator is given in the left panel of Fig. 1. This shows the imaginary part of $\mathcal{G}(k=0,z)$ for a SCGF self-energy obtained from a CDBonn NN interaction. The cut of $\mathcal{G}$ across the real axis is determined by the spectral function, $\text{Im}\,\{\mathcal{G}(k,\omega_-) - \mathcal{G}(k,\omega_+)\} = \mathcal{A}(k,\omega)$. As expected, this extension to the complex plane does not yield a pole.

To recover Eq. (2) from the real frequency Fourier transform, we supposed that one could deform the integration contour from slightly above the real axis to just include the pole in the lower half-plane. The discontinuity of $\mathcal{G}$ across the real axis, however, precludes us from doing so. Instead, one could eliminate the discontinuity by providing a functional continuation, $\tilde{\mathcal{G}}$, which is continuous across the real axis. This continuation will have a pole in the lower half-plane. It is this pole, *i.e.* that of$\tilde{G}$ rather than that of $\mathcal{G}$, that can be associated to a qp.

The propagator in the complex energy plane fulfills a Dyson equation:

$$\mathcal{G}(k,z) = \frac{1}{z - \frac{k^2}{2m} - \Sigma(k,z)}\,, \tag{7}$$

with a self-energy, Σ, extended to complex energies in analogy with the complex extension of the propagator:

$$\Sigma(k,z) \equiv \int \frac{d\omega}{2\pi} \frac{\gamma(k,\omega)}{z-\omega}\,. \tag{8}$$

The function $\gamma(k,\omega)$ is proportional to the imaginary part of the self-energy near the real axis. The complex plane self-energy has the same analytic structure as the propagator, *i.e.* it is an analytic function except for the real axis. A way to obtain $\tilde{\mathcal{G}}$, is to use the complex Dyson equation, Eq. (7), with Σ replaced by the following extrapolation of the self-energy into the lower half-plane, $\tilde{\Sigma}$:

$$\tilde{\Sigma}(k,z) \equiv \begin{cases} \Sigma(k,z), & \text{Im}\ z > 0 \\ \Sigma^*(k,z), & \text{Im}\ z \leq 0 \end{cases}\,. \tag{9}$$

This choice eliminates the discontinuity of Σ across the real energy axis and produces a pole in the lower half-plane propagator. This procedure seems to be valid in the case of the retarded propagator in nuclear matter, but one cannot exclude a different outcome in other cases.

The right plot in Fig. 1 shows the imaginary part of $\tilde{\mathcal{G}}(k=0,z)$. $\tilde{\mathcal{G}}$ is analytic across the real axis, but develops an isolated pole in the lower half-plane. The position of this pole is given by a zero in the denominator of the complex propagator, Eq. (7). This gives rise to the complex equation:

$$z(k) = \frac{k^2}{2m} + \text{Re}\tilde{\Sigma}(k,z(k)) + i\text{Im}\ \tilde{\Sigma}(k,z(k))\,. \tag{10}$$

The solution, $z(k) = \varepsilon(k) + i\Gamma(k)$, yields the *fully dressed* qp spectrum and inverse lifetime. As we have access to the SCGF self-energies in the complex plane via Eq. (8), we are able to compute numerically the dressed spectra and lifetimes for different momenta, densities and temperatures. Note that, in the most general case, the solution to the previous equation need not be unique. We have not found any signature of multiple solutions in our nuclear matter calculations.

Journal of Physics: Conference Series **427** (2013) 012009 doi:10.1088/1742-6596/427/1/012009

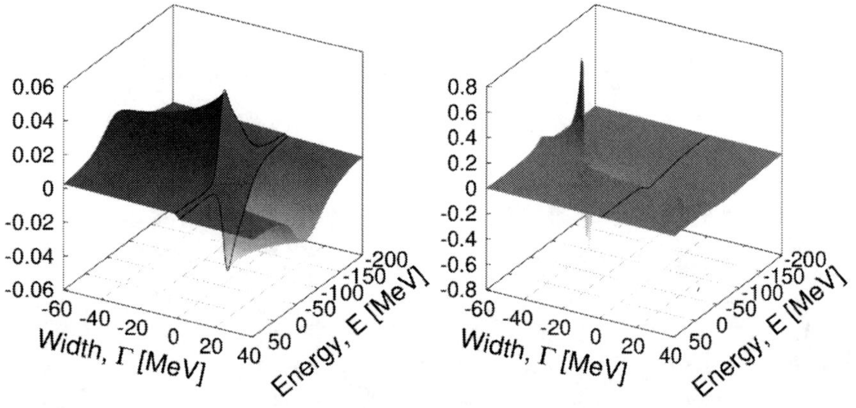

Figure 1. Imaginary part of the propagator in the complex energy plane, $z = E + i\Gamma$, at zero momentum for a CDBonn self-energy at $\rho = 0.16$ fm^{-3} and $T = 5$ MeV. The left plot corresponds to the usual propagator, $\mathcal{G}(k, z)$, while the right plot represents $\tilde{\mathcal{G}}(k, z)$. Solid lines show the imaginary part of the propagator just above/below the real axis, $\pm\mathcal{A}(k, \omega)/2$.

Our strategy can therefore be divided in four steps. First, we perform a ladder SCGF calculation of the self-energy along the real axis. We then extend the self-energy to the complex plane using its Lehmann representation, Eq. (9). The knowledge of $\Sigma(k, z)$ in the upper half plane is then used to find $\tilde{\Sigma}$ in the lower half-plane. Finally, using Eq. (10), one can determine the qp properties and hence the mean-free path. In the next section, we will discuss the result of these calculations for symmetric nuclear matter.

2.2. *Other approaches*

Previous calculations have relied on solving Eq. (10) using successive approximations for the dependence on the imaginary part of the complex energy of $\tilde{\Sigma}$ [18, 19, 21]. In the following, the different orders in the approximation are referred to as *renormalizations*, in accordance to the usual solid state nomenclature [31]. At the lowest order, known as *first renormalization*, one completely neglects the dependence on the imaginary part of z. This provides the usual definition of a qp:

$$\varepsilon_1(k) = \frac{k^2}{2m} + \text{Re}\tilde{\Sigma}(k, \varepsilon_1(k)) \,, \tag{11}$$

$$\Gamma_1(k) = \text{Im}\ \tilde{\Sigma}(k, \varepsilon_1(k)) \,, \tag{12}$$

which coincides with the peak of the spectral function. A *second renormalization* qp pole is obtained by expanding the self-energy around $z_1(k)$ to first order in the imaginary part of z,

$$\varepsilon_2(k) = \varepsilon_1(k) - \text{Im}\ \tilde{\Sigma}(k, \varepsilon_1(k))\ \text{Im}\ \frac{1}{1 - \tilde{\Sigma}'(z_1(k))} \,, \tag{13}$$

$$\Gamma_2(k) = \Gamma_1(k)\ \text{Re}\frac{1}{1 - \tilde{\Sigma}'(z_1(k))} \,. \tag{14}$$

In the context of nuclear physics, it has generally been assumed that the dependence of $\tilde{\Sigma}$ on the imaginary part of z is soft and can be ignored in the previous derivatives [18]. This gives

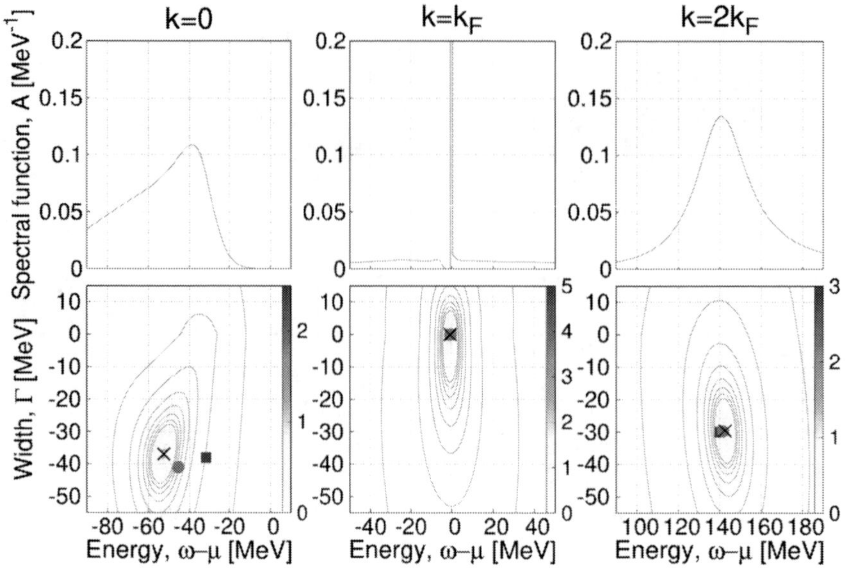

Figure 2. Upper panels: spectral function at $\rho = 0.16$ fm^{-3} and $T = 0$ MeV for the CDBonn interaction. Lower panels: absolute value of $\tilde{G}$ in the same conditions. The fully dressed pole is indicated by a cross, while the circle (square) show the position of the first (second) renormalization quasi-particle.

rise to a slightly different expression for the qp pole:

$$\varepsilon_{2'}(k) = \varepsilon_1(k) \tag{15}$$

$$\Gamma_{2'}(k) = \Gamma_1(k)\frac{1}{1 - \mathrm{Re}\tilde{\Sigma}'(\varepsilon_1(k))} . \tag{16}$$

As we shall see, this approximation is well justified only above k_F. Note that the previous expressions do not need any evaluation of self-energies in the complex plane. Access to real-energy self-energies, as those provided by Brueckner-Hartree-Fock, is enough to compute qp properties under the previous assumptions.

Once the qp spectrum and inverse lifetimes are known, one can go ahead and compute the mean-free path. Note, however, that this requires a consistent determination of ε, Γ and v, via Eq. (3). For the nuclear physics renormalization, Eqs. (15) and (16), the prefactor on the inverse lifetime is the inverse of the $\omega-$mass. The group velocity involves the full effective mass, $\frac{m^*}{m} = \frac{m_\omega}{m}\frac{m_k}{m}$. As a consequence, the mean-free path is only renormalized by the $k-$mass, $\lambda_{2'}(k) = \frac{m}{m_k}\lambda_0$, with respect to the uncorrected mean-free path, $\lambda_0(k) = k/[2m \,\mathrm{Im}\, \Sigma(k, \varepsilon_1(k)_+)]$. In this case, one can interpret the $k-$mass as a non-locality correction to the lowest order mean-free path. Within the nuclear physics community, this is a well-known correction that increases the mean-free path towards correct values [18, 19, 21].

3. Results

The upper panels of Figure 2 show the SCGF spectral function, as a function of energy, for three different characteristic momenta ($k = 0, k_F$ and $2k_F$). These have been obtained from a $T = 0$ self-energy based on the CDBonn interaction [27]. The lower panels give the absolute value of the continued propagator. One can clearly see, from the contour levels, that $\tilde{\mathcal{G}}$ develops a pole

Journal of Physics: Conference Series **427** (2013) 012009 doi:10.1088/1742-6596/427/1/012009

in the lower half-plane. The location of this fully dressed pole is consistent with the numerical solution of Eq. (10), shown with a cross. Differences between this pole and the first or second renormalization predictions (corresponding to the circle and square, respectively) are visible at $k = 0$. At and above the Fermi surface, discrepancies disappear and the fully dressed pole coincides with the first and second renormalizations. This points towards a very soft dependence of Σ on the imaginary part of z for $k \geq k_F$. The contour lines, which are basically aligned along the imaginary axis, validate this picture. Let us highlight that, at the Fermi surface itself, the calculation yields a quasi-particle with zero width, providing a verification of Fermi liquid theory [34].

Nuclear many-body calculations are subject to uncertainties associated to the underlying NN interaction as well as to the approximation scheme itself. These uncertainties are explored in Fig. 3. The upper panel displays the mean-free path of a nucleon in homogeneous nuclear matter at a density of $\rho = 0.16$ fm^{-3} and zero temperature, obtained with a CDBonn self-energy supplemented with an Urbana-type 3BF [27]. As expected, we find that the largest discrepancies between different approximations occur for hole energies, below -20 MeV, in a region where λ is already relatively small. In contrast, above 50 MeV, all approximations give similar results, except for λ_0 (dashed line), which is not corrected for non-locality and thus should not be taken as a realistic prediction. $\lambda_{2'}$ (dashed-dotted line) is only somewhat larger than λ_0 because of the small m_k associated to the SCGF results. Note that the classical kinetic theory prediction, $\lambda \sim (\rho\sigma_{np})^{-1}$ (dotted lines), is well below all quantum in-medium mean-free paths. The latter flatten at high energies, and remain constant, at a value of around $4 - 5$ fm.

In contrast, the lower panel of Fig. 3 focuses on the NN interaction dependence of our results. All the mean-free paths here are obtained from the full pole in the complex plane. The $T = 0$ mean-free path with 3BF (solid line) is slightly larger than that obtained without 3BFs (dashed). We also study the importance of the two-body NN interactions at a temperature of $T = 5$ MeV. CDBonn results are shown in a dotted-dashed line, whereas Av18 results correspond to the double-dotted-dashed line. The largest differences are observed, again, at negative energies. Overall, the effect of changing two-body NN interactions is as large as that of switching 3BFs on or off. If one takes the spread between different lines as an estimate of theoretical uncertainties, these amount to less than 1 fm beyond 50 MeV. Note that this uncertainty is of the same order of that obtained from experimental estimates [37, 38] and that the theoretical prediction agrees well with those.

Fig. 4 focuses on the temperature and density dependence of the mean-free path in nuclear matter obtained from the full complex plane. This demonstrates that our method can be applied in a wide range of conditions. The upper panel displays the temperature dependence of λ for a typical density of a nuclear interior, $\rho = 0.16$ fm^{-3}. The effect of temperature is relevant in an area of about 20 MeV around the Fermi surface. By switching on temperature, the mean-free path at the Fermi surface immediately becomes finite due to thermal damping. The damping effect becomes larger as temperature increases and thus the value of λ at the Fermi surface, $\omega = \mu$, decreases as temperature increases. These results indicate that finite temperature calculations provide reliable estimates of zero-temperature properties away from the Fermi surface. In other words, thermal effects do not dominate the very high and low energy behaviour of the mean-free path.

The density dependence of the mean-free path, displayed in the lower panel of Fig. 4, is more pronounced at all energies. Near the Fermi surface, an increase in density leads to an increase of λ. This is expected based on degeneracy arguments. A system at larger density is more degenerate and thus closer to the zero-temperature case, where λ diverges at the Fermi surface. At an energy of around 70 MeV this tendency is reversed, though, and λ tends to decrease as the density grows. This behaviour is more intuitive, but it remains to be seen whether the $1/\rho$ dependence predicted by classical transport theory is valid in the fully correlated, quantum

Journal of Physics: Conference Series **427** (2013) 012009 doi:10.1088/1742-6596/427/1/012009

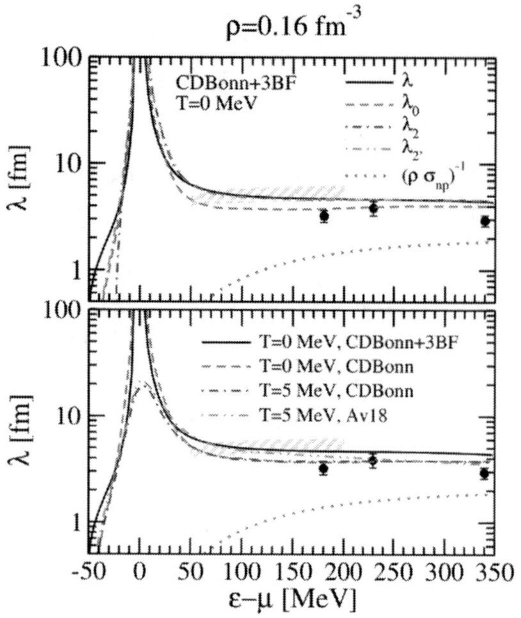

Figure 3. Mean-free path of a nucleon in nuclear matter as a function of energy. Upper panel: results obtained with a CDBonn+3BF self-energy at $T = 0$ MeV. The different approximations are commented on in the text. Lower panel: results obtained from the fully dressed pole for different NN forces and two different temperatures. The shaded band and solid dots correspond to the experimental results of Refs. [37, 38], respectively.

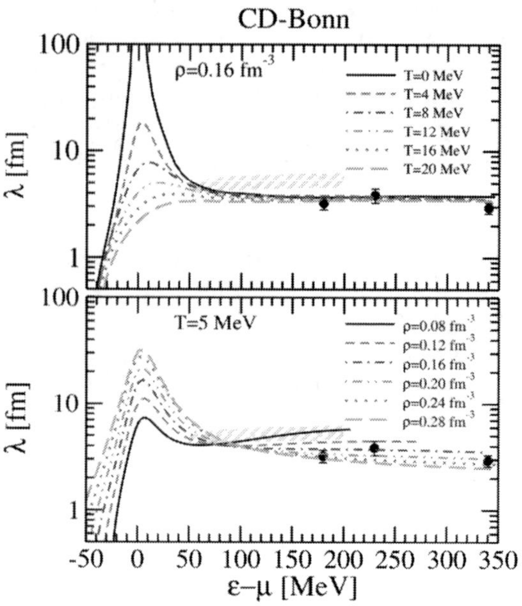

Figure 4. Mean-free path of a nucleon in nuclear matter as a function of energy. Upper panel: temperature dependence obtained with a CDBonn self-energy at a constant density of $\rho = 0.16$ fm^{-3}. Lower panel: density dependence obtained with a CDBonn self-energy at a constant temperature of $T = 4$ MeV.

case. The low density results cover the band of experimental data of Ref. [37], which has been extracted phenomenologically from finite nuclei. On average, these have a typical density which is lower than 0.16 fm^{-3} and thus this prediction meshes well with experimental estimates.

4. Conclusions and outline

We have devised a new method to obtain the mean-free path of a nucleon in the medium. The method involves the extension of Green's functions techniques into the complex plane and the determination of quasi-particle properties from a pole in the lower half-plane. Our approach provides a validation of previously used approximations by taking into account the full dependence on the imaginary part of the energy. In the nuclear medium, the calculation of the pole within this method provides similar mean-free paths as those obtained with earlier, cruder calculations. The renormalization induced by this procedure is specially relevant for hole properties. With all many-body corrections properly implemented, we obtain a mean-free path of around $4 - 5$ fm at saturation density and energies above 50 MeV.

The nucleon mean-free path lies at the heart of several theoretical and experimental considerations. Our future work will systematically assess the density, temperature and isospin asymmetry dependence of the mean-free path. In addition, we plan to compute other transport coefficients using our ladder self-energies. Bulk viscosities are particularly important in the context of neutron stars. Ultimately, we aim at producing a consistent set of transport properties that can be validated thoroughly against pulsar observations.

Acknowledgments

This work has been supported by a Marie Curie Intra European Fellowship within the 7th Framework programme, STFC grants ST/I005528/1 & ST/J000051/1 and the Espace de Structure Nucléaire Théorique (ESNT, CEA Saclay).

References

[1] Baldo M 1999 *Nuclear Methods and the Nuclear Equation of State* International Review of Nuclear Physics, Vol. 8 (World Scientific (Singapore))
[2] Lattimer J M and Prakash M 2007 *Phys. Rep.* **442** 109 (*Preprint* 0612440)
[3] Steiner A W, Lattimer J M and Brown E F 2010 *Astrophys. J.* **722** 33–54
[4] Özel F, Baym G and Güver T 2010 *Phys. Rev. D* **82** 101301
[5] Balescu R 1975 *Equilibrium and Nonequilibrium Statistical Mechanics* (*Wiley-Interscience publication* vol 76) (Wiley)
[6] Page D, Prakash M, Lattimer J and Steiner A 2011 *Phys. Rev. Lett.* **106** 081101
[7] Ho W, Andersson N and Haskell B 2011 *Phys. Rev. Lett.* **107** 101101
[8] Cutler C and Lindblom L 1987 *Astrophys. J.* **314** 234
[9] Wambach J, Ainsworth T and Pines D 1993 *Nucl. Phys. A* **555** 128–150
[10] Benhar O and Valli M 2007 *Phys. Rev. Lett.* **99** 232501
[11] Haskell B, Degenaar N and Ho W C G 2012 *MNRAS* **424** 93–103
[12] Benhar O, Polls A, Valli M and Vidaña I 2010 *Phys. Rev. C* **81**(2) 024305
[13] Abrikosov A A and Khalatnikov I M 329 *Rep. Prog. Phys.* **22** 329
[14] Economou E N 2006 *Green's Functions in Quantum Physics* 3rd ed Springer Series in Solid-State Sciences (Springer)
[15] Kadanoff L P and Baym G 1962 *Quantum Statistical Mechanics* First ed (New York: Benjamin)
[16] Mahan G D 1990 *Many-particle Physics* Second ed (NY: Plenum Press)
[17] Enss T, Haussmann R and Zwerger W 2011 *Ann. Phys.* **326** 770–796
[18] Negele J and Yazaki K 1981 *Phys. Rev. Lett.* **47** 71–74
[19] Fantoni S, Friman B L and Pandharipande V R 1981 *Phys. Lett. B* **104** 89–91
[20] Sinha B 1983 *Phys. Rev. Lett.* **50** 91–94
[21] Mahaux C 1983 *Phys. Rev. C* **28** 1848–1849
[22] Glavanakov I V 2004 *Physics of Atomic Nuclei* **67** 1580–1595
[23] Li G, Machleidt R and Zhuo Y 1993 *Phys. Rev. C* **48** 1062–1068
[24] Zuo W, Bombaci I and Lombardo U 1999 *Phys. Rev. C* **60** 024605
[25] Sammarruca F 2008 *Phys. Rev. C* **77** 047301
[26] Rios A, Polls A and Dickhoff W H 2009 *Phys. Rev. C* **79** 18 (*Preprint* 0904.2183)
[27] Somà V and Bożek P 2008 *Phys. Rev. C* **78** 054003
[28] Fradkin E S 1969 *Nucl. Phys.* **12** 466–484
[29] Engelsberg S and Schrieffer J R 1963 *Phys. Rev.* **131** 993–1008
[30] Eiguren A and Ambrosch-Draxl C 2008 *Phys. Rev. Lett.* **101** 036402
[31] Eiguren A, Ambrosch-Draxl C and Echenique P 2009 *Phys. Rev. B* **79** 245103
[32] Frick T and Müther H 2003 *Phys. Rev. C* **68** 034310
[33] Dickhoff W H and Barbieri C 2004 *Prog. Part. Nucl. Phys.* **52** 377–496
[34] Dickhoff W H and Van Neck D 2005 *Many-body theory exposed!* 1st ed (World Scientific)
[35] Abrikosov A A, Gorkov L P and Dzyaloshinskii I Y 1965 *Quantum Field Theoretical Methods in Statistical Physics* 2nd ed (Pergamon Press)
[36] Baym G and Mermin N D 1961 *J. Math. Phys.* **2** 232
[37] Nadasen A, Schwandt P, Singh P P, Jacobs W W, Bacher A D, Debevec P T, D K M and Meek J T 1981 *Phys. Rev. C* **23** 1023
[38] Renberg, P U and Measday, D F and Pepin, M and Schwaller, P and Favier, B and Richard-Serre C 1972 *Nucl. Phys. A* **183** 81

Journal of Physics: Conference Series **427** (2013) 012010 doi:10.1088/1742-6596/427/1/012010

Towards a nonequilibrium Green's function description of nuclear reactions

A. Rios[1], **B. Barker**[2] **and P. Danielewicz**[2]

[1]Department of Physics, University of Surrey, Guildford, Surrey, GU2 7XH, United Kingdom
[2]National Superconducting Cyclotron Laboratory and Department of Physics and Astronomy, Michigan State University, East Lansing, MI 48824-1321, USA

E-mail: a.rios@surrey.ac.uk

Abstract. Semiclassical time-dependent approaches are nowadays able to describe energetic central collisions of heavy isotopes with minimal assumptions. Simulations in the mean-field picture have provided insight on both nuclear structure and low-energy many-body reaction mechanisms. In this context, nonequilibrium Greens functions techniques have the potential to improve the description of the time evolution of nuclear systems by introducing effects beyond the mean-field. We describe a first attempt to use the Kadanoff-Baym dynamics for one-dimensional nuclear systems, with a particular emphasis on the process of correlation build-up.

1. Introduction

Nonequilibrium Green's function (NGF) techniques [1, 2, 3, 4] represent a powerful tool to describe the evolution of correlated quantum many-body systems. The study of the Kadanoff-Baym (KB) dynamics in different quantum many-body systems is an on-going challenge for theorists. The first direct implementations of NGF in homogeneous nuclear systems saw light almost 30 years ago [2]. Posterior advances include, among others, the extension to homogeneous electronic systems [5], small atoms [6] and molecules [7], quantum dots [8] and, more recently, the study of double excitations [9, 10]. Within nuclear physics, NGF techniques have been mostly used for derivations rather than exploited directly [11, 12, 4]. With so few quantitative studies on the impact of correlations in the time evolution of nuclear systems, one might wonder whether nuclear reactions simulations based on the mean-field picture or on semiclassical approaches are valid at all. To clarify these issues, we would like to develop nuclear reaction simulations beyond the Hartree-Fock approximation. Green's functions out of equilibrium, as described by the two-time Kadanoff-Baym approach, are an ideal tool for this purpose.

Our interest on the many-body aspects of nuclear dynamics ties well with the recent advances in nuclear structure theory. Effective field theories based on the chiral expansion together with renormalization techniques have provided access to perturbative QCD-inspired nuclear interactions [13]. As a direct consequence, first principles many-body calculations of medium-mass and heavy nuclei are becoming possible. These are based on a variety of methods, from coupled cluster [14] to equilibrium Green's function techniques [15]. The extension of similar *ab initio* methods to the realm of central nuclear collisions is an exciting prospect, as these have been mostly studied within phenomenological approaches.

On a more general scale, the advancement of dynamical quantum many-body methods is necessary for our understanding of systems that range from ultracold atoms [16] to nuclei [17] or atoms [6]. Simulations with NGF techniques can provide a systematic understanding of correlation effects via conserving approximations [18]. Moreover, different approximations may account for the effects of different types of correlations. The description, however, remains of a single-particle type and other many-body effects are averaged out. Not surprisingly, in the semiclassical limit the KB equations yield the Boltzmann equation (BE). In the few body limit, KB dynamics seem to run into some trouble, possibly associated to the small number of degrees of freedom [19].

Historically, just a handful of methods have been developed to describe central nuclear reactions. Few of these are general enough to be employed in generating practical predictions. The time-dependent Hartree-Fock (TDHF) method has been exhaustively employed in describing low-energy reactions [20]. Nowadays, TDHF simulations can be performed in full 3D and involve nuclei as heavy as uranium [21]. However, the validity of TDHF requires correlations to play a negligible role in the dynamics [12]. At low energies, one might argue that the role of correlations is minimized by antisymmetrization. Conversely, one would expect correlations to dominate at higher energies, where the Pauli principle weakens.

In the semiclassical limit, the two-time Kadanoff-Baym (KB) equations reduce to a single-time, Boltzmann equation (BE). The BE has been extensively used to analyze reactions at intermediate and high-energies [22, 23]. This approach deals with the increasing complexity of the reaction by using simplifying assumptions on an average, single-particle level. Because of their semiclassical nature, however, these descriptions have remained genuinely disconnected from the quantum methods employed for nuclear structure. More importantly, there is no systematic way of improving upon the BE-type approaches.

Correlations can lead to a fast thermalization of the occupation of single-particle states and to enhanced stopping, compared to what can be found in TDHF [24, 25, 26]. Similarly, quantum effects may dominate in high-density regions and thus affect reaction observables. Specifically aiming at increasing incident energies, it is therefore important to develop a quantal approach to central nuclear reactions that, besides the mean-field effects, can also incorporate correlations. The NGF approach that we try to develop can also remove some of the serious limitations of both TDHF and the BE. Unfortunately, even at the single-particle level, NGF techniques require handling vast amounts of information that can easily overwhelm the capabilities of computing systems, rendering this approach impractical.

Here we summarise our first attempts to implement NGF's to idealised, one-dimensional (1D) nuclear systems. The study of 1D nuclear systems (nuclear slabs) is academic, but offers an initial benchmark for ideas involved in the time evolution. Furthermore, 1D calculations provide simple and clear visualisations. First, we describe the application of mean-field dynamics in the context of nuclear collisions. In this case, the use of Green's functions techniques illustrates the importance of off-diagonal matrix elements in the space representation for the dynamics of central nuclear collisions. Further, we explore 1D simulations of the time evolution of correlated systems within the second order Born approximation.

2. Kadanoff-Baym equations

We describe the dynamics of correlated nuclear systems using the KB equations. Within this approach, the time evolution can incorporate different levels of correlations. For certain intrinsically-consistent many-body approximations to the self-energy, one can show that the time evolution induced by the self-consistent KB equations preserves the conservation laws obeyed by the system as a whole [18, 2]. We will use both the mean-field approximation and the second-order, direct Born approximation, which do preserve conservation laws.

The KB equations are nominally derived without any particular assumption on the physical system under consideration, hence their generality. They describe the time evolution of the so-called Wightman functions, which are expectation values of products of operators pertaining to different times. Those containing single-particle information are related to the expectation values of products of two single-particle annihilation and creation operators at different time arguments,

$$G^<(x_1,t_1;x_2,t_2) = i\left\langle \hat{a}^\dagger(x_2,t_2)\,\hat{a}(x_1,t_1)\right\rangle , \tag{1}$$

$$G^>(x_1,t_1;x_2,t_2) = -i\left\langle \hat{a}(x_1,t_1)\,\hat{a}^\dagger(x_2,t_2)\right\rangle . \tag{2}$$

The expectation values, $\langle\cdot\rangle \equiv \mathrm{Tr}\{\hat{\rho}_i\,\cdot\}$, are taken with respect to an initial density matrix, $\hat{\rho}_i$ at $t = t_0$. We consider initial uncorrelated states, even though the theory can incorporate correlations in the initial state [2].

The KB equations for 1D systems, governing the evolution of Green's functions in their arguments,

$$\left\{ i\hbar\frac{\partial}{\partial t_1} + \frac{\hbar^2}{2m}\frac{\partial^2}{\partial x_1^2} \right\} G^{\lessgtr}(\mathbf{1}\mathbf{1}') = \int \mathrm{d}x_{\bar{1}}\, \Sigma_{HF}(\mathbf{1}\bar{\mathbf{1}})\, G^{\lessgtr}(\bar{\mathbf{1}}\mathbf{1}') + \int_{t_0}^{t_1} \mathrm{d}\bar{\mathbf{1}}\, \Sigma^+(\mathbf{1}\bar{\mathbf{1}})\, G^{\lessgtr}(\bar{\mathbf{1}}\mathbf{1}') + \int_{t_0}^{t_{1'}} \mathrm{d}\bar{\mathbf{1}}\, \Sigma^{\lessgtr}(\mathbf{1}\bar{\mathbf{1}})\, G^-(\bar{\mathbf{1}}\mathbf{1}') , \tag{3}$$

$$\left\{ -i\hbar\frac{\partial}{\partial t_{1'}} + \frac{\hbar^2}{2m}\frac{\partial^2}{\partial x_{1'}^2} \right\} G^{\lessgtr}(\mathbf{1}\mathbf{1}') = \int \mathrm{d}x_{\bar{1}}\, G^{\lessgtr}(\mathbf{1}\bar{\mathbf{1}})\, \Sigma_{HF}(\bar{\mathbf{1}}\mathbf{1}') + \int_{t_0}^{t_1} \mathrm{d}\bar{\mathbf{1}}\, G^+(\mathbf{1}\bar{\mathbf{1}})\, \Sigma^{\lessgtr}(\bar{\mathbf{1}}\mathbf{1}') + \int_{t_0}^{t_{1'}} \mathrm{d}\bar{\mathbf{1}}\, G^{\lessgtr}(\mathbf{1}\bar{\mathbf{1}})\, \Sigma^-(\bar{\mathbf{1}}\mathbf{1}') , \tag{4}$$

follow from considerations of the equations of motion for creation and destruction operators [1]. The simplified notation $\mathbf{1} = (x_1, t_1)$ has been introduced. The retarded (+) and advanced (-) functions are defined according to $F^\pm(\mathbf{1},\mathbf{2}) = F^\delta(\mathbf{1},\mathbf{2}) \pm \Theta\left[\pm(t_1 - t_2)\right]\left[F^>(\mathbf{1},\mathbf{2}) - F^<(\mathbf{1},\mathbf{2})\right]$, with F^δ standing for a possible singular contribution at $t_1 = t_2$. The generalized self-energy $\Sigma(\mathbf{1},\mathbf{2})$ introduces interaction effects on the time evolution and also describes excitation processes within the system [1, 2].

The self-energy in the previous equations has been separated into two different components. The first one involves the Hartree-Fock (HF) contribution, $\Sigma_{HF}(\mathbf{1},\mathbf{2})$, which accounts for the instantaneous, one-body interaction of the considered particle with the mean-field produced by the other particles of the system. The term involving $\Sigma^{\lessgtr}$ describes time-dependent excitation processes, beyond the mean-field changes. Such terms account for the effect of correlations on the time evolution and need to be included for a complete description of nuclear reactions.

1D calculations are rather economical in terms of CPU and storage. Let us consider the difficulties associated to storing density matrices (or Green's functions) in D dimensions without explicit spin or isospin degrees of freedom. A uniform mesh of size N_x in each direction will yield matrices of size N_x^{2D}. Fairly sparse meshes with $N_x \sim 10$ are already computationally demanding at present. For 1D this is not an issue and we will present results with $N_x \sim 100$, that can be carried out straightforwardly. In addition to spatial variables, NGF requires a suitable infrastructure to keep track of the time degree of freedom, as memory effects come into play. From Eq. (3), one can easily see that the time derivative of the Green's function $G^<$ at times t_1 and $t_{1'}$ depends on the Green's functions and self-energies at all the previous times t, $t_0 < t < t_1$ and $t_0 < t < t_{1'}$, via the time integrals on the r.h.s. Consequently, to find a solution of the KB equations, one must keep track of all the previous time-steps. For a uniform mesh of N_t points in time, the $G^{\lessgtr}$ functions thus require handling $N_x^{2D} \times N_t^2$ matrices. For $D > 1$, this becomes a major concern in the numerical implementation of the KB equations [27].

3. Mean-field approximation

The KB equations simplify substantially when the correlation effects, described in terms of $\Sigma^{\lessgtr}$, are neglected. In that case, the evolution equations for $G^<$ and $G^>$ can be decoupled. Since the single-particle observables are more straightforwardly expressed in terms of $G^<$, we concentrate on the evolution of the latter. Thanks to the instantaneous nature of Σ_{HF}, the set of equations for the time-diagonal elements of the Wightman function, $t_1 = t_{1'}$, can also be closed and there is no need to consider explicitly time off-diagonal terms. Note, in particular, that this reduces the dimensionality of the problem to N_x^{2D}, as one can update $G^<$ at every time step and there is no need to keep explicit track of expectation values involving the past.

The evolution equations simplify even more when assuming a negligible range for the nucleon-nucleon interactions. In that case, the evolution equation becomes:

$$i\hbar\frac{\partial}{\partial t}G^<(x_1,x_{1'};t) = \left[-\frac{\hbar^2}{2m}\left\{\frac{\partial^2}{\partial x_1^2} - \frac{\partial^2}{\partial x_{1'}^2}\right\} + U(x_1,t) - U(x_{1'},t)\right] G^<(x_1,x_{1'};t)\,, \quad (5)$$

This can easily be linked to TDHF by expanding the Wightman functions in terms of single-particle states [17]. For the same initial state and mean-field parametrization, both approaches yield identical dynamics. Our goal here is not so much to provide an accurate description of 1D nuclear mean-field dynamics, but rather to explore its implementation in terms of Green's functions.

In the following, we shall assume a spin-isospin saturated system and thus suppress the (degenerate) spin and isospin indices. In the strict 1D interpretation of the 1D calculations, the density of nucleons is $n_1(x,t) = \nu\,\rho(x,x;t)$, where $\nu = 4$ represents the spin-isospin degeneracy of the system, and the nucleon number is $A = \nu N_s$. N_s is the number of shells filled in the calculation. One can connect this 1D interpretation to 3D calculations by assuming that matter is uniform in the y and z directions. With this, the density in the 3D interpretation, becomes proportional to the 1D density, $n(x,t) = \xi\, n_1(x,t)$,s with $\xi = \sqrt{\frac{5}{3}}\left(\frac{\pi n_0^2}{6\nu^2}\right)^{1/3}$ [17]. The advantage of developing a 3D interpretation is that it allows one to employ well-known 3D mean-field parameterizations for 1D calculations. We choose a simple nuclear mean-field parametrization for our calculations,

$$U(t) = \frac{3}{4}\,t_0\, n(x,t) + \frac{2+\sigma}{16}\,t_3\,[n(x,t)]^{1+\sigma}\,, \quad (6)$$

with the parameters fitted to the saturation properties of nuclear matter [17]. For such a local mean-field, the time evolution of the density matrix can be implemented numerically in a rather straightforward way using the Split Operator Method (SOM) [17, 28]. This involves a continuous switching between configuration space and momentum space which is optimally accomplished using Fast Fourier Transforms (FFT) [28]. In practical terms, we use an implementation of the SOM involving a uniform mesh in the spatial dimension, with periodic boundary conditions at the edges.

3.1. Preparation of the initial state

Before discussing the dynamics at the mean-field level, we will comment on the construction of initial states. The initial state for a reaction (or any other dynamical process) should be a ground state within the same approximation scheme as used in the dynamics. This guarantees that observables for the ground state should do not change with time. One way of preparing the initial system in a consistent manner is through an imaginary-time evolution [11, 4], employing analogous approximations for the self-energy within the imaginary- and real-time domains. Naturally, it may be worthwhile to explore alternative ways of preparing the initial state. If we restrict ourselves, for practical purposes, to dynamics in the real time, an obvious choice is that of an adiabatic switching.

At time $t \to -\infty$, the system may be assumed to represent the ground state for a known (or solvable) external potential, U_0. Here we will limit ourselves to a 1D Harmonic Oscillator (HO), which yields a relatively good starting point for nuclear physics purposes [29]. Provided that the interaction is switched on slowly enough while the external HO is steadily eliminated, the system may be evolved to the interacting ground state [30]. In the absence of level crossings or phase transitions, the procedure should be generally valid. We show in the following examples of this at the Hartree-Fock level. Further, we will present preliminary calculations of analogous results within the two-time dynamics at the direct Born approximation.

With the adiabatic switching, the single-particle potential acquires an explicit time dependence, and we adopt

$$U_t = F(t-\tau_0)\,U_0 + [1 - F(t-\tau_0)]\,U\,, \tag{7}$$

where U is our mean-field parametrization and $U_0 = \frac{1}{2}\,m\,\Omega^2\,x^2$ is the precursor HO potential. Ω can be chosen to minimize the evolution of geometric characteristics foes nuclear slabs [17]. The switching function F above is equal to 1 at some initial argument t_i (< 0) and to 0 at some final argument t_f (> 0) when the switching over is completed. This switching function is constructed in terms of a monotonically *decreasing* function f,

$$F(t) = \frac{f(t) - f(t_f)}{f(t_i) - f(t_f)}\,, \tag{8}$$

for the argument values $t_i \leq t \leq t_f$. In the switching function for Eq. (7), we have most often employed

$$f(t) = \frac{1}{1 + e^{t/\tau}} \tag{9}$$

within Eq. (8). Here, τ represents a transition time that, for the sake of adiabaticity of the switching, should be longer than any characteristic time of the system. Notably, whenever $f(t_i) \simeq 1$ and $f(t_f) \simeq 0$, such as for $|t_{i,f}| \gg \tau$ in the case of (9), then the two functions f and F practically coincide,

$$F(t) \simeq f(t)\,. \tag{10}$$

Slower adiabatic switchings should provide better approximations to the mean-field ground states. In the top panel of Fig. 1, we show the evolution of the energy per nucleon for a HO slab initiated at time $t = -1000\,\text{fm}/c$. We concentrate on a single case, with $N_s = 2$ HO shells filled (*i.e.* $A = 8$ nucleons in the 1D interpretation). We consider different transition times, τ, that characterise the adiabaticity of the transition. For any of the employed transition times $\tau \geq 5\,\text{fm}/c$, the energy evolves to a value very close to that for the static Hartree-Fock solution (see inset). Judging the quality of the adiabatic transition on the basis of the energy alone is potentially treacherous, though, as the energy is quadratic in the deviation of wave function from the ground state. In principle, final states for the adiabatic evolution might be found, with a wavefunction poorly approximating that of the mean-field ground-state, but giving an energy close to the ground-state value.

Alternatively, the total density, n, is linear in the deviation of the wavefunction from the ground-state, so it may provide a better measure of the wavefunction quality. With this in mind, in the bottom panel of Fig. 1, we show the evolution of the size of the slab, defined as twice its average extent in 1D. Indeed, for $\tau = 5$ and $10\,\text{fm}/c$, slab sizes exhibit significant

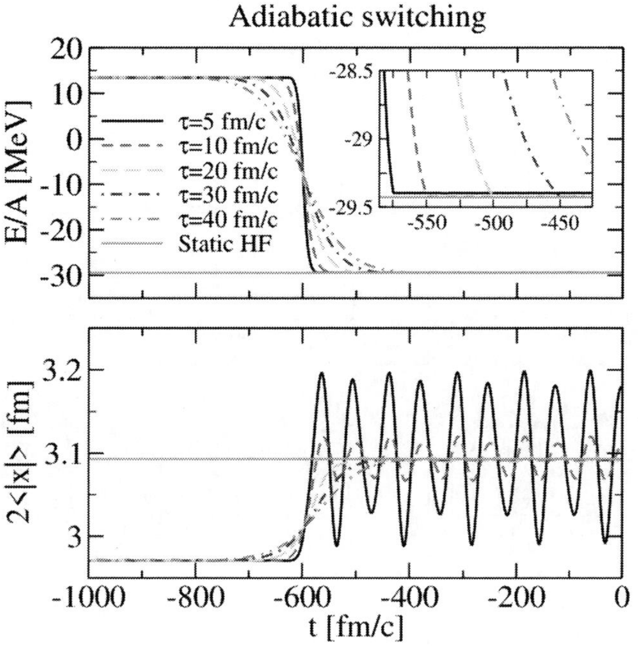

Figure 1. Time evolution of the energy per particle (upper panel) and the size of the slab (lower panel) when adiabatically switching from a starting HO configuration to a final mean-field state. Different values of the transition time τ are considered. The inset in the top panel shows a magnified portion of the time evolution of the energy. For reference, mean-field results from static Hartree-Fock solutions are shown as straight solid lines.

oscillations in the final state, indicating that the ground state has not been yet satisfactorily reached. Instead, the final state in the dynamics is an excited state of the final hamiltonian and thus fluctuates over time. For $\tau \gtrsim 30\,\text{fm}/c$, the oscillations become insignificant, with the slab sizes practically coinciding with that of the static Hartree-Fock solution.

In a nuclear reaction, the construction of the initial state involves two steps. First, the ground state of the two reacting systems has to be built. Second, the two initial states have to be boosted so as to collide with each other. The boost can be accomplished through a simple multiplication by momentum phase factors [31]:

$$G_1^<(x_1, x_{1'}) = \mathrm{e}^{iPx_1/\hbar}\, G_0^<(x_1, x_{1'})\, \mathrm{e}^{-iPx_{1'}/\hbar}\,, \qquad (11)$$

where $G_1^<$ corresponds to a moving slab at momentum per nucleon P, while $G_0^<$ is that of an idle slab. As a consequence of the boost, the kinetic energy for the slab increases by the amount $\Delta K/A = P^2/2m$. The net Wightman function of the colliding system is constructed as a sum of functions for two countermoving slabs, $G^<(x_1, x_{1'}) = G_1^<(x_1, x_{1'}) + G_2^<(x_1, x_{1'})$. We will here concentrate on symmetric collisions, with the second slab being a symmetric reflection of the first. Our conclusions, however, are general and broadly valid also for asymmetric reaction.

3.2. *Collisions of slabs*

The relatively simple 1D model of nuclear collisions discussed here demonstrates a surprisingly rich range of phenomena [32, 33, 17]. Qualitatively different physical processes are observed within the model when changing the center-of-mass (CM) energy for the reactions. At low

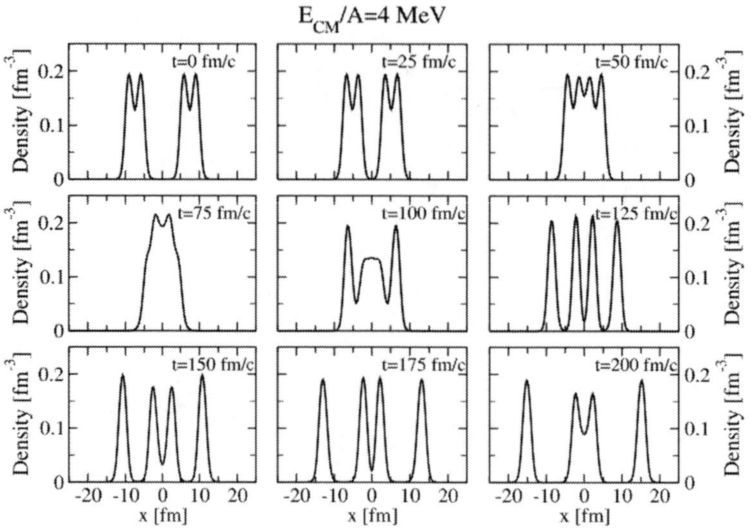

Figure 2. Evolution of the CM density profile for a collision of two $A = 8$ slabs at the collision energy of $E_{CM}/A = 4\,\mathrm{MeV}$.

collision energies ($E_{CM}/A \sim 0.1 - 0.5\,\mathrm{MeV}$), the nuclear slabs fuse into one compound slab that remains excited for long times. For intermediate energies ($E_{CM}/A \sim 0.5 - 15\,\mathrm{MeV}$), a fusion process is observed, followed by a break-up into a number of smaller slabs. Higher reaction energies ($E_{CM}/A > 15\,\mathrm{MeV}$) yield a pile-up of density at the system center, followed by a violent break-up phase with the formation of a low-density neck. The process is reminiscent of multifragmentation in nuclear reactions.

We will not explore this wide phenomenology here, but rather focus on one specific case that will highlight the importance of space off-diagonal matrix elements in $G^<(x, x')$ specifically. Let us concentrate on the collision energy range immediately above fusion reactions. When the two colliding slabs fuse, the compound system is too excited in energy to remain bound and the system undergoes a fission process into three fragments. Because of the symmetry of this specific collision, the final state is formed by a central fragment and two outgoing, symmetric slabs. Figure 2 provides the snapshots for the time evolution of one such collision at $E_{CM}/A = 4\,\mathrm{MeV}$. Following the fusion of the original slabs at $t \sim 50\,\mathrm{fm}/c$, two density peaks form at the edges of the compound system and subsequently emerge from the central region to opposite sides. The central region eventually recontracts and oscillates, as it is a highly excited state of an $A \sim 8$ system. The $A \sim 4$ fragments remain excited too, as evidenced in the small changes with time of their central density.

3.3. Off-diagonal structure of the density matrix

At equal arguments in a specific representation, the function $-iG^<$ of Eq. (1) yields the single-particle density in that representation. At different arguments, the off-diagonal matrix elements reflect existing correlations between magnitudes and phases of single-particle wavefunctions. As we have discussed earlier, the task of following all the elements in 3D is likely to overwhelm present computer storage capabilities. Nevertheless, the quantities of most direct physical interest, including densities, tend to be associated with either diagonal values of the functions or values close to the diagonal [1]. In the specific physical cases where the matrix elements that lie sufficiently far from the diagonal can be ignored, NGF calculations would become possible

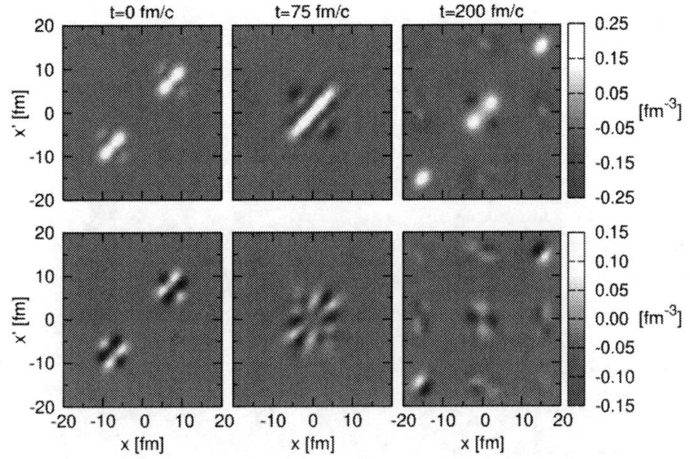

Figure 3. Intensity plots representing the real (upper panels) and the imaginary part (lower panels) of the scaled density matrix, $\xi\nu\,\rho(x,x';t)$, for a collision at $E_{CM}/A = 4\,\text{MeV}$.

due to a drastic reduction in dimensionality. We argue that central nuclear reactions are good examples of this, as the late stages involve limited phase coherence between different elements.

To examine and quantify the importance of elements far from the diagonal in space representation, let us explore the off-diagonal structure of $-iG^<$ in the same reaction presented in Fig. 2. The upper (lower) panels of Fig. 3 show intensity plots for the imaginary (real) parts of $G^<(x,t;x',t)$ for three different times in the reaction, chosen to represent the initial, overlap and late instants. The diagonal of the upper panels, $x = x'$, coincide with the density profiles displayed in Fig. 2. To understand off-diagonal structures, it is useful to think of the expansion of the Wightman function in terms of natural orbitals,

$$-iG^<(x_1,t_1;x_{1'},t_{1'}) = \sum_\alpha n_\alpha \phi_\alpha(x_1,t_1)\,\phi^*_\alpha(x_{1'},t_{1'})\,. \tag{12}$$

The wavefunctions of the initial slabs are confined to within the size of the slab and, consequently, $-iG^<$ will have a limited, square support in the (x,x') plane. The left panels of Fig. 3 show this confinement around the two initial slabs. Changes in sign as one moves away from the diagonal are attributable to the boost factors.

At this collision energy, once the compound slab is formed (central panel), it remains in contact for a too short time for the single-particle orbitals to spread out evenly over the size of the compound system. In consequence, the square-like support for the density matrix of the compound system is only approximate. The two outgoing fragments and the central compound system are clearly identified as blobs near the diagonal, $x = x'$. In addition, cross-correlations develop between individual fragments, as signaled by the patches of significant values of the density matrix far away from the $x = x'$ diagonal. As an example, the structures at $x \sim 0$ and $x' \sim 15\,\text{fm}$ for $t = 200\,\text{fm}/c$ in Fig. 3, both for the real and imaginary parts of the matrix, signal the overlap of single-particle states between the central and the outgoing fragments.

The discussed correlation patches may be understood in terms of the fragmentation of natural orbitals. Obitals can have components in each fragment rather than being split among them. Consequently, if the two fragments are in positions x_1 and x_2, when taking the product in Eq. (12) one will find non-zero elements of $G^<$ around (x_1,x_2). Physically, this is a reflection of

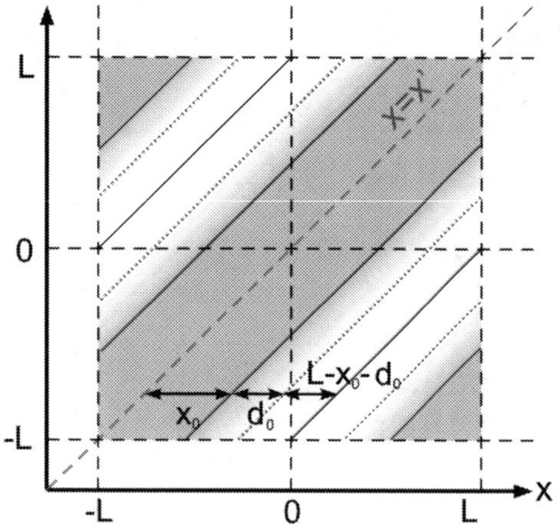

Figure 4. The elimination procedure of off-diagonal elements in terms of superoperators eliminates off-diagonal elements in the white region, $|x - x'| > x_0 + d_0$.

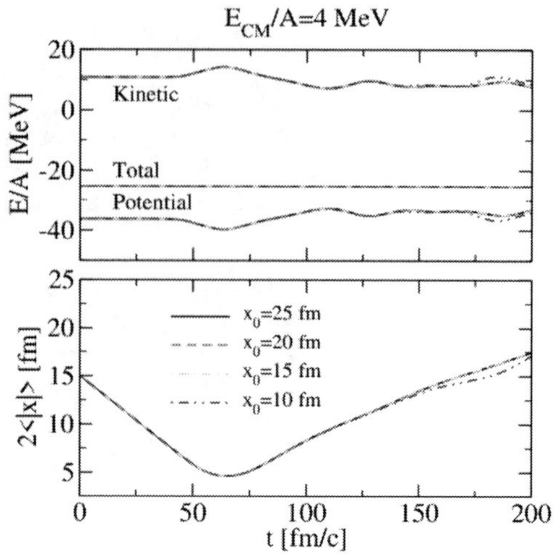

Figure 5. Time evolution of the energy per particle (upper panel) and of the system extent (lower panel) in a collision at $E_{CM}/A = 4\,\text{MeV}$ for different values of the cutoff parameter x_0 at fixed $d_0 = 2$ fm.

the fact that, during the breakup, the nucleons from the original orbitals have finite probabilities of ending up in different fragments. The amplitudes for those possibilities maintain a phase relationship leading to correlations in the density matrix. The entanglement of the internal wavefunctions is thus the sole reason for the persistence of the far-away off-diagonal patches in the density matrix. The real and imaginary parts for those structures are comparable in magnitude, see Fig. 3. This points to the involvement of significant relative phases, as expected from the significant difference in momentum and position. Note that these entanglement correlations persist for fragments that are 30 fm apart for the $t = 200\,\text{fm}/c$ panels of Fig. 3.

Physics beyond the mean-field dynamics (short-range correlations, particle and gamma decays, etc.) are likely to introduce additional decoherence between the separating fragments, beyond that stemming only from different mean-field orbitals. Correspondingly, higher values for the off-diagonal elements are likely to persist more in a mean-field approach than in any simulation with correlations. As a corollary, if one finds out that the far off-diagonal elements may be disregarded within the mean-field dynamics, then it should be even more possible to disregard such elements in more realistic approaches.

We have quantified the importance of off-diagonal matrix elements using two different methods. First, we have employed an *ad hoc* procedure based on super-operators that eliminate off-diagonal elements in the time evolution [17]. Fig. 4 sketches the spatial structure of the super-operator. White areas, beyond $|x - x'| > x_0 + d_0$, are artificially eliminated at each time step. The remaining blue (grey) areas are left unaffected. Note that the structure is chosen to preserve the periodicity in (x, x') space. Fixing $d_0 = 2$ fm, one can progressively eliminate more and more off-diagonal matrix elements by decreasing x_0. For the break-up reaction explored here, the results of this procedure are shown in Fig. 5. Neither the total energy nor its kinetic or potential components (upper panel) seem to be affected by these cuts. The extent of the system, presented in the bottom panel, is also unaffected, pointing towards a good average description of the density distribution in the system. Overall, the elimination procedure we have devised indicates that off-diagonal elements play a very small feedback effect on the dynamics.

The off-diagonal suppression procedure we have just presented, though, does not directly involve any substantial savings in terms of storage or CPU time. The most substantial savings can be achieved by using an alternative spatial discretization scheme in a rotated coordinate frame. This rotation is the computational equivalent of a switch to Wigner space. With this, one could control the spareness and lengths of the mesh in the relative coordinate while keeping the average (physical) direction unaffected. Our preliminary calculations indicate that this implementation reproduces well the original dynamics while simultaneously reducing substantially the total numerical effort.

4. Two-time Kadanoff-Baym evolution

We turn now our attention to beyond mean-field dynamics as obtained from the Kadanoff-Baym dynamics. We use the second order approximation in the propagator expansion. The so-called direct Born self-energy in momentum space reads:

$$\Sigma^{\lessgtr}(p,t;p',t') = \int \frac{dp_1}{2\pi}\frac{dp_2}{2\pi} V(p-p_1)V(p'-p_2)G^{\lessgtr}(p_1,t;p_2,t')\Pi^{\lessgtr}(p-p_1,t;p'-p_2,t') \tag{13}$$

$$\Pi^{\lessgtr}(p,t;p',t') = \int \frac{dp_1}{2\pi}\frac{dp_2}{2\pi} G^{\lessgtr}(p_1,t;p_2,t')G^{\gtrless}(p_2-p',t';p_1-p,t)\,, \tag{14}$$

In this exploratory study, we have not included the exchange term despite the fact that it may induce significant effects. As it is common in nuclear physics, we use two different interactions for the mean-field self-energy and the Born term. The Hartree-Fock self-energy of Eq. (6) has a zero-range as well as a density dependence that effectively accounts for three-body correlations. The attractive component of the mean-field provides stability. The Born term requires a finite range interaction and we choose the (time-independent) form:

$$V(p) = V_0\sqrt{\pi}\eta^2 p^2 e^{-\frac{\eta^2 p^2}{4}} \tag{15}$$

with $V_0 = 40$ MeV and $\eta = 3$ fm. In coordinate space representation, this interaction qualitatively reproduces the characteristic short-range repulsion and long-range attraction of the nucleon-nucleon interaction. Note that the p^2 dependence at low momenta is necessary to avoid divergences in the 1D scattering cross section.

The solution of the KB equations, Eqs. (3) and (4), is a computationally demanding task in the two-time plane [27], all the more so for inhomogeneous systems [6]. We advance the solution of these following the algorithms presented in Ref. [34]. In analogy to the mean-field case, we use Fast-Fourier transforms to switch from real to momentum space thus allowing for quick phase-factor multiplications and calculation of convolution integrals. At each time step, we demand self-consistency of the density before we proceed further on in time. Results presented here correspond to a mesh of $N_x = 40$ points with $N_t = 150$ time-steps.

In analogy to the mean-field case, we want to explore the potential of adiabatic switching to build initially correlated ground states. To that end, we follow a two-step process. First, we build a mean-field ground state using the same adiabatic switching, from a HO to a mean-field ground state, described in Sec. 3.1. Further, we turn on adiabatically the Born self-energy. Because of the two-time nature of the time propagation, the adiabatic switch-on should also develop a two-time structure. The origin of this double time dependence can be found in Eq. (13), which can be formulated also for time-dependent interactions. Replacing the time independent interaction, $V(p)$, by the adiabatically evolving one, $\tilde{V}(p,t) = (1-F(t))V(p)$, the self-energy becomes directly proportional to the product $(1-F(t))\times(1-F(t'))$. The adiabatically switched-on self-energies, however, feed back to the propagator via the KB evolution and are also sensitive to associated memory effects. Consequently, the time dependence of the self-energy can be more complex than that given by the product of adiabatic factors as time proceeds.

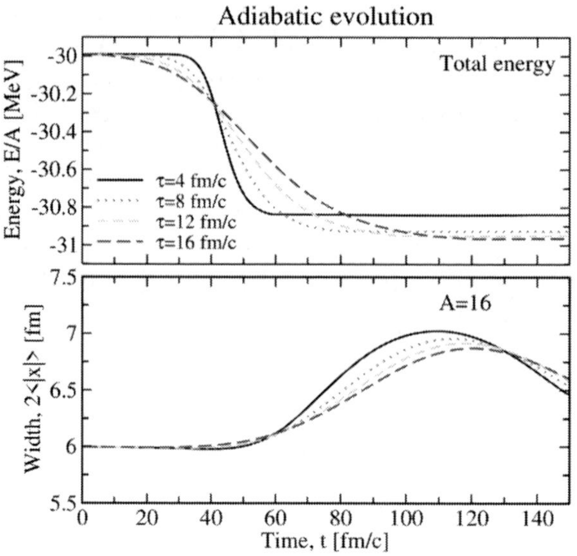

Figure 6. Time evolution of the energy per particle (upper panel) and the size of the slab (lower panel), when adiabatically switching from a starting mean-field configuration to a final correlated state. Different values of the transition time τ are considered.

Fig. 6 shows preliminary results for the correlated adiabatic switching on for a slab with $N_s = 4$, *i.e.* $A = 16$ in the 1D interpretation. The upper panel shows the total energy of the system. The switch-on provides an attractive contribution to the energy of the order of ~ 1 MeV. Note that, due to numerical limitations in this preliminary implementation, we have had to restrict the time evolution to relatively short timescales. As a direct consequence, the range of adiabatic switchings is also restricted. The final energy that is reached as τ increases does change for all the values of τ explored in the figure, which suggests that we have not yet reached adibatically the ground state of the final system. With correlations, however, the switching needs to be slow compared to characteristic correlation times [35]. The success of the switching procedure should be judged again with the degree to which the final state is stationary.

The lower panel of Fig. 6 focuses on the time evolution of the slab's size in the adiabatic evolution. When correlations are turned on, the slab swells in size. With the present time extent, we can only resolve an initial oscillation in slab size. A rapid damping of this oscillation is expected as τ increases and the evolution becomes more and more adiabatic. Naively, one expects the final state obtained in these dynamical calculations to be the same as the equivalent ground state obtained in imaginary time. We plan to carry on a detailed analytical and numerical comparison in the near future.

Note that, by introducing correlations in the dynamics, the occupation probabilities of the single-particle states, n_α, gradually move away from a Fermi step function to a correlated shape. Net energy is gained at the cost of populating states with higher single-particle energies. The redistribution of the state populations provides damping mechanisms in nuclear reactions. Their microscopic description is one of the major goals of our long term project.

5. Conclusions

The mean-field time evolution provides a first qualitative guidance on the dynamics of nuclear slabs. In terms of numerical implementation, it has allowed us to develop an understanding of initial state dynamics through adiabatic switching. This procedure provides good results when implemented with standard TDHF codes. At the mean-field level, the implementation of the dynamics within the NGF theory introduces a double-argument description of the system. Correlations between spatial off-diagonal elements are generated in a nuclear collision whenever single-particle orbitals can be spread across different fragments (or, in the case of multifragmentation, among the whole system).

We have also developed the tools to simulate correlated dynamics by implementing the KB equations numerically. In this first, preliminary study, we have explored the potential of these techniques to yield correlated ground states. To this aim, we have performed an adiabatic switching of the Born self-energy on top of Hartree-Fock results. These indicate that relatively large timescales are needed for the system to thermalize. Such long times are as of now inaccessible due to numerical limitations. The hope is that these limitations can be overcome in the near future by further optimization and parallelization of our numerical codes.

In spite of their considerable general potential in describing quantal time-dependent systems, NGF techniques have been, so far, underutilized in nuclear physics. Specific potential applications within nuclear physics include the description of large-amplitude collective motion and of central nuclear collisions. The techniques could yield a seamless description for nuclear structure [36]. Quite more generally, a more thorough understanding of the interplay between spatial dynamics, temporal evolution and beyond mean-field correlations in quantum many-body systems is a challenging prospect. The overall consistent picture provided by NGF's seems to be the best candidate to address these issues rigorously.

Acknowledgments

This work was supported by the National Science Foundation, under Grant No. PHY-1068571, by a Marie-Curie Intra-European Fellowship within the 7th European Community Framework Programme and by STFC grants ST/I005528/1 & ST/J000051/1.

References

[1] Kadanoff L P and Baym G 1962 *Quantum Statistical Mechanics* (New York: Benjamin)
[2] Danielewicz P 1984 *Ann. Phys. (N.Y.)* **152** 239
[3] Botermans W and Malfliet R 1990 *Phys. Rept.* **198** 115
[4] Köhler H S 1995 *Phys. Rev. C* **51** 3232
[5] Kwong N H, Bonitz M, Binder R and Köhler H S 1998 *Phys. Stat. Sol. B* **206** 197
[6] Dahlen N E and van Leeuwen R 2007 *Phys. Rev. Lett.* **98** 153004
[7] Balzer K, Bauch S and Bonitz M 2010 *Phys. Rev. A* **82** 033427
[8] Balzer K, Bonitz M, van Leeuwen R, Dahlen N E and Stan A 2009 *Phys. Rev. B* **79** 245306
[9] Balzer K, Hermanns S and Bonitz M 2012 *EPL* **98** 67002
[10] Säkkinen N, Manninen M and van Leeuwen R 2012 *New J. Phys.* **14** 013032
[11] Danielewicz P 1984 *Ann. Phys. (N.Y.)* **152** 305
[12] Tohyama M 1987 *Phys. Rev. C* **36** 187–192
[13] Bogner S, Furnstahl R and Schwenk A 2010 *Prog. Part. Nucl. Phys.* **65** 94–147
[14] Hagen G, Papenbrock T, Dean D J and Hjorth-Jensen M 2008 *Phys. Rev. Lett.* **101** 092502
[15] Dickhoff W H and Barbieri C 2004 *Prog. Part. Nucl. Phys.* **52** 377
[16] Gasenzer T 2009 *Eur. Phys. J. ST* **168**(1) 89–148
[17] Rios A, Barker B, Buchler M and Danielewicz P 2011 *Ann. Phys.* **326** 1274–1319
[18] Baym G 1962 *Phys. Rev.* **127** 1391
[19] Puig von Friesen M, Verdozzi C and Almbladh C O 2009 *Phys. Rev. Lett.* **103** 176404
[20] Negele J W 1982 *Rev. Mod. Phys.* **54** 913
[21] Golabek C and Simenel C 2009 *Phys. Rev. Lett.* **103**(4) 042701
[22] Bertsch G F and Das Gupta S 1988 *Phys. Rept.* **160** 189–233
[23] Bonasera A, Gulminelli F and Molitoris J 1994 *Phys. Rep.* **243** 1
[24] Köhler H S 1980 *Nucl. Phys. A* **343** 315
[25] Grangé P, Richert J, Wolschi G and Weidenmüller H A 1981 *Nucl. Phys. A* **356** 260
[26] Tohyama M 1985 *Phys. Lett. B* **163** 14
[27] Köhler H S, Kwong N H and Yousif H A 1999 *Comp. Phys. Comm.* **123** 123

[28] Feit M D, Fleck J A and Steiger A 1982 *J. Comp. Phys.* **47** 412
[29] Ring P and Schuck P 1980 *The Nuclear Many-Body Problem* (Berlin: Springer)
[30] Messiah A 1999 *Quantum Mechanics* (NY: Dover Publications)
[31] Thouless D J and Valatin J G 1962 *Nucl. Phys.* **31** 211
[32] Bonche P, Koonin S and Negele J W 1976 *Phys. Rev. C* **13** 1226
[33] Köhler H S and Nilsson B S 1984 *Nucl. Phys. A* **417** 541
[34] Stan A, Dahlen N E and van Leeuwen R 2009 *Jour. Chem. Phys.* **130** 224101
[35] Morawetz K and Kohler H 1999 *Eur. Phys. J. A* **4** 291
[36] Dickhoff W and van Neck D 2005 *Many-Body Theory Exposed!* (London: World Scientific)

Non-equilibrium renormalised contacts for transport in nanodevices with interaction: a quasi-particle approach

H. Ness, L. K. Dash
Department of Physics, University of York, Heslington, York YO10 5DD, UK
European Theoretical Spectroscopy Facility (ETSF)

Abstract. We present an application of a new formalism to treat the quantum transport properties of fully interacting nanoscale junctions We consider a model single-molecule nanojunction in the presence of two kinds of electron-vibron interactions. In terms of the electron density matrix, one interaction is diagonal in the central region and the second off-diagonal between the central region and the left electrode. We use a non-equilibrium Green's function technique to calculate the system's properties in a self-consistent manner. The interaction self-energies are calculated at the Hartree-Fock level in the central region and within a dynamical mean-field-like approach for the crossing interaction. Our calculations are performed for different transport regimes ranging from the far off-resonance to the quasi-resonant regime, and for a wide range of parameters. They show that a non-equilibrium (i.e. bias dependent) dynamical (i.e. energy dependent) renormalisation is obtained for the contact between the left electrode and the central region in the form of a non-equilibrium renormalisation of the lead embedding potential. The conductance is affected by the renormalisation of the contact: the amplitude of the main resonance peak is modified as well as 'the lineshape of the first vibron side-band.

1. Introduction

The theory of quantum electronic transport in nano-scale devices has evolved rapidly over the past decade, as advances in experimental techniques have made it possible to measure transport properties down to single-molecule devices. The development of accurate theoretical methods for the description of quantum transport at the single-molecule level is essential for continued progress in research areas as such molecular electronics, spintronics, and thermoelectrics [1].

One of the longstanding problems in quantum charge transport is the establishment of a theoretical framework which allows for quantitatively accurate predictions of conductance from first principles. The need for methods going beyond the conventional approaches, based on equilibrium electronic structure calculations combined with Landauer-like elastic scattering, has been clear for a number of years. Only recently have more advanced methods to treat electronic interaction appeared (for example see Refs. [2, 3, 4]).

Alternative frameworks to deal with the steady-state or time-dependent transport are given by many-body perturbation theory, based on the non-equilibrium (NE) Green's function (GF) formalism: in these approaches, the interactions and (initial) correlations are taken into account by using conserving approximations for the many-body self-energy [5-14].

Other interactions, such as electron-phonon coupling, also play an important role in single-molecule quantum transport. These interactions are more important in nanoscale systems, as the electronic probability density is concentrated in a small region of space and thus normal screening mechanisms are ineffective. Such interactions are also crucial in inelastic electron tunneling spectroscopy. Such a technique constitutes an important basis for spectroscopy of molecular junctions, yielding insight into the vibrational modes (single molecule phonons called vibrons) and ultimately the atomic structure of the junction [15].

There have been many theoretical investigations focusing on the effect of electron-vibron coupling in molecular- and atomic-scale wires [16-51]. In all these studies, the interactions have always been considered to be present in the central region (i.e. the molecule) only, with the latter connected to two non-interacting terminals. Interactions are also assumed not to cross at the contracts between the central region and the leads. When electronic interactions are present throughout the system, as within density-functional theory calculations, they are treated only at the mean-field level and do not allow for any inelastic scattering events. However, such approximations are valid only in a very limited number of practical cases. The interactions, in principle, exist throughout the entire system.

In a recent paper we derived a general expression for the current in nano-scale junctions with interaction present everywhere in the system [52]. The importance of such extended interactions in nano-scale devices has also been addressed, for electron-electron interaction, in recently developed approaches such as Refs. [2, 53].

In the present paper, we apply our formalism [52] to a specific model of single-molecule nanojunctions. We focus on a model system in the presence of electron-vibron interaction both within the molecule *and* between the molecule and one of the leads. We adopt a quasiparticle-like approach to treat the crossing interactions (i.e. there are some restrictions on the components of the self-energy for the crossing interaction). We show how the interaction crossing at one interface of the molecular nanojunctions affects the transport properties by renormalising the coupling at the interface in a dynamical and bias-dependent manner.

The paper is organised as follows: In Sec. 2, we briefly recall the main result of our expression for the current in fully interacting systems. In Sec. 3, we present the model Hamiltonian for the system which includes two kinds of electron-vibron interaction: a Holstein-like Hamiltonian combined with a Su-Schrieffer-Heeger-like Hamiltonian. In this section, we also describe how the corresponding self-energies are calculated and the implications of such approximations on the current expression at the left and right interfaces. In Sec. 4, we show how the non-equilibrium dynamical renormalisation affects the generalised embedding potential of the left (L) lead, and how in turn this affects the conductance of the nanojunction. We finally conclude and discuss extensions of the present work in Sec. 5.

2. General theory for quantum transport

We consider a two-terminal device, made of three regions left-central-right, labelled $L-C-R$, in the steady-state regime. In such a device the interaction—which we specifically leave undefined (e.g. electron-electron or electron-phonon)—is assumed to be well described in terms of the single-particle self-energy $\Sigma^{\rm MB}$ and spreads throughout the entire system.

We use a compact notation for the Green's function G and the self-energy Σ matrix elements $M(\omega)$. They are annotated M_C (M_L or M_R) for the elements in the central region C (left L, right R region respectively), and M_{LC} (or M_{CL}) and M_{RC} (or M_{CR}) for the elements between region C and region L or R. There are no direct interactions between the two electrodes, i.e. $\Sigma^{\rm MB}_{LR/RL} = 0$.

In Refs. [52, 54], we showed that for a finite applied bias V the steady-state current $I_L(V)$

flowing through the left LC interface is given by:

$$I_L = \frac{e}{\hbar}\int \frac{d\omega}{2\pi}\mathrm{Tr}_{\{C\}}\left[G^r_C\tilde{\Upsilon}^{L,l}_C + G^a_C(\tilde{\Upsilon}^{L,l}_C)^\dagger + G^<_C(\tilde{\Upsilon}^L_C - (\tilde{\Upsilon}^L_C)^\dagger)\right] \\ +\mathrm{Tr}_{\{L\}}\left[\Sigma^{\mathrm{MB},>}_L G^<_L - \Sigma^{\mathrm{MB},<}_L G^>_L\right] \tag{1}$$

where the $\tilde{\Upsilon}^X_C$ quantities are

$$\begin{aligned} \tilde{\Upsilon}^L_C(\omega) &= \Sigma^a_{CL}(\omega)\ \tilde{g}^a_L(\omega)\ \Sigma^r_{LC}(\omega), \\ (\tilde{\Upsilon}^L_C)^\dagger &= \Sigma^a_{CL}\ \tilde{g}^r_L\ \Sigma^r_{LC}, \\ \tilde{\Upsilon}^{L,l}_C &= \Sigma^<_{CL}\,(\tilde{g}^a_L - \tilde{g}^r_L)\,\Sigma^r_{LC} + \Sigma^r_{CL}\ \tilde{g}^<_L\ \Sigma^r_{LC}. \end{aligned} \tag{2}$$

By definition $\Sigma_{LC}(\omega) = V_{LC} + \Sigma^{\mathrm{MB}}_{LC}(\omega)$ (similarly for the CL components) where $V_{LC/CL}$ are the nominal coupling matrix elements between the L and C regions. $\tilde{g}^x_L(\omega)$ are the GF of the region L renormalised by the interaction *inside* that region, where $x = r, a, <$ stands for the retarded, advanced and lesser GF components respectively. For example, for the advanded and retarded components, we have $(\tilde{g}^{r/a}_L(\omega))^{-1} = (g^{r/a}_L(\omega))^{-1} - \Sigma^{\mathrm{MB},r/a}_L(\omega)$ where all quantities are defined only in the subspace L.

The second trace in Eq. (1) corresponds to inelastic events induced by the interaction in the L lead. At equilibrium, because of the detailed balance equation $\Sigma^> G^< = \Sigma^< G^>$, this term vanishes. At non-equilibrium, this is generally not the case. However, when a local detailed balance equation holds, i.e. the system is locally in a (quasi)equilibrium state, this terms vanishes since one recovers locally $\Sigma^{\mathrm{MB},>}G^< = \Sigma^{\mathrm{MB},<}G^>$. Hence, in practice, we do not have to worry about the infinite sum in the trace $\mathrm{Tr}_{\{L\}}[...]$ since there will always be a region/boundary in the left lead L beyond which the system is at (quasi)equilibrium.

The first trace in the current equation Eq. (1) corresponds to a generalisation of the result of Meir and Wingreen [55]. It encompasses the cases for which the interactions are present in the three L, C, R regions as well as in between the L/C and C/R regions. It also bears some resemblance to the expression derived by Meir and Wingreen [55] when written as:

$$I^{\mathrm{MW}}_L = \frac{e}{\hbar}\int \frac{d\omega}{2\pi}\mathrm{Tr}_{\{C\}}\left[G^r_C\Sigma^{L,<}_C + G^a_C(\Sigma^{L,<}_C)^\dagger + G^<_C(\Sigma^{L,a}_C - \Sigma^{L,r}_C)\right]. \tag{3}$$

where we use the standard definitions $\Sigma^{L,<}_C = -(\Sigma^{L,<}_C)^\dagger = V_{CL}\ g^<_L\ V_{LC} = \mathrm{i}f_L\Gamma_L$ and $\Sigma^{L,a}_C - \Sigma^{L,r}_C = V_{CL}(g^a_L - g^r_L)V_{LC} = \mathrm{i}\Gamma_L$.

By comparing Eq. (1) and Eq. (3), we can see that the quantities $\tilde{\Upsilon}_{LC}$ ($\tilde{\Upsilon}^\dagger_{LC}$) and $\tilde{\Upsilon}^l_{LC}$ are now playing the role of the L lead self-energy Σ^a_L (Σ^r_L) and $\Sigma^<_L$ respectively when the interactions cross at the LC interface.

3. Model for the interaction

3.1. Hamiltonians

We consider a single-molecule junction in the presence of electron-vibron interaction both inside the central region and crossing at the contacts. We further concentrate on a model for the central region which consists of a single molecular level coupled to a single vibrational mode. A full description of our methodology, for the interaction inside the region C, is provided in Refs. [56, 57, 58]. Moreover, we consider that some electron-vibron interaction exists also at one contact (the left L electrode for instance). This model typically corresponds to an experiment for a molecule chemisorbed onto a surface (the left electrode) with a tunneling barrier to the right R lead.

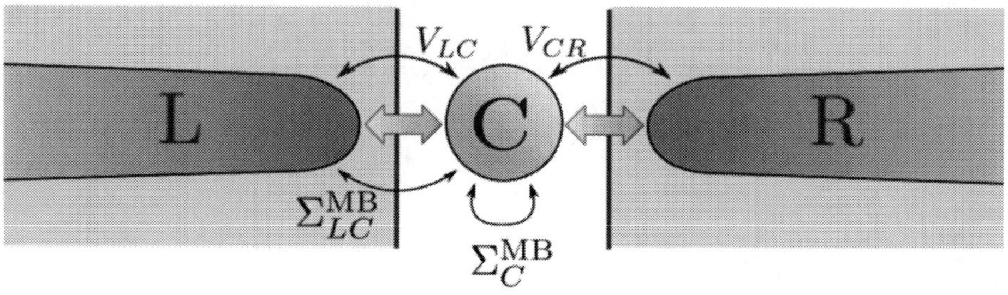

Figure 1. (Color online) Schematic representation of a central scattering region C connected to the left L and right R electrodes. Interactions are given by the coupling of the region C to the L (R) electrode ($V_{LC/CL}$ and $V_{RC/CR}$), and by the many-body effects within the central region (Σ_C^{MB}) and at the LC interface ($\Sigma_{LC}^{\mathrm{MB}}$). Σ_C^{MB} corresponds to the coupling of an electron with the vibron mode ω_0 (with coupling strength γ_0), and $\Sigma_{LC}^{\mathrm{MB}}$ corresponds to the coupling of a hopping electron with another vibron mode ω_A (with coupling strength γ_A).

In the following model, we consider two kinds of electron-vibron coupling: a local coupling in the sense of a Holstein-like coupling of the electron charge density with a internal degree of freedom of vibration inside the central region, and an off-diagonal coupling in the sense of a Su-Schrieffer-Heeger-like coupling [59, 17] to another vibration mode involving the hopping of an electron between the central C region and the L electrode. A schematic representation of the molecular junction is given in Figure 1.

The Hamiltonian for the region C is

$$H_C = \varepsilon_0 d^\dagger d + \hbar\omega_0 a^\dagger a + \gamma_0(a^\dagger + a)d^\dagger d, \tag{4}$$

where $d^\dagger$ (d) creates (annihilates) an electron in the molecular level ε_0. The electron charge density in the molecular level is coupled to the vibration mode of energy ω_0 via the coupling constant γ_0, and $a^\dagger$ (a) creates (annihilates) a vibration quantum in the vibron mode ω_0. The central region C is nominally connected to two (left and right) one-dimensional tight-binding chains via the hopping integral t_{0L} and t_{0R}. The corresponding electrode $\alpha = L, R$ self-energy is $\Sigma_\alpha^r(\omega) = t_{0\alpha}^2 e^{ik_\alpha(\omega)}/\beta_\alpha$ with the dispersion relation $\omega = \varepsilon_\alpha + 2\beta_\alpha \cos(k_\alpha(\omega))$ where ε_α and β_α are the tight-binding on-site and off-diagonal elements of the electrode chains.

To describe the electron-vibron interaction existing across the left contact, we consider that the hopping integral t_{0L} is actually dependent on some generalised coordinate X. The latter represents either the displacement of the centre-of-mass of the molecule or of some chemical group at the end of the molecule link to the L electrode. At the lowest order, the matrix element can be linearised as $t_{0L}(X) = t_{0L} + t'_{0L}X$. Hence the hopping of an electron from the C region to the L region (and *vice versa*) is coupled to a vibration mode (of energy ω_A) via the coupling constant γ_A (itself related to t'_{0L}). The corresponding Hamiltonian is given by

$$H_{LC} = \gamma_A(b^\dagger + b)(c_L^\dagger d + d^\dagger c_L) + \omega_A b^\dagger b, \tag{5}$$

where $b^\dagger$ (b) creates (annihilates) a vibration quantum in the vibron mode ω_A, the generalised coordinate is $X = \sqrt{\hbar/(2m_A\omega_A)}(b^\dagger + b)$, and $c_L^\dagger$ (c_L) creates (annihilates) an electron in the level ε_L of the L electrode.

The Hamiltonians Eq. (4) and Eq. (5) are used to calculate the corresponding electron self-energies at different orders of the interaction γ_0 and γ_A using conventional non-equilibrium diagrammatic techniques [56, 58].

Furthermore, at equilibrium, the whole system has a single and well-defined Fermi level $\mu^{\rm eq}$. A finite bias V, applied across the junction, lifts the Fermi levels as $\mu_{L,R} = \mu^{\rm eq} + \eta_{L,R} eV$. The fraction of potential drop [60] at the left contact is η_L and $\eta_R = \eta_L - 1$ at the right contact, with $\mu_L - \mu_R = eV$ and $\eta_L \in [0, 1]$.

3.2. Self-energies for the interactions

In this paper, we consider different approximations to the treat the interaction inside the central region and the interaction crossing at the left contact. First of all, calculating exactly the corresponding interaction self-energies is a tremendous task since, in principle, they depend on both the phonon progator and the electron Green's functions in all the different parts of the system. Hence, for the first application of our formalism, we proceed step by step in terms of the increasing complexity of the interaction; and we limit ourselves to approximations for the self-energies that are well known and well controlled.

The electron-vibron self-energies in the central region C are calculated within the self-consistent Born approximation (i.e. diagrams with one phonon line). The details of these calculations have been reported elsewhere [56, 58]. For the crossing interaction, we consider a mean-field-like approach for the electron-vibron coupling at the LC interface. Within such an approximation, we can understand the results of the calculations in terms of renormalisation of the LC contact.

Furthermore considering the crossing interaction occuring only at one interface permit us to check and test the consistency of our formalism. Indeed, with no interaction crossing at the right CR interface, the current I_R is given by the conventional Meir and Wingreen formula, i.e. Eq. (3) for the CR interface. In order to have a consistent formalism, we need to have current conservation, and such a constraint is best tested with crossing interaction at only one interface. The corresponding results are shown in detail in 4.1.

Within mean-field-like approximations, the effects of the crossing interaction correspond to a renormalisation of the coupling in a static or a dynamical manner. The corresponding self-energies have only retarded and advanced components $\Sigma^{r/a}_{LC}$ and $\Sigma^{r/a}_{CL}$. The extra inelastic effects included in the components $\Sigma^{\gtrless}_{CL}$ are neglected altogether.

Hence the Υ^X_C quantities defined in Eq. (2) become:

$$\begin{aligned} \tilde{\Upsilon}^L_C &= \Sigma^a_{CL}\, g^a_L\, \Sigma^r_{LC}, \\ (\tilde{\Upsilon}^L_C)^\dagger &= \Sigma^a_{CL}\, g^r_L\, \Sigma^r_{LC}, \\ \tilde{\Upsilon}^{L,l}_C &= \Sigma^r_{CL}\, g^<_L\, \Sigma^r_{LC}, \\ (\tilde{\Upsilon}^{L,l}_C)^\dagger &= -\Sigma^a_{CL}\, g^<_L\, \Sigma^a_{LC}. \end{aligned} \tag{6}$$

with $\Sigma^{r/a}_{LC} = V_{LC} + \Sigma^{{\rm MB},r/a}_{LC}$.

Correspondingly, the generalised embedding potentials for the left contact defined as $Y^{L,x}_C = [\Sigma^r_{CL} g_L \Sigma^r_{LC}]^x$ [52, 54] with $x = r, a, \gtrless$, are now given by

$$\begin{aligned} Y^{L,r}_C(\omega) &= \Sigma^r_{CL} g^r_L \Sigma^r_{LC}, \\ Y^{L,a}_C(\omega) &= \Sigma^a_{CL} g^a_L \Sigma^a_{LC}, \\ Y^{L,<}_C(\omega) &= \Sigma^r_{CL} g^<_L \Sigma^a_{LC}. \end{aligned} \tag{7}$$

In the static limit, the mean-field approach leads to the Hartree expression for the electron-vibron self-energies at the LC interface:

$$\Sigma^{{\rm MB},r/a}_{LC} = -2\frac{\gamma^2_A}{\omega_A}\langle n_{LC}\rangle, \tag{8}$$

where

$$\langle n_{LC}\rangle = -i\int \frac{d\omega}{2\pi}\, G^{<}_{LC}(\omega). \tag{9}$$

The self-energy $\Sigma^{\mathrm{MB},r/a}_{LC}$ is independent of the energy ω and leads to a static (nonetheless bias-dependent) renormalisation of the nominal coupling $V_{CL} = V_{LC} = t_{0L}$ between the L and C regions. This NE renormalisation will induce, among other effects, a bias-dependent modification of the broadening of the spectral features of the C region. We have analysed these effects in details in Ref.[51]. The renormalisation is such that the amplitude of the current is reduced in comparison with the current values obtained when the interaction is present only in the central region. The NE static renormalisation of the contact is highly non-linear and non-monotonic in function of the applied bias, and the larger effects occur at applied biases corresponding to resonance peaks in the dynamical conductance. The conductance is also affected by the NE renormalisation of the contact, showing asymmetric broadening around the resonance peaks and some slight displacement of the peaks at large bias in function of the coupling strengh γ_A.

Beyond the static limit, we can develop a dynamical mean-field-like approach for the electron-vibron coupling following Ref. [62]. The retarded self-energy containing all orders of the electron-vibron coupling is expressed as a continued fraction as shown analytically in Refs. [63, 64]. We have already used such an approach to study the transport properties of organic molecular wires which are dominated by the propagation of polarons [16, 17] or solitons [18]. The expression for the corresponding self-energy $\Sigma^{r}_{LC}(\omega)$ is given by:

$$\Sigma^{r}_{LC}(\omega) = \cfrac{\gamma_A^2}{G^{r}_{LC}(\omega-\omega_A)^{-1} - \cfrac{2\gamma_A^2}{G^{r}_{LC}(\omega-2\omega_A)^{-1} - \cfrac{3\gamma_A^2}{G^{r}_{LC}(\omega-3\omega_A)^{-1} - \ldots}}} \tag{10}$$

where G^{r}_{LC} is the retarded component of the off-diagonal GF between the central region and the left lead L. Its closed expression is obtained from the corresponding Dyson equation $G^{r}_{LC} = [g\Sigma G]^{r}_{LC}$:

$$G^{r}_{LC}(\omega) = \left[[g^{r}_{L}]^{-1} - \Sigma^{r}_{LC}\left[[\tilde{g}^{r}_{C}]^{-1} - Y^{R,r}_{C}\right]^{-1}\Sigma^{r}_{CL}\right]^{-1}\Sigma^{r}_{LC}\left[[\tilde{g}^{r}_{C}]^{-1} - Y^{R,r}_{C}\right]^{-1}, \tag{11}$$

with $\tilde{g}^{r}_{C}(\omega) = [\omega - \varepsilon_0 - \Sigma^{\mathrm{MB},r}_{C}(\omega)]^{-1}$ and $Y^{R,r}_{C}(\omega) = V_{CR}\, g^{r}_{R}(\omega) V_{RC}$.

Such a dynamical mean-field-like approach, which corresponds to a quasi-particle approach since the self-energy $\Sigma^{r/a}_{LC/CL}$ can be incorporated into a Schrödinger-like equation, is expected to affect the transport properties via the non-equilibrium dynamical renormalisation of the contacts, i.e. through the generalised embedding potentials $Y^{L,x}_{C}(\omega)$ as well as through the corresponding $\Upsilon^{x}_{C}(\omega)$ quantities.

Finally, note that for the lowest-order expansion, the self-energy Σ^{r}_{LC} takes a simple form:

$$\Sigma^{r}_{LC}(\omega) = \gamma_A^2 G^{r}_{LC}(\omega - \omega_A)\ . \tag{12}$$

4. Results

We have perfomed calculations, in a self-consistent manner, for many different values of the Hamiltonian parameters. We present below the most characteristic results for different transport regimes and for different coupling strengths γ_A, while the interaction in the region C is taken to be in the intermediate coupling regime $\omega_0 = 0.2$, $\gamma_0 = 0.15$, i.e. $\gamma_0/\omega_0 = 0.75$.

The nominal couplings between the central region and the electrodes $t_{0L,R}$, before NE renormalisation, are not too large, so that we can discriminate clearly between the different

Journal of Physics: Conference Series **427** (2013) 012011 doi:10.1088/1742-6596/427/1/012011

vibron side-band peaks in the spectral functions. The values chosen for the parameters are typical values for realistic molecular junctions [65, 58]. In the following the current is given in units of charge per time, the conductance in units of the quantum of conductance $G_0 = 2e^2/\hbar$ and the bias V and the embedding potential Y_C in natural units of energy where $e = 1$ and $\hbar = 1$.

4.1. Conserving approximation

One of the most important physical conditions that our formalism needs to fulfil is the constraint of current conservation. We use conserving approximations to calculate the interaction self-energies in the central region C and an quasi-particle approximation for crossing interaction at the left interface.

With our choice for $\Sigma^r_{LC}(\omega)$ given by Eq. (10), we find that the left lead generalised embedding potential $Y^{L,r}_C(\omega)$ is renormalised by the crossing interaction. This is clearly seen in Figure 2, which shows the imaginary part of $Y^{L,r}_C$ for different transport regimes, different coupling at the LC interface and for both equilibrium and non-equilibrium conditions.

At equilibrium, and in the absence of contact renormalisation, the imaginary part of $Y^{L,r}_C$ is simply the imaginary part of the conventional L lead self-energy $\Sigma^r_L(\omega) = t^2_{0L} e^{ik_L(\omega)}/\beta_L$ and corresponds to a semi-elliptic functions with non-zero values within the energy range $-2\beta_L \leq \omega \leq +2\beta_L$ (see dashed lines in figure 2). The spectral weight of $\Im m\ \Sigma^r_L$ is given by $\int d\omega\ \Im m\ \Sigma^r_L(\omega) = t^2_{0L} \int d\omega\ \Im m\ g^r_L(\omega) = -\pi t^2_{0L}$.

In the presence of the renormalised L contact, we obtained a strong deviation from the semi-elliptic function as shown by the (red) thin lines and (black) thick lines in Figure 2, which correspond respectively to the equilibrium (bias $V = 0$) and the non-equilibrium (bias $V = 1.0$) conditions. This result indicates a strong reduction of the available transport channels in the renormalised L lead embedding potential for regions of energy where $\Sigma^{r/a}_{LC/LC}(\omega)$ has non-zero value.

However, a very important property is that the total spectral weight of $\Im m\ \Sigma^r_L$ is conserved for all the calculations we have performed, i.e. for all the different transport regimes, coupling strength γ_A and ω_A at the LC interface and all applied bias. We find that in all the cases $\int d\omega\ \Im m\ Y^{L,r}_C(\omega) = -\pi t^2_{0L}$ (with maximum deviation of 0.05% arising from numerical errors in the results shown in Figure 2). So in this sense, we can say that the approximation for $\Sigma^{r,a}_{LC/LC}$ is conserving.

Correspondingly, we have also carefully checked that the current is conserved for all the calculations presented in the present paper, i.e. that $|I_L + I_R|/|I_L| \sim |I_L + I_R|/|I_R| < 10^{-5}$, with the current I_L at the LC interface is given by expression Eq. (1) and the current I_R at the CR interface does not contain any contact renormalisation, and hence is given by a Meir and Wingreen like expression Eq. (3).

4.2. Dynamical non-equilibrium renormalisation

The dynamical renormalisation of the L lead embedding potential $Y^{L,r}_C$ presents features (dips) at some energies. Qualitatively speaking, there is a form of correlation between them and the features that exist in the spectral function of the central region $A_C(\omega) = (G^a_C(\omega) - G^r_C(\omega))/2\pi i$.

Figure 3 shows the spectral function $A_C(\omega)$, rescaled by some factor, for a set of parameters corresponding to panel (a) in figure 2. The bottom panel of figure 3 shows the imaginary part of $Y^{L,r}_C(\omega)$ from which the semi-elliptic background has been subtracted, i.e. $\Delta Y^{L,r}_C = Y^{L,r}_C - \Sigma^r_L$. Qualitatively speaking, both quantities present similar features, shifted in energy by an amount corresponding to ω_A ($\omega_A = 0.2$ in the calculations) as expected from the definition of $\Sigma^{r/a}_{LC/LC}(\omega)$.

One can then expect a maximum effect of the dynamical renormalisation in an integrated quantity such as the current when the features in $Y^{L,r}_C(\omega)$ coincide with those in the spectral

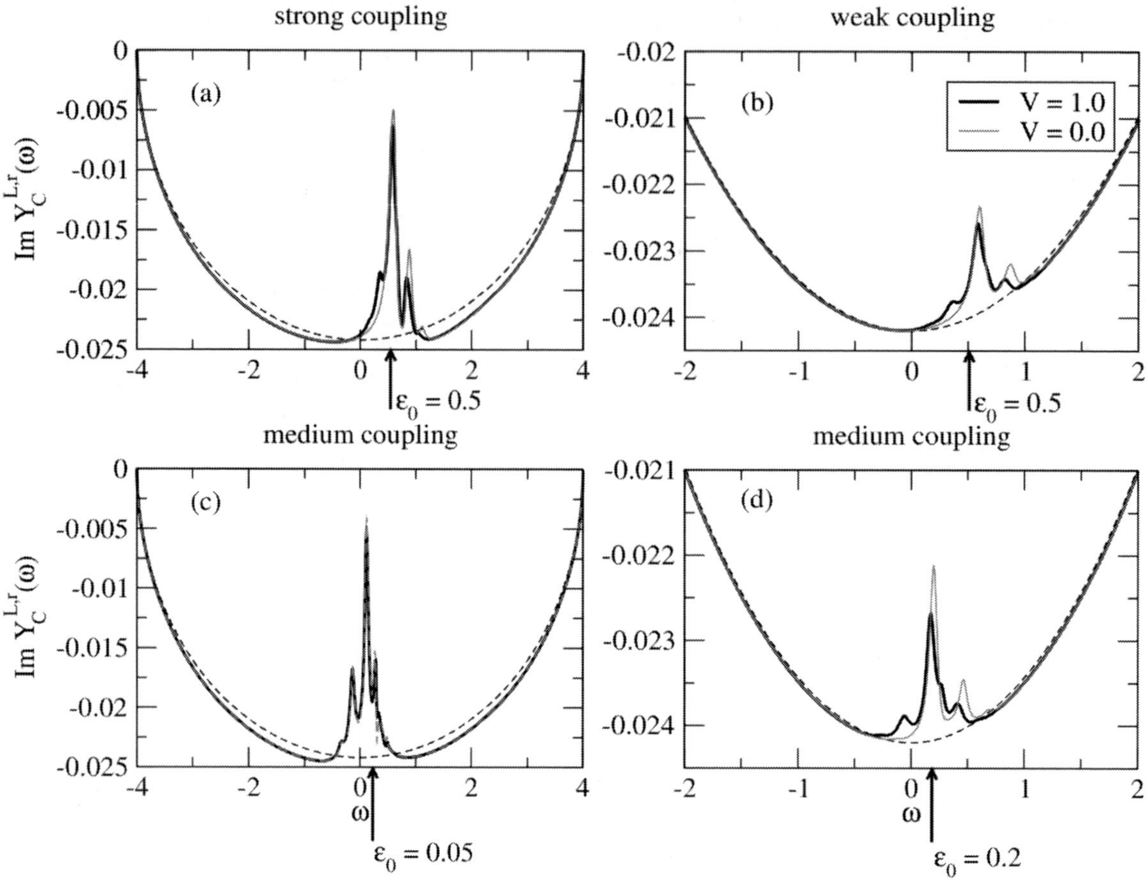

Figure 2. (Color online) Imaginary part of the left-lead generalised embedding potential $Y_C^{L,r}(\omega)$. The dashed lines represent the L lead spectral function in the absence of renormalisation at the LC interface (i.e. semi-elliptic density of states). The thin (red) lines are the spectral functions $\Im m Y_C^{L,r}(\omega)$ at equilibrium ($V = 0$). The thick (black) lines are the non-equilibrium spectral functions $\Im m Y_C^{L,r}(\omega)$ at finite bias ($V = 1.0$). These show a strong reduction of the spectral functions on energy range of $\Delta\omega \approx 1$ around ε_0. Panel (a) Off-resonant regime $\varepsilon_0 = 0.5$ and strong coupling at the interface $\omega_A = 0.20$, $\gamma_0 = 0.28$. (b) Off-resonant regime $\varepsilon_0 = 0.5$ and weak coupling at the interface $\omega_A = 0.20$, $\gamma_0 = 0.07$. (c) Resonant regime $\varepsilon_0 = 0.05$ and medium coupling at the interface $\omega_A = 0.20$, $\gamma_0 = 0.14$. (d) Intermediate regime $\varepsilon_0 = 0.2$ and medium coupling at the interface $\omega_A = 0.10$, $\gamma_0 = 0.07$. The other parameters are $\omega_0 = 0.20$, $\gamma_0 = 0.15$, $t_{0R} = t_{0L} = 0.22$, $\beta_\alpha = 2.0$, $\epsilon_\alpha = 0.0$, $\eta_V = 1$.

function $A_C(\omega)$ within a given bias energy-window.

We now consider the modification of the dynamical conductance $G(V) = dI/dV$ induced by the crossing interaction $\Sigma_{LC/LC}^{r/a}(\omega)$ for different transport regimes and for different crossing interaction strength. Figures 4 and 5 show $G(V)$ for three different transport regimes for weak to strong crossing interaction strength γ_A, and for two values of ω_A ($\omega_A = 0.2$ for Fig. 4 and $\omega_A = 0.1$ Fig. 5).

The dynamical renormalisation of the left contact slightly affects the main resonance peak in the conductance, and more importantly the lineshape of the first vibron side-band peak above the main conductance peak. The most important effects are obtained for the strong crossing

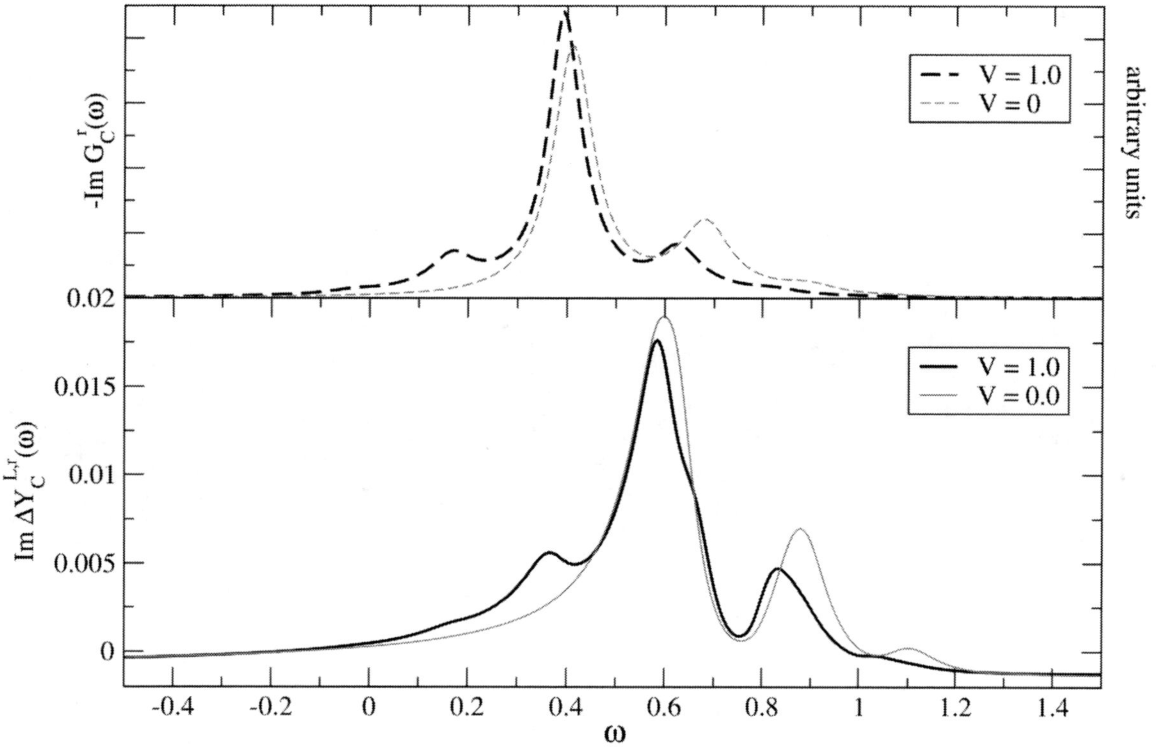

Figure 3. (Color online) Rescaled spectral function proportional to $-\Im m G_C^r$ for the central region (top panel) and imaginary part of the L lead generalised embedding potential relative to the non-renormalised L lead self-energy: $\Delta Y_C^{L,r} = Y_C^{L,r} - \Sigma_L^r$ (bottom panel). The calculations are performed for the same parameters as in panel (a) of Figure 2 and for both zero bias and finite bias $V = 1$. Here $\mu_R = 0$ and $\mu_L = V$. The qualitative correlations between the features in $\Delta Y_C^{L,r}$ and in the spectral function of the C region $-\Im m G_C^r$ are clearly observed.

interaction strength $\gamma_A = 0.28$. Even if the ratio γ_A/ω_A is conserved between the calculations shown in Figure 4 and Figure 5, the absolute value of γ_A is the crucial quantity that governs the effects of the dynamical renormalisation. Furthermore, for the different sets of parameters we have considered, our calculations show that the lower order expansion for Σ_{LC}^r (see Eq. ()) provides a good approximation to the results obtained with a larger number of levels in the continued fraction expansion of Σ_{LC}^r (see panel (a) in fig.).

We now check the effects of varying the nominal coupling $t_{0\alpha}$ on the conductance. The results are shown in Figure 6 for the off-resonant transport regime with and without dynamical renormalisation. With renormalisation of the left contact, the conductance values decrease with increasing coupling $t_{0\alpha}$ to the leads while the conductance peaks broaden. This is seen more clearly in the right panel of Figure 6 where the rescale conductance $G(V)/G_{\max}$ is plotted such that the main resonance peak has an amplitude of 1.

In the presence of dynamical renormalisation of the L contact, the main conductance peak follows the same behaviour as in the absence of renormalisation. However, one observes a highly non-linear behaviour of the modification of the first vibron side-band peak. The complete and detailed understanding of such modification is rather complex, and beyond the scope of the present paper. However it is strongly related to the new features in the renormalised L lead embedding potential which depart from the non-interacting case.

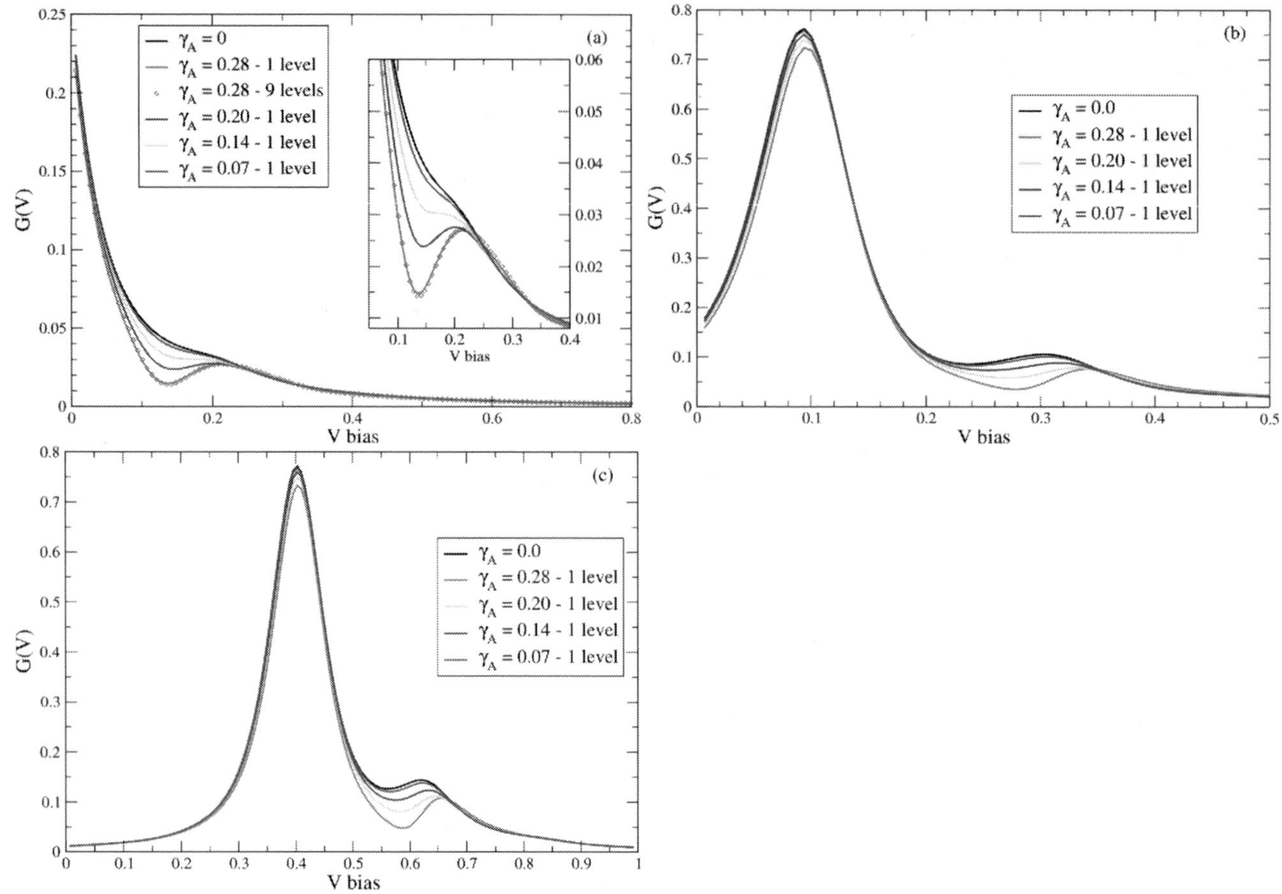

Figure 4. (Color online) Dynamical conductance $G(V)$ for different transport regimes and different coupling strength γ_A, and $\omega_A = 0.20$. Panel (a) Resonant regime $\varepsilon_0 = 0.05$. Panel (b) Off-resonant regime $\varepsilon_0 = 0.2$. Panel (c) Off-resonant regime $\varepsilon_0 = 0.5$. The non-equilibrium dynamical renormalisation of the LC contact affects both the main conductance peak and the first vibron side-band peak. The other parameters are $\omega_0 = 0.20$, $\gamma_0 = 0.15$, $t_{0R} = t_{0L} = 0.22$, $\beta_\alpha = 2.0$, $\epsilon_\alpha = 0.0$, $\eta_V = 1$.

5. Conclusion

We have studied the transport properties through a two-terminal nanoscale device with interactions present not only in the central region but also with interaction crossing at the interface between the left lead and the central region. To calculate the current for such a fully-interacting system, we have used our recently developed quantum transport formula [52] based on the NEGF formalism. As a first practical application, we have considered a prototypical single-molecule nanojunction with electron-vibron interaction. In terms of the electron density matrix, the interaction is diagonal in the central region for the first vibron mode and off-diagonal between the central region and the left electrode for the second vibron mode. The interaction self-energies are calculated in a self-consistent manner using the lowest order Hartree-Fock-like diagram in the central region and a quasi-particle (dynamical mean-field-like) approach for the crossing interaction. Our calculations were performed for different transport regimes ranging from the far off-resonance to the quasi-resonant regime, and for a wide range of parameter values.

They show that, for this model, we obtain a non-equilibrium (i.e. bias dependent) dynamical

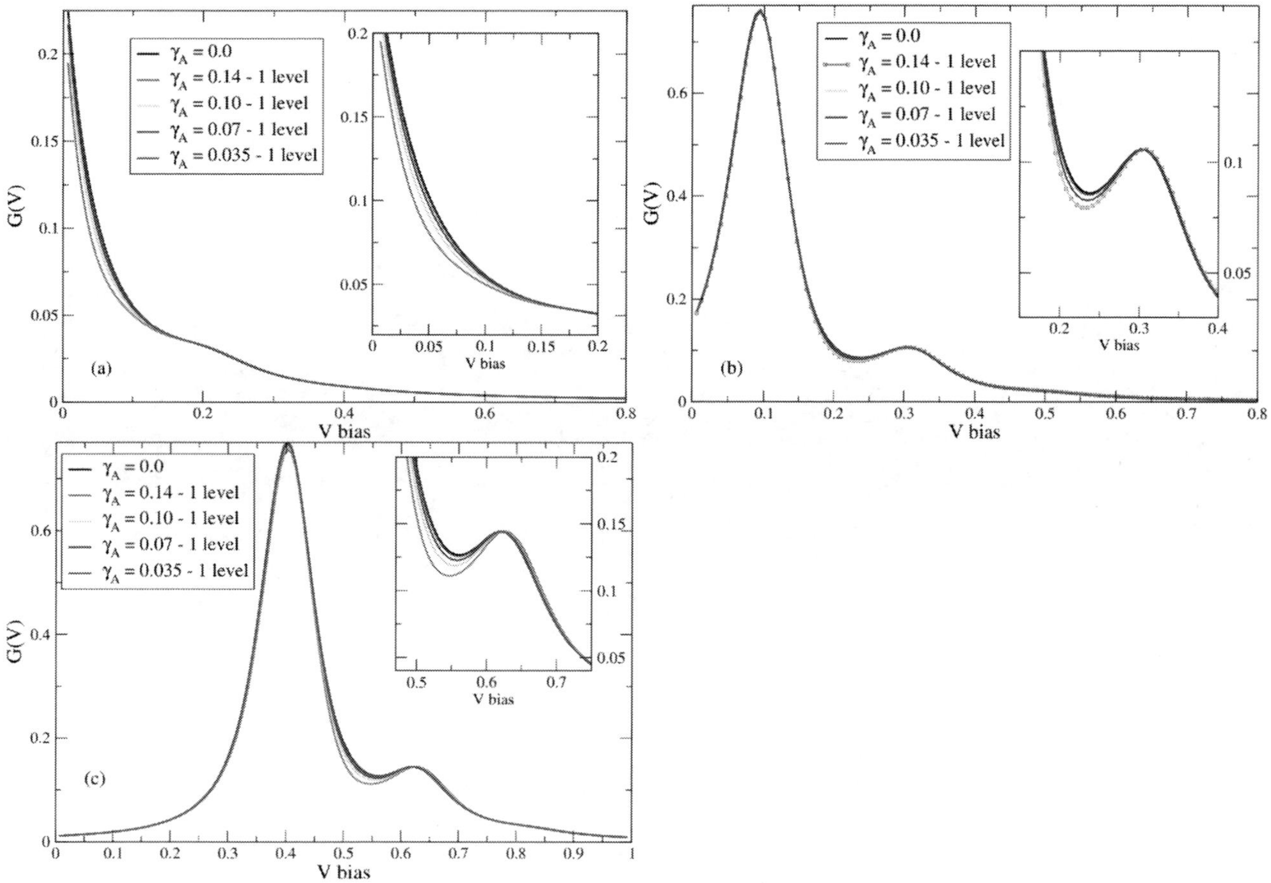

Figure 5. (Color online) Dynamical conductance $G(V)$ for different transport regimes and different coupling strength γ_A, and $\omega_A = 0.10$. Panel (a) Resonant regime $\varepsilon_0 = 0.05$. Panel (b) Off-resonant regime $\varepsilon_0 = 0.2$. Panel (c) Off-resonant regime $\varepsilon_0 = 0.5$. The non-equilibrium dynamical renormalisation of the LC contact affects both the main conductance peak and the first vibron side-band peak. The other parameters are $\omega_0 = 0.20$, $\gamma_0 = 0.15$, $t_{0R} = t_{0L} = 0.22$, $\beta_\alpha = 2.0, \epsilon_\alpha = 0.0$, $\eta_V = 1$.

(i.e. energy dependent) renormalisation of the generalised embedding potential of the L left lead. The renormalisation is such that the amplitude of the corresponding spectral function is reduced around the molecular level energy ε_0 over an energy range roughly equal to the energy support of the spectral density of the central region C. This corresponds a reduction of the number of transport channels in the left lead at given energy, even if the total spectral weight of the L embedding potential is conserved.

The non-equilibrium dynamical renormalisation of the L contact is highly non-linear and non-monotonic in function of the applied bias, and the larger effects occur at applied bias for which features are present in both the spectral function of the central region C and the generalised embedding potential of the L lead. The conductance is affected by the NE renormalisation of the contact: the amplitude of the main resonance peak is modified as well as the lineshape of the first vibron side-band peak.

Finally, extensions of the present study are now considered to go beyond the quasi-particle approach for the crossing interaction self-energy by including as well as the lesser and greater

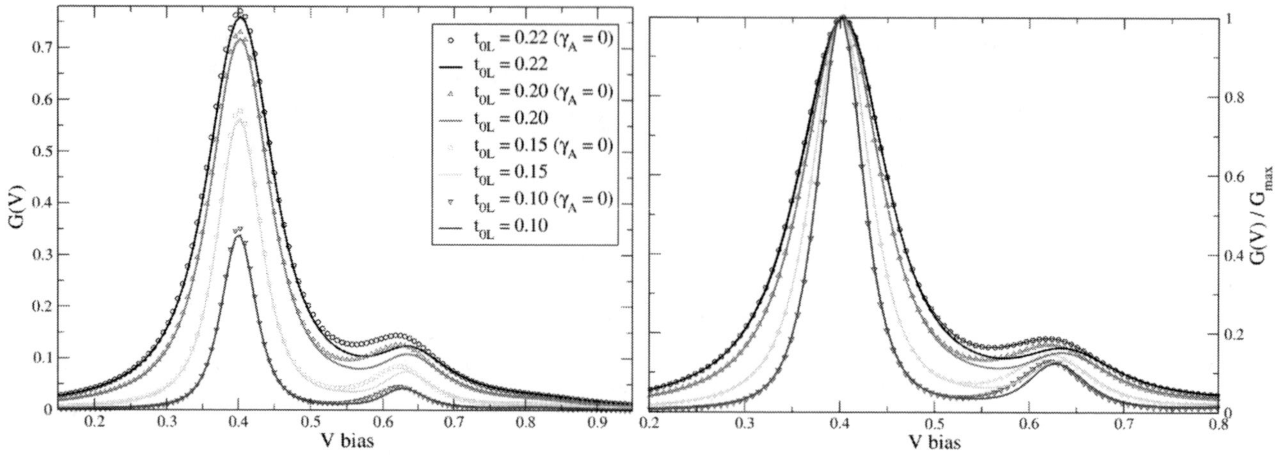

Figure 6. (Color online) (Left Panel) Dynamical conductance $G(V)$ for the off-resonant transport regime ($\varepsilon_0 = 0.50$) and different norminal coupling strength $t_{0R} = t_{0L}$ to the leads. (Right Panel) Dynamical conductance $G(V)$ rescaled such that the amplitude of the main resonance peak is 1 quantum of conductance. The other parameters are $\gamma_A = 0.2$, $\gamma_A = 0.14$ (unless otherwise indicated), $\omega_0 = 0.20$, $\gamma_0 = 0.15$, $\beta_\alpha = 2.0, \epsilon_\alpha = 0.0$, $\eta_V = 1$.

components of the LC interface self-energy. This should lead to dynamical NE renormalisation of the contact involving inelastic scattering processes.

References

[1] Widawsky J R, Darancet P, Neaton J B and Venkataraman L 2012 *Nano Letters* **12** 354
[2] Strange M, Rostgaard C, Häkkinen H and Thygesen K S 2011 *Physical Review B* **83** 115108
[3] Rangel T, Ferretti A, Trevisanutto P E, Olevano V and Rignanese G M 2011 *Phys. Rev. B* **84** 045426
[4] Darancet P, Ferretti A, Mayou D and Olevano V 2007 *Phys. Rev. B* **75** 075102
[5] Baym G 1962 *Physical Review* **127** 1391
[6] von Barth U, Dahlen N E, van Leeuwen R and Stefanucci G 2005 *Physical Review B* **72** 235109
[7] van Leeuwen R, Dahlen N E, Stefanucci G, Almbladh C O and von Barth U 2006 *Lecture Notes in Physics* **706** 33
[8] Kita T 2010 *Progress of Theoretical Physics* **123** 581
[9] Tran M T 2008 *Physical Review B* **78** 125103
[10] Myöhänen P, Stan A, Stefanucci G and van Leeuwen R 2008 *EuroPhysics Letters* **84** 67001
[11] Myöhänen P, Stan A, Stefanucci G and van Leeuwen R 2010 *Journal of Physics: Conference Series* **220** 012017
[12] Perfetto E, Stefanucci G and Cini M 2010 *Physical Review Letters* **105** 156802
[13] Velický B, Kalvová A and Špička V 2010 *Physical Review B* **81** 235116
[14] von Friesen M P, Verdozzi C and Almbladh C O 2009 *Physical Review Letters* **103** 176404
[15] Arroyo C R, Frederiksen T, Rubio-Bollinger G, Vélez M, Arnau A, Sánchez-Portal D and Agraït N 2010 *Physical Review B* **81** 075405
[16] Ness H and Fisher A J 1999 *Physical Review Letters* **83** 452
[17] Ness H, Shevlin S A and Fisher A J 2001 *Physical Review B* **63** 125422
[18] Ness H and Fisher A J 2002 *Europhysics Letters* **57** 885
[19] Mii T, Tikhodeev S and Ueba H 2003 *Physical Review B* **68** 205406
[20] Montgomery M J, Hoekstra J, Sutton A P and Todorov T N 2003 *Journal of Physics: Condensed Matter* **15** 731
[21] Troisi A, Ratner M A and Nitzan A 2003 *Journal of Chemical Physics* **118** 6072
[22] Chen Y C, Zwolak M and di Ventra M 2005 *Nano Letters* **4** 1709
[23] Lorente N and Persson M 2000 *Physical Review Letters* **85** 2997
[24] Frederiksen T, Brandbyge M, Lorente N and Jauho A P 2004 *Physical Review Letters* **93** 256601
[25] Galperin M, Ratner M A and Nitzan A 2004 *Nano Letters* **4** 1605

[26] Mitra A, Aleiner I and Millis A J 2004 *Physical Review B* **69** 245302
[27] Pecchia A, di Carlo A, Gagliardi A, Sanna S, Frauenhein T and Gutierrez R 2004 *Nano Letters* **4** 2109
[28] Chen Z, Lü R and Zhu B 2005 *Physical Review B* **71** 165324
[29] Ryndyk D A and Keller J 2005 *Physical Review B* **71** 073305
[30] Sergueev N, Roubtsov D and Guo H 2005 *Physical Review Letters* **95** 146803
[31] Viljas J K, Cuevas J C, Pauly F and Häfner M 2005 *Physical Review B* **72** 245415
[32] Yamamoto T, Watanabe K and Watanabe S 2005 *Physical Review Letters* **95** 065501
[33] Cresti A, Grosso G and Parravicini G P 2006 *Journal of Physics: Condensed Matter* **18** 10059
[34] Han J E 2006 *Physical Review B* **73** 125319
[35] Troisi A and Ratner M A 2006 *Nano Letters* **6** 1784
[36] de la Vega L, Martín-Rodero A, Agraït N and Levy-Yeyati A 2006 *Physical Review B* **73** 075428
[37] Frederiksen T, Paulsson M, Brandbyge M and Jauho A P 2007 *Physical Review B* **75** 205413
[38] Galperin M, Ratner M A and Nitzan A 2007 *Journal of Physics: Condensed Matter* **19** 103201
[39] Ryndyk D A and Cuniberti G 2007 *Physical Review B* **76** 155430
[40] Schmidt B B, Hettler M H and Schön G 2007 *Physical Review B* **75** 115125
[41] Troisi A, Beebe J M, Picraux L B, van Zee R D, Stewart D R, Ratner M A and Kushmerick J G 2007 *Proceedings of the National Academy of Sciences of the USA* **104** 14255
[42] Asai Y 2008 *Physical Review B* **78** 045434
[43] Benesch C, Čížek M, Klimeš J, Kondov I, Thoss M and Domcke W 2008 *Journal of Physical Chemistry C* **112** 9880
[44] Paulsson M, Frederiksen T, Ueba H, Lorente N and Brandbyge M 2008 *Physical Review Letters* **100** 226604
[45] Egger R and Gogolin A O 2008 *Physical Review B* **77** 113405
[46] Monturet S and Lorente N 2008 *Physical Review B* **78** 035445
[47] McEniry E J, Frederiksen T, Todorov T N, Dundas D and Horsfield A P 2008 *Physical Review B* **78** 035446
[48] Schmidt B B, Hettler M H and Schön G 2008 *Physical Review B (Condensed Matter and Materials Physics)* **77** 165337
[49] Secker D, Wagner S, Ballmann S, Hartle R, Thoss M and Weber H B 2011 *Physical Review Letters* **106** 136807
[50] Härtle R, Butzin M, Rubio-Pons O and Thoss M 2011 *Physical Review Letters* **107** 046802
[51] Ness H and Dash L K 2012 *Physical Review Letters* **108** 126401
[52] Ness H and Dash L K 2011 *Physical Review B* **84** 235428
[53] Perfetto E, Stefanucci G and Cini M 2012 *Physical Review B* **85** 165437
[54] Ness H and Dash L K 2012 *Journal of Physics A: Mathematical and Theoretical* **45** 195301
[55] Meir Y and Wingreen N S 1992 *Physical Review Letters* **68** 2512
[56] Dash L K, Ness H and Godby R W 2010 *Journal of Chemical Physics* **132** 104113
[57] Ness H, Dash L K and Godby R W 2010 *Physical Review B* **82** 085426
[58] Dash L K, Ness H and Godby R W 2011 *Physical Review B* **84** 085433
[59] Heeger A J, Kivelson S, Schrieffer J R and Su W P 1988 *Review of Modern Physics* **60** 781
[60] Datta S, Tian W D, Hong S H, Reifenberger R, Henderson J I and Kubiak C P 1997 *Physical Review Letters* **79** 2530
[61] Ness H and Dash L K 2012 *submitted*
[62] Ciuchi S, de Pasquale F, Fratini S and Feinberg D 1997 *Physical Review B* **56** 4494–4512
[63] Ness H 2006 *Journal of Physics: Condensed Matter* **18** 6307
[64] Cini M and D'Andrea A 1988 *Journal of Physics C: Solid State Physics* **21** 193
[65] Dash L K, Ness H, Verstraete M and Godby R W 2012 *Journal of Chemical Physics* **136** 064708

Nonequilibrium Green's functions in the study of heat transport of driven nanomechanical systems

Liliana Arrachea and Bruno Rizzo

Departamento de F'ısica e IFIBA, Facultad de Ciencias Exactas y Naturales, Universidad de Buenos Aires. Pabellón I, Ciudad Universitaria (1428) Buenos Aires, Argentina

E-mail: lili@df.uba.ar

Abstract. We review a recent theoretical development based on non-equilibrium Green's function formalism to study heat transport in nanomechanical devices modeled by phononic systems of coupled quantum oscillators driven by ac forces and connected to phononic reservoirs. We present the relevant equations to calculate the heat currents flowing along different regions of the setup, as well as the power developed by the time-dependent forces. We also present different strategies to evaluate the Green's functions exactly or approximately within the weak driving regime. We finally discuss the different mechanisms in which the ac driving forces deliver the energy. We show that, besides generating heat, the forces may operate exchanging energy as a quantum engine.

1. Introduction

The Green's function formalism provided the theoretical framework for many studies of electron transport in mesoscopic devices. Since the pioneer work by Caroli and coworkers, [1] the celebrated generalized Landauer-Büttiker formula by Meir and Wingreen [2] and the generalization to setups driven by the time-dependent voltages and fields of Refs. [3, 4], several interesting applications can be found in the literature. Details of these historical developments and some important aspects of the subsequent progress have been reviewed in papers of this conference series [5, 6].

In the last years, transport experiments carried out in the presence of ac driving fields and the study of the related pumping phenomena [7–13] motivated the development of strategies to formulate and efficiently solve the corresponding Dyson's equations in theoretical models for these setups. A rigorous formulation was presented in Refs. [14] for the problem of a mesoscopic ring threaded by a time-dependent magnetic flux. Such treatment was afterwards adapted to study particle transport in models of quantum pumps consisting in quantum dots where local ac voltages are applied, with [15] and without many-body interactions [16]. In the latter case, an explicit relation between the Green's function and the Floquet scattering matrix was presented in Ref. [17].

Microscopic models and the framework provided by non-equilibrium Green's functions also offer the adequate basis for the study of mechanisms for heat generation, energy transport and quantum refrigeration by electrons in mesoscopic structures under ac driving, [18] as well as the possibility of characterizing these processes by defining effective temperatures [19]. In the context of phononic systems, J-S Wang and coworkers proposed recently in Refs. [20–22] a treatment analogous to that of Ref. [1] to describe thermal transport induced by a temperature gradient under stationary conditions. The present contribution is devoted to review a recent related development, which focuses on the study of heat transport by phonons under ac driving [23]. One of the motivations is to investigate cooling mechanisms in these systems. This is because refrigeration by means of manipulating phonons, rather than electrons, has the appealing feature of being suitable for insulating as well as metallic systems. On the other hand, cooling nanomechanical systems is a very active field of research [24–30] and it is interesting to explore alternative mechanisms.

In a recent work,[31] it was shown that when a barrier (or perturbation) moves along a cavity connecting two gases of acoustic phonons, a sizable amount of heat can be transferred from the colder to the hotter gas. When the process is repeated as a cycle, the result is refrigeration until the colder reservoir reaches a minimum temperature determined by geometrical parameters of the device and the velocity at which the barrier travels through the cavity. That mechanism was afterward analyzed in the framework of a microscopic model which was solved by means of a nonequilibrium Green's functions treatment analogous to the one proposed to study heat transport in ac driven electron systems. In the subsequent sections we briefly review that theoretical treatment. We also present some approximations to study the problem in the limit of weak driving. Finally, we show that, besides refrigeration, another interesting mechanism suggested to take place in quantum pumps [18] and quantum capacitors [32], can be also realized in phononic systems. Namely, the exchange of energy between the driving fields.

2. Background

2.1. Microscopic Model

The standard microscopic model for acoustic phonons consists of masses coupled by springs. The one-dimensional (1d) version of this model became popular in the context of heat transport since the pioneer work by Fermi, Pasta, and Ulam [33] and several subsequent studies devoted to understand the violation of the Fourier law in these systems [20–22, 34–38]. Fourier law implies a linear drop of the local temperature along a system connected between reservoirs at different temperatures. It is the counterpart of Ohm law in the context of heat transport, and it is nowadays rather natural to accept that it should not be expected to hold through a finite chain of harmonic oscillators, along which heat propagates coherently. In fact, a similar effect is known to take place in the behavior of the local voltage along a mesoscopic structure placed between two reservoirs at different chemical potentials [39–41]. In these systems, the voltage drop takes place at the contacts and the concept of "contact resistance" was introduced to characterize this feature, while it vanishes along the structure through which electrons propagate coherently. It is remarkable that the studies on the origin of the lack of the Fourier law in phononic systems and the introduction of the concept of the contact resistance in electron systems followed parallel routes and the connection between these two phenomena has not been mentioned even in review articles [37, 38]. More recently, heat pumping has been analyzed through a two-level system driven by an ac field coupled to phononic baths at different temperatures, [42] between semi-infinite harmonic chains with a time-dependent coupling, [43] and in anharmonic molecular junctions in contact to phononic baths [44].

We consider the one-dimensional (1d) system sketched in Fig. 1, with a central finite chain of N molecules with identical masses m_c coupled by springs of constant k_0. This system is connected, at certain sites l_α, to semi-infinite chains of masses m_α coupled by springs of constant k_α. These chains play the role of reservoirs, which we assume being kept at temperatures T_α.

$$H = \sum_{\alpha=1}^{L} H_\alpha + H_c(t) + H_{\text{cont}}. \tag{1}$$

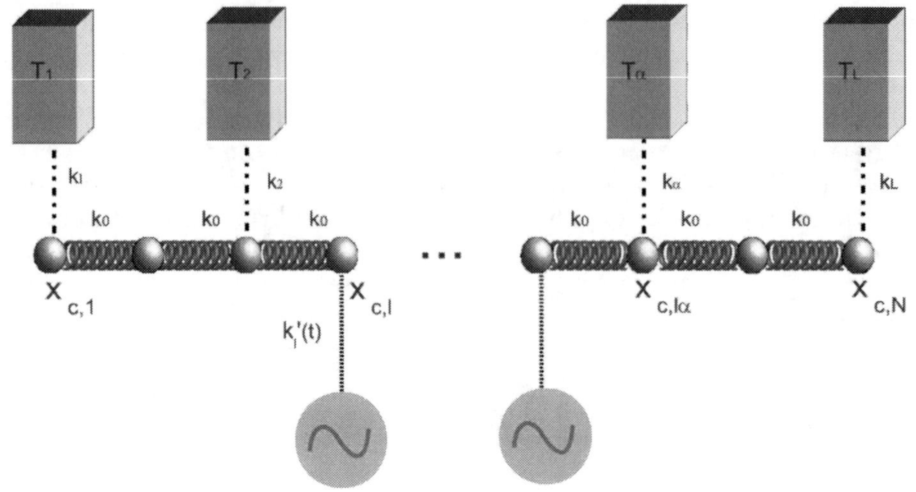

Figure 1. Sketch of the setup containing a central chain of acoustic phonons represented by a 1d lattice of atoms with equal masses connected by springs. This system is in contact to several reservoirs at different temperatures and affected by driving forces acting locally at the different positions of the central chain.

The first term H_α represents the reservoirs; the second $H_c(t)$ describes the central chain. We assume that external time-dependent perturbations are applied at this system. The Hamiltonian reads

$$H_c(t) = \sum_{l=1}^{N} \frac{p_{c,l}^2}{2m_c} + \sum_{l=1}^{N-1} \frac{k_0}{2}(x_{c,l} - x_{c,l+1})^2 + \sum_{l=1}^{N} \frac{k_l'(t)}{2} x_{c,l}^2, \tag{2}$$

where the last term of this expression describes the time dependent perturbation, which is represented by time-dependent elastic forces at different positions of the central chain. It is useful to define two characteristic frequencies in the problem: the frequency $\omega_c = \sqrt{k_0/m_c}$, and the driving frequency Ω_0.(We shall work on units such that $\hbar = k_B = 1$). The contact between the central chain and the reservoirs is described by the Hamiltonian

$$H_{\text{cont}} = \sum_{\alpha=1}^{L} \frac{k_\alpha}{2}(x_{\alpha,1} - x_{c,l_\alpha})^2. \tag{3}$$

Notice that the site $j = 1$ of each reservoir couples to the site l_α of the central region. The reservoir Hamiltonians are given by

$$H_\alpha = \sum_{j=1}^{N_\alpha} \left[\frac{p_{\alpha,j}^2}{2m_\alpha} + \frac{k_\alpha}{2}(x_{\alpha,j} - x_{\alpha,j+1})^2 \right]. \tag{4}$$

In the limit of semi-infinite chains, $N_\alpha \to \infty$, it is convenient to express the degrees of freedom of the reservoirs in terms of normal modes. For open boundary conditions, this corresponds to performing the following transformation,

$$x_{\alpha,l} = \sqrt{\frac{2}{N_\alpha + 1}} \sum_{n=0}^{N_\alpha} \sin(q_n^\alpha l)\, x_{\alpha,n} \tag{5}$$

and

$$p_{\alpha,l} = \sqrt{\frac{2}{N_\alpha + 1}} \sum_{n=0}^{N_\alpha} \sin(q_n^\alpha l)\, p_{\alpha,n}, \tag{6}$$

where

$$q_n^\alpha = \frac{n\pi}{N_\alpha + 1}, \qquad n = 0, \ldots, N_\alpha. \tag{7}$$

The corresponding Hamiltonians transform into

$$H_\alpha = \sum_n \left\{ \frac{p_{\alpha,n}^2}{2m_\alpha} + \frac{k_\alpha}{2} \left[1 - \cos(q_n^\alpha)\right] x_{\alpha,n}^2 \right\}. \tag{8}$$

2.2. Energy balance

Following a procedure similar to that used in Ref. [18], we consider the evolution in time of the energy density stored at a given elementary volume and write the corresponding conservation equation. Since we are treating a 1d lattice, and our Hamiltonians involve only nearest-neighbors terms, the minimum volume that we consider is one that encloses nearest-neighbors sites, like the left box of Fig. 2. If we denote by $dE_{l,l+1}/dt$ the total energy flow exiting and entering a volume of the central chain without connection to reservoirs, the equation for the conservation of the energy for reads:

$$\frac{dE_{l,l+1}}{dt} = J_{l+1}^Q(t) - J_l^Q(t) + P_l(t) + P_{l+1}(t), \tag{9}$$

where $J_l^Q(t)$ denotes the incoming energy current from site $l-1$ towards the site l and $J_{l+1}^Q(t)$ the outgoing flow from site $l+1$ to $l+2$. We denote with a positive sign the flows pointing from left to right. These quantities are evaluated from

$$\frac{dE_{l,l+1}}{dt} = -i\langle [H, H_{l,l} + H_{l+1,l+1} + H_{l,l+1}] \rangle + \frac{d[H_l + H_{l+1}](t))}{dt}, \tag{10}$$

where $H_{l,l'}$ are the matrix elements of $H_c(t)$ on the sites l, l'. The explicit calculation cast

$$J_l^Q = \frac{k_0}{m_c} \langle p_{c,l} x_{c,l-1} \rangle, \tag{11}$$

for the current and

$$P_l(t) = \frac{\partial k_l'(t)}{\partial t} \langle x_{c,l}^2 \rangle, \tag{12}$$

for the power developed by the force applied at the site l.

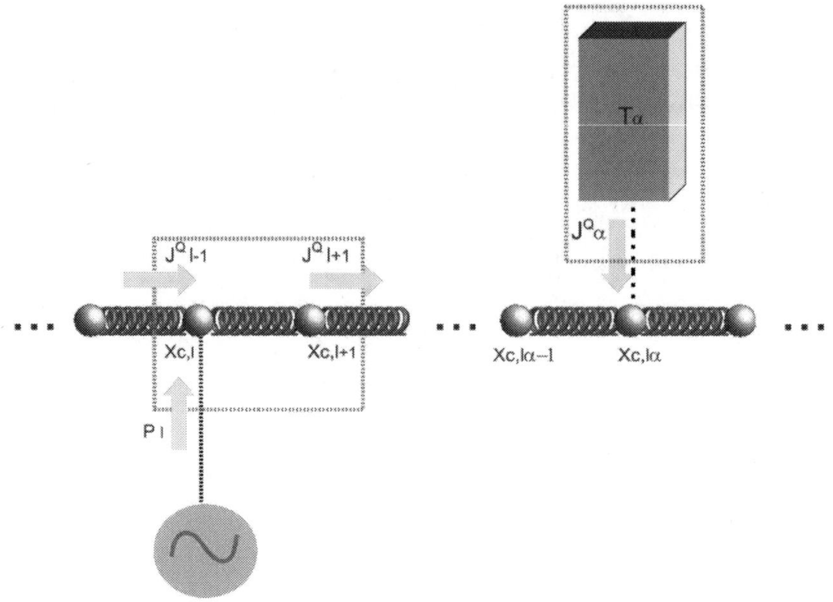

Figure 2. Sketch of the setup indicating the construction employed to derive the expressions for the heat currents as well as the poser developed by the external forces.

2.2.1. Heat flow into reservoirs We now focus on the right box of the sketch of Fig. 2. The variation in time of the energy stored in the reservoir α is

$$\frac{dE_\alpha(t)}{dt} \quad = \quad -i\langle[H, H_\alpha]\rangle = J_\alpha^Q(t), \tag{13}$$

where we assume that no driving force is acting at the contact site of the reservoir. The current $J_\alpha^Q(t)$ flows along the contact between the site l_α of the central chain and the reservoir α and it is assumed to be positive when it enters the reservoir. Since we are dealing with phonons, this energy flow is a pure heat flow and its explicit expression reads

$$J_\alpha^Q(t) = \sum_n \frac{k_\alpha}{m_c}\sqrt{\frac{2}{N_\alpha+1}}\sin(q_n^\alpha)\langle x_{\alpha,n}\, p_{c,l_\alpha}\rangle. \tag{14}$$

Conservation of energy implies that the *total* average of the power invested by the external fields is dissipated into the reservoirs at a rate

$$\sum_{\alpha=1}^{L} \overline{J}_\alpha^Q = \sum_{l=1}^{N} \overline{P}_l, \tag{15}$$

where

$$\overline{P}_l = \frac{1}{\tau}\int_0^\tau dt\, P_l(t), \tag{16}$$

with $\tau = 2\pi/\Omega_0$, is the power developed by the external fields at the site l averaged over one period, while

$$\overline{J}^Q_\alpha = \frac{1}{\tau}\int_0^\tau dt\, J^Q_\alpha(t), \tag{17}$$

is the dc component of the heat current at the reservoir α. Notice that a net amount of work has to be invested in order to pump heat from one reservoir to the other. In the next sections we show that carrying out an analysis similar to that developed in Ref. [18] leads to the conclusion that the rate at which the total work done by the external fields is dissipated as heat flowing into the reservoirs is proportional to Ω_0^2.

3. Non-equilibrium Green's function approach

3.1. Relevant Green's functions

The theory of non-equilibrium Green's functions, which was developed independently by L. P. Kadanoff and G. Baym [45], Schwinger [46] and Keldysh [47] has been employed many times in the context of electron transport, as mentioned in the introduction. Details can be found in that literature. In the present case, the general procedure is completely analogous. The relevant Green's functions are the bigger and lesser functions

$$\begin{aligned} D^>(1,1') &= i\langle X_1(t_1)X_{1'}(t_{1'})\rangle, \\ D^<(1,1') &= i\langle X_{1'}(t_{1'})X_1(t_1)\rangle, \end{aligned} \tag{18}$$

where $X(1) = \sqrt{m_1}x_1(t_1)$ is a phononic operator and $1, 1'$ is a schematic notation that labels the spacial and time coordinates. The corresponding retarded and advanced Green's functions read

$$D^R(1,1') = \theta(t_1 - t_{1'})[D^>(1,1') - D^<(1,1')] \tag{19}$$
$$D^A(1,1') = \theta(t_{1'} - t_1)[D^<(1,1') - D^>(1,1')], \tag{20}$$

where the four Green functions satisfy the following relations $D^A(1,1') = [D^R(1',1)]^*$, $D^<(1,1') = -[D^>(1',1)]^*$ and $D^R(1,1') - D^A(1,1') = D^>(1,1') - D^<(1,1')$.

3.2. Dyson's equation

We now turn to derive the Dyson's equations to evaluate the Green's functions in our problem. The operators x and p satisfy canonical commutation relations:

$$[x_l, x_{l'}] = 0, \qquad [x_l, p_{l'}] = i\delta_{l,l'}. \tag{21}$$

For coordinates l, l' within the central chain, the lesser and greater Green's functions read

$$D^<_{l,l'}(t,t') = im_c\langle x_{c,l'}(t')x_{c,l}(t),\rangle \tag{22}$$
$$D^>_{l,l'}(t,t') = im_c\langle x_{c,l}(t)x_{c,l'}(t')\rangle, \tag{23}$$

while the retarded Green's function is

$$D^R_{l,l'}(t,t') = -i\theta(t-t')m_c\langle[x_{c,l}(t), x_{c,l'}(t')]\rangle. \tag{24}$$

Now we proceed to the calculation of the equation of motion of the retarded Green's function. To this end we must take the second derivative respect to t', and use the Eherenfest's theorem. The result is

$$-\left[\frac{\partial^2}{\partial t'^2} + 2\frac{k_0}{m_c} + \frac{k'_{l'}(t)}{m_c}\right] D^R_{l,l'}(t,t') + \frac{k_0}{m_c}\left[D^R_{l,l'-1}(t,t') + D^R_{l,l'+1}(t,t')\right] \tag{25}$$
$$= \delta_{l,l'}\delta(t-t') + \frac{k_\alpha}{m_c}\left[D^R_{l,l_\alpha}(t,t')\delta_{l',l_\alpha} - D^R_{l,\alpha_1}(t,t')\delta_{l',l_\alpha}\right], \tag{26}$$

where l_α refers to the site of the central chain connected to reservoir α and α_1 the first site of the same reservoir. In turn, the integral representation for the Dyson's equation corresponding to the function D^R_{l,α_1} reads

$$D^R_{l,\alpha_1}(t,t') = -\frac{k_\alpha}{m_\alpha}\sum_{n=0}^{N_\alpha}\frac{2}{N_\alpha+1}\sin^2(q_n^\alpha)\int_{-\infty}^{+\infty} dt_1 D^R_{i,l_\alpha}(t,t_1)d^R_{q_n^\alpha}(t_1,t'), \tag{27}$$

where

$$d^R_{q_n^\alpha}(\omega) = \frac{1}{2E_{\alpha,n}}\left[\frac{1}{\omega+i\eta-E_{\alpha,n}} - \frac{1}{\omega+i\eta+E_{\alpha,n}}\right], \tag{28}$$

with $\eta = 0^+$ stands for the equilibrium Green's function within the uncoupled reservoir α. If we replace the Eq.(27) into Eq.(25), we can define the self energy as

$$\Sigma^R_\alpha(t,t') = \frac{k_\alpha^2}{m_c m_\alpha}\sum_{n=0}^{N_\alpha}\frac{2}{N_\alpha+1}\sin^2(q_n^\alpha)d^R_{q_n^\alpha}(t,t'), \tag{29}$$

(a more explicit expression is presented in Appendix A), and Eq.(25) can be written in the simpler matrix form

$$-\partial_{t'}^2\hat{D}^R(t,t') + \hat{D}^R(t,t')\hat{M}(t') - \int dt_1 \hat{D}^R(t,t_1)\hat{\Sigma}^R(t_1,t') = \delta(t-t'), \tag{30}$$

with $\hat{M}(t) = \hat{M}^{(0)} + \hat{M}^{(1)}(t)$, where

$$M^{(0)}_{l,l'} = \begin{cases} \frac{k_0}{m_c}(2\delta_{l,l'} - \delta_{i',i\pm1}) + \frac{k_\alpha}{m_c}\delta_{l',l_\alpha}\delta_{l,l'}, & 1<l<N, \\ (\frac{k_0}{m_c}+\frac{k_1}{m_c})\delta_{l,l'} - \frac{k_0}{m_c}\delta_{l',l+1}, & i=1, \\ (\frac{k_0}{m_c}+\frac{k_N}{m_c})\delta_{l,l'} - \frac{k_0}{m_c}\delta_{l',l-1}, & l=N, \end{cases} \tag{31}$$

$$M^{(1)}_{l,l'}(t) = \frac{k_l'(t)}{m_c}\delta_{l,l'}, \tag{32}$$

and $\Sigma^R_{l,l'}(t,t') = \sum_{\alpha=1}^L \delta_{l,l'}\delta_{l,l_\alpha}\Sigma^R_\alpha(t,t')$.

On the another hand, the Dyson's equation for the lesser function is obtained by recourse to Langreth's rules

$$D^<_{l,l'}(t,t') = \sum_\alpha \int dt_1 \int dt_2\, D^R_{l,l_\alpha}(t,t_1)\,\Sigma^<_\alpha(t_1-t_2)D^A_{l_\alpha,l'}(t_2,t'), \tag{33}$$

and

$$D^<_{l,q_n^\alpha}(t,t') = -\frac{k_\alpha}{m_\alpha}\sum_n\sqrt{\frac{2}{N_\alpha+1}}\sin(q_n^\alpha)\int dt_1[D^R_{l,l_\alpha}(t,t_1)\,d^<_{q_n^\alpha}(t_1-t') + D^<_{l,l_\alpha}(t,t_1)\,d^A_{q_n^\alpha}(t_1-t')], \tag{34}$$

where Eq. (33) corresponds to coordinates along the central chain while Eq. (34) corresponds to one coordinate along the chain and the other on the reservoir. The advanced Green's function of the uncoupled reservoir is $d^A_{q_n^\alpha}(t_1,t') = [d^R_{q_n^\alpha}(t',t_1)]^*$, while the corresponding lesser function reads

$$d^<_{q_n^\alpha}(\omega) = \frac{i\pi n_\alpha(\omega)}{E_{\alpha,n}}\left[\delta(\omega-E_{\alpha,n}) - \delta(\omega+E_{\alpha,n})\right], \tag{35}$$

with $n_\alpha(\omega) = 1/(e^{\omega/T_\alpha}-1)$ being the Bose-Einstein distribution function, which depends on the temperature T_α of the reservoir α. In terms of the spectral function defined in Appendix A, the Fourier transform of the lesser Green's function reads

$$\Sigma^<_\alpha(\omega) = i n_\alpha(\omega)\Gamma_\alpha(\omega). \tag{36}$$

Journal of Physics: Conference Series **427** (2013) 012012 doi:10.1088/1742-6596/427/1/012012

3.3. Solution of Dyson's equation

To obtain the retarded function of the central region, $\hat{D}^R$, one has to solve the differential equations (30). For the case of terms that depend harmonically on time, it is convenient to follow the strategy proposed in Refs. [14, 16] for electron systems and recently adapted to phonons in Ref. [23]. We summarize the main steps in the present section. It is convenient to perform the Fourier transform with respect to the "delayed time" in the retarded Green's function

$$\hat{D}^R(t,\omega) = \int_{-\infty}^{+\infty} dt' e^{i(\omega+i0^+)(t-t')} \hat{D}^R(t,t'), \tag{37}$$

and to represent the time-dependent perturbation of the Hamiltonian in terms of its Fourier expansion

$$\hat{M}^{(1)}(t) = \sum_{k=-K,k\neq 0}^{K} \hat{M}_k^{(1)} e^{-ik\Omega_0 t}. \tag{38}$$

Substituting in the Dyson's equation (30) results

$$\hat{D}^R(t,\omega) = \hat{D}^{(0)}(\omega) + \sum_{k\neq 0,-\infty}^{+\infty} e^{-ik\Omega_0 t} \hat{D}^R(t,\omega+k\Omega_0)\hat{M}_k^{(1)}\hat{D}^{(0)}(\omega), \tag{39}$$

where

$$\hat{D}^{(0)}(\omega) = \left[\omega^2 \hat{I} - \hat{M}^{(0)} - \Sigma^R(\omega)\right]^{-1}, \tag{40}$$

corresponds to the stationary component of the retarded Green's function of the central chain connected to the reservoirs but without the effect of the time-dependent perturbations. Notice that because of the periodic structure of the time dependent part of the Hamiltonian, the retarded Green's function is also periodic in time with period τ. For this reason it is useful and natural to represent this function in terms of a Fourier series as follows

$$\hat{D}^R(t,\omega) = \sum_{k=-\infty}^{+\infty} e^{-ik\Omega_0 t} \mathcal{D}(k,\omega) \tag{41}$$

where the functions $\mathcal{D}(k,\omega)$ are also known as Floquet components. The exact solution for the set of coupled linear equations Eq.(39) can be obtained numerically by following the procedure introduced in Ref. [48] for fermionic systems driven by ac potentials. Alternatively, it is possible to solve these equations approximately in the limits of weak amplitudes $\hat{M}_k^{(1)}$ of the time-dependent perturbations and in the limit of low frequencies Ω_0. We briefly present these approximate treatments in the next subsections.

3.4. Perturbative solution of Dyson's equation

In most cases, the set of linear equations of Eq.(39) must be solved numerically. For this reason, it is convenient to carry out a systematic expansion in powers of $\hat{M}^{(1)}$, to obtain analytical expressions. When these amplitudes are small (compared with the energies of the Hamiltonian independent of time), a perturbative solution in these parameters is a good approximation. Assuming that the external perturbations $\hat{M}^{(1)}(t)$ contain K harmonics, as expressed in Eq. (38), the Green's function evaluated up to the 1st order in the amplitudes $\hat{M}_k^{(1)}$ is

$$\hat{D}^R(t,\omega) \sim \sum_{k=-K}^{K} e^{-ik\Omega_0 t} \hat{\mathcal{D}}(k,\omega), \tag{42}$$

with

$$\hat{\mathcal{D}}(0,\omega) = \hat{D}^{(0)}(\omega), \tag{43}$$

$$\hat{\mathcal{D}}(\pm k,\omega) = \hat{D}^{(0)}(\omega \pm k\Omega_0)\hat{M}^{(1)}_{\pm k}\hat{D}^{(0)}(\omega), \quad k = 1,\ldots,K. \tag{44}$$

3.5. Low driving frequency solution of Dyson's equation
Let us now consider the Dyson equation in Eq. (39) in the limit of low driving frequency Ω_0.

1st order A solution exact up to $\mathcal{O}(\Omega_0)$ can be obtained by expanding Eq. (39) as follows:

$$\begin{aligned}\hat{D}(t,\omega) \sim\ & \hat{D}^{(0)}(\omega) + \hat{D}(t,\omega)\hat{M}^{(1)}(t)\hat{D}^{(0)}(\omega) + \\ & i\partial_\omega \hat{D}(t,\omega)\frac{d\hat{M}^{(1)}(t)}{dt}\hat{D}^{(0)}(\omega).\end{aligned} \tag{45}$$

We define the frozen Green function

$$\hat{D}_f(t,\omega) = \left[\hat{D}^{(0)}(\omega)^{-1} - \hat{M}^{(1)}(t)\right]^{-1}, \tag{46}$$

in terms of which the exact solution of the Dyson equation at $\mathcal{O}(\Omega_0)$ reads

$$\hat{D}^{(1)}(t,\omega) = \hat{D}_f(t,\omega) + i\partial_\omega \hat{D}_f(t,\omega)\frac{d\hat{M}^{(1)}(t)}{dt}\hat{D}^{(0)}(\omega). \tag{47}$$

2nd order To obtain the solution exact up to $\mathcal{O}(\Omega_0^2)$, we consider the following expansion of equation Eq. (39):

$$\begin{aligned}\hat{D}(t,\omega) \sim\ & \hat{D}^{(0)}(\omega) + \hat{D}(t,\omega)\hat{M}^{(1)}(t)\hat{D}^{(0)}(\omega) \\ & + i\partial_\omega \hat{D}(t,\omega)\frac{d\hat{M}^{(1)}(t)}{dt}\hat{D}^{(0)}(\omega) \\ & - \frac{1}{2}\partial_\omega^2 \hat{D}(t,\omega)\frac{d^2\hat{M}^{(1)}(t)}{dt^2}\hat{D}^{(0)}(\omega).\end{aligned} \tag{48}$$

The solution exact up to $\mathcal{O}(\Omega_0^2)$ is

$$\begin{aligned}\hat{D}^{(2)}(t,\omega) =\ & \hat{D}_f(\omega) + i\partial_\omega \hat{D}^{(1)}(t,\omega)\frac{d\hat{M}^{(1)}(t)}{dt}\hat{D}_f(\omega) \\ & -\frac{1}{2}\partial_\omega^2 \hat{D}_f(t,\omega)\frac{d^2\hat{M}^{(1)}(t)}{dt^2}\hat{D}_f(\omega).\end{aligned} \tag{49}$$

One can obtain this Green function numerically by first discretizing the time in the interval $0 \le t \le \tau$, then solving Eq. (49) for each one time, and finally evaluating the Fourier transform to obtain the Floquet representation with which the dc component of the heat current can be calculated from Eq. (55).

4. dc heat current

In this section, we present the steps to calculate the net heat current flowing into a given reservoir. We start by writing the time-dependent heat current given in Eq.(14) as follows

$$J_\alpha^Q(t) = \sum_n k_\alpha \sqrt{\frac{2}{N_\alpha+1}}\sin(q_n^\alpha)\lim_{t\to t'}\frac{\partial}{\partial t'}\langle x_{\alpha,n}(t')x_{l_\alpha}(t)\rangle \tag{50}$$

$$= -\sum_n \frac{k_\alpha}{m_c}\sqrt{\frac{2}{N_\alpha+1}}\sin(q_n^\alpha)\lim_{t\to t'}\mathrm{Re}\{i\frac{\partial}{\partial t'}D^<_{l_\alpha,q_n^\alpha}(t,t')\}. \tag{51}$$

We now substitute Eq.(34) and the representation of Eq.(41) to write the dc component of the above current as

$$\mathcal{J}_\alpha^Q(t) = -\sum_n k_{n,\alpha} \int_{-\infty}^{+\infty} \frac{d\omega}{2\pi} \mathrm{Re}\{\omega \mathcal{D}_{l_\alpha,l_\alpha}(0,\omega) d_{q_n^\alpha}^<(\omega) + \sum_{\beta=1}^{L} \sum_{k=-\infty}^{+\infty} (\omega + k\Omega_0) |\mathcal{D}_{l_\alpha,l_\beta}(k,\omega)|^2 \Sigma_\beta^<(\omega) d_{q_n^\alpha}^A(\omega + k\Omega_0)\} \tag{52}$$

where $k_{n,\alpha} = \left(\frac{k_\alpha^2}{m_\alpha m_c}\right) \frac{2}{N_\alpha+1} \sin^2(q_n^\alpha)$. Finally we obtain

$$\mathcal{J}_\alpha^Q = \int_{-\infty}^{+\infty} \frac{d\omega}{2\pi} [\omega 2\mathrm{Im}\{\mathcal{D}_{l_\alpha,l_\alpha}(0,\omega)\} \Gamma_\alpha(\omega) n_\alpha(\omega) + \sum_{\beta=1}^{L} \sum_{k=-\infty}^{+\infty} (\omega + k\Omega_0) |\mathcal{D}_{l_\alpha,l_\beta}(k,\omega)|^2 \Gamma_\beta(\omega) \Gamma_\alpha(\omega + k\Omega_0) n_\beta(\omega)]. \tag{53}$$

An alternative representation to this current can be obtained after recasting the first term of the above equation by recourse to the identity

$$\mathcal{D}_{l,l'}(k,\omega) - \mathcal{D}_{l',l}^*(-k,\omega + k\Omega_0) = -i \sum_{k'} \sum_{\alpha} \mathcal{D}_{l,l_\alpha}(k + k', \omega - k'\Omega_0) \times \Gamma_\alpha(\omega - k'\Omega_0) \mathcal{D}_{l',l_\alpha}^*(k', \omega - k'\Omega_0). \tag{54}$$

This leads to the following equation for the dc heat current,

$$\mathcal{J}_\alpha^Q = \sum_{\beta=1}^{L} \sum_{k=-\infty}^{+\infty} \int_{-\infty}^{+\infty} \frac{d\omega}{2\pi} (\omega + k\Omega_0) [n_\beta(\omega) - n_\alpha(\omega + k\Omega_0)] \Gamma_\alpha(\omega + k\Omega_0) \Gamma_\beta(\omega) |\mathcal{D}_{l_\alpha,l_\beta}(k,\omega)|^2. \tag{55}$$

This equation constitutes a generalization to the case of ac driving of a Landauer-Büttiker formula for the heat transport in phononic systems (see Refs. [20–22])

$$J_\alpha^Q = \int_{-\infty}^{+\infty} \frac{d\omega}{2\pi} \omega \sum_{\beta=1}^{L} \mathcal{T}_{\alpha\beta}(\omega) [n_\beta(\omega) - n_\alpha(\omega)], \tag{56}$$

being $\mathcal{T}_{\alpha\beta}(\omega) = |D_{l_\alpha,l_\beta}^R(\omega)|^2 \Gamma_\alpha(\omega) \Gamma_\beta(\omega)$, the transmission function between the reservoirs α and β. Equation (55) indicates that a net heat current may exist even in the absence of a temperature difference between the reservoirs. Such a flow is a consequence of the pumping with the ac forces and contains, in general, an incoming component which accounts for the work done by the ac fields and which is dissipated into the reservoirs. As shown in Ref. [23] under certain conditions, a net current between the reservoirs can be established. This current can go against a temperature gradient, therefore allowing for cooling.

5. Mean power developed by the external forces

We now turn to analyze another interesting aspect of the heat transport, which are the mechanisms used by the driving forces to transfer a net amount of energy to the phononic system. For sake of simplicity we restrict our analysis to the case of driving forces that oscillate with a single harmonic component and a phase-lag, of the form

$$k_l'(t) = K_0 \cos(\Omega_0 t + \delta_l). \tag{57}$$

The mean power developed by the force acting on the site l corresponds to calculating the dc component of the time-dependent power of Eq.(12). In terms of Green's functions can be written as follows

$$\overline{P}_l = \frac{-i}{\tau m_c} \int_0^\tau dt \frac{\partial k_l'(t)}{\partial t} D_{l,l}^<(t,t). \tag{58}$$

According to the Dyson's equation (33) and in terms of the Floquet-Fourier representation of the Green's function (41) the above equation reads

$$\overline{P}_l = -\frac{\Omega_0}{m_c} \sum_\alpha \sum_{k,k'} \int_{-\infty}^{+\infty} \frac{d\omega}{2\pi} M_{k'}^{(1)} k' \mathcal{D}_{l,l_\alpha}(k,\omega) \Sigma_\alpha^<(\omega) \mathcal{D}_{l,l_\alpha}^*(k+k',\omega), \tag{59}$$

For the case of ac forces of the form (57) this equation results

$$\overline{P}_l = \frac{\Omega_0 K_0}{2m_c} \sum_k \sum_{\alpha=1}^{L} \int_{-\infty}^{+\infty} \frac{d\omega}{2\pi} i n_\alpha(\omega) [\mathcal{D}_{l,l_\alpha}(k,\omega)\Gamma_\alpha(\omega)\mathcal{D}_{l,l_\alpha}^*(k-1,\omega)e^{i\delta_l} \tag{60}$$

$$-\mathcal{D}_{l,l_\alpha}(k,\omega)\Gamma_\alpha(\omega)\mathcal{D}_{l,l_\alpha}^*(k+1,\omega)e^{-i\delta_l}]. \tag{61}$$

A simple analytical result can be derived for weak driving, in which case we can use the perturbation treatment of Sec. 3.4. This leads to a result for the mean power which is exact up to $\mathcal{O}(K_0^2)$ and reads

$$\overline{P}_l \sim \frac{\Omega_0 K_0^2}{2\pi m_c} \sum_{j=1}^{N} \int_{-\infty}^{+\infty} d\omega \mathrm{Im}\left\{[\cos(\delta_l - \delta_j) - i\sin(\delta_l - \delta_j)]\right. \tag{62}$$

$$\left.\left[\gamma_{l,j}(\omega) D_{l,j}^{(0)}(\omega - \Omega_0) + \gamma_{l,j}^*(\omega) D_{l,j}^{(0)*}(\omega + \Omega_0)\right]\right\}, \tag{63}$$

with

$$\gamma_{l,j}(\omega) = \sum_{\alpha=1}^{L} n_\alpha(\omega) D_{l,l_\alpha}^{(0)*}(\omega)\Gamma_\alpha(\omega)\mathcal{D}_{j,l_\alpha}^{(0)}(\omega). \tag{64}$$

It is particularly interesting to analyze this result in the limit of low driving frequency Ω_0. To this end we expand the previous equation up to second order in this parameter. In the ensuing result we find that in this limit the mean power contains two different kinds of components $\overline{P}_l = \overline{P}_l^{ex} + \overline{P}_l^{dis}$, being

$$\overline{P}_l^{ex} = -\frac{\Omega_0 K_0^2}{\pi m_c} \sum_{j=1}^{N} \sin(\delta_l - \delta_j) \int_{-\infty}^{+\infty} d\omega \mathrm{Re}\left\{\gamma_{l,j}(\omega) D_{l,j}^{(0)}(\omega)\right\}, \tag{65}$$

$$\overline{P}_l^{dis} = -\frac{\Omega_0^2 K_0^2}{\pi m_c} \sum_{j=1}^{N} \cos(\delta_l - \delta_j) \int_{-\infty}^{+\infty} d\omega \mathrm{Im}\left\{\gamma_{l,j}(\omega) \frac{\partial D_{l,j}^{(0)}(\omega)}{\partial\omega}\right\}, \tag{66}$$

where the first component is $\propto \Omega_0$ and it is the leading contribution for low driving frequency. Since $\sin(\delta_l - \delta_j)$ is an antisymmetric function, this term satisfies

$$\sum_{l=1}^{N} \overline{P}_l^{ex} = 0, \tag{67}$$

indicating that some of the forces deliver, while others receive a net amount of energy. This is the essential ingredient for this setup to have an operational regime typical of an engine. In the present case, it is particularly appealing The fact that the exchanged energy is of mechanical nature. The second component, which is $\propto \mathcal{O}(\Omega_0^2)$, is non-vanishing and accounts for the conservation of the energy expressed in Eq. (15). This component, thus, describes the net amount of energy that is dissipated into the reservoirs in the form of heat. It is important to notice that for a driving like (57), the exchange component is non-vanishing only in the case of a driving with a phase-lags $\delta_l - \delta_j \neq n\pi$. Another possibility to get finite values for $\overline{P}_l^{ex}$ is to consider local forces with different amplitudes. It can be in general shown that a multiple-parametric driving is a necessary condition to have the exchange mechanism in the present setup.

6. Summary and conclussions

We have considered a simple model to study heat transport in nanomechanical systems. It consists in the usual model for acoustic phonons coupled to phononic reservoirs and generalized to include the effect of a time-dependent perturbation in the form of elastic forces acting at different places of the structure. We have reviewed a theoretical treatment based on non-equilibrium Green's functions, which is analogous to the one previously proposed for electron systems under ac-driving. We have derived the equations for the heat current along different regions of the setup and shown how can be evaluated in terms of the Green's functions. We have also derived expressions to evaluate the power developed by the external ac forces. In the limit of weak driving amplitudes and low frequencies, we have analytically shown that the mean power done by each force, contains a component that represents the heat dissipated into the reservoirs. For a driving with multiple parameters, corresponding for instance to driving forces oscillating with a phase lag, there is an additional component, which corresponds to energy that can be exchanged between the different forces. This component dominates for low enough driving frequencies and enables an operational regime of the setup as a quantum engine. A similar mechanism has been identified in driven electron systems like quantum pumps [18] and in coupled quantum capacitors [32]. The present case has the appealing feature of involving mechanical work.

The present theoretical approach sets a general framework for the study of heat transport by vibrational degrees of freedom in driven systems. It can be adapted, generalized and improved in many directions, to analyze several interesting situations. In particular, we have considered phononic reservoirs, but the present treatment can be easily adapted to deal with photonic reservoirs. We have considered coupled atoms, but this treatment can be easily generalized to consider coupled molecules, which have internal vibrational degrees of freedom. Improvements to model realistic setups with many vibrational modes can also be straightforwardly implemented. Extensions of the present treatment to consider the anharmonic effects of Refs. [20–22] with ac driving are also possible. To finalize, it is interesting to mention that a family of 1d interacting models of cold atoms can be reduced to a Luttinger Hamiltonian, which is basically a model of harmonic oscillators. Dynamical evolution of such systems in quenching protocols including coupling to reservoirs have been recently considered [49]. The present scheme could be useful to study such systems under ac driving.

Acknowledgments

LA thanks R. Capaz, C. Chamon and E. Mucciolo for many discussions. We acknowledge support from CONICET, UBACyT and MINCYT through PICT, Argentina.

Appendix A. Self-energy corresponding to the coupling to a semi-infinite chain of acoustic phonons

The explicit evaluation of the sum over the normal modes entering the expression of the self energy (29) can be expressed in terms of a spectral density $\Gamma_\alpha(\omega)$ as follows

$$\Sigma_\alpha^R(t,t') = \int_{-\infty}^{+\infty} \frac{d\omega}{2\pi} e^{-i\omega(t-t')} \int_{-\infty}^{+\infty} \frac{d\omega'}{2\pi} \frac{\Gamma_\alpha(\omega')}{\omega - \omega' + i\eta}, \tag{A.1}$$

with $\eta > 0$ and

$$\begin{aligned}\Gamma_\alpha(\omega) &= \lim_{N_\alpha \to \infty} \frac{2\pi k_\alpha^2}{(N_\alpha+1) m_c m_\alpha} \sum_{n=0}^{N_\alpha} \sin^2(q_n^\alpha) \frac{1}{E_{\alpha,n}} \\ &\quad \times [\delta(\omega - E_{\alpha,n}) - \delta(\omega + E_{\alpha,n})] \\ &= \text{sgn}(\omega) \frac{k_\alpha}{m_c} \Theta\Big(1 - (\frac{k_\alpha - m_\alpha \omega^2}{k_\alpha})^2\Big) \\ &\quad \times \sqrt{1 - \left(\frac{k_\alpha - m_\alpha \omega^2}{k_\alpha}\right)^2},\end{aligned} \tag{A.2}$$

being $E_{\alpha,n} = \sqrt{k_\alpha(1-\cos q_n^\alpha)/m_\alpha}$. Notice that the Fourier transform of the self energy is perfectly define because is an equilibrium function.

References

[1] Caroli C, Combescot R, Nozieres P and Saint-James D 1971 *J. Phys. C: Solid State Phys.* **4** 916

[2] Meir Y and Wingreen N S 1992 *Phys. Rev. Lett.* **68** 2512

[3] Pastawski H M 1992 *Phys. Rev. B* **46** 4053

[4] Jauho A P, Wingreen N S and Meir Y 1994 *Phys. Rev. B* **50** 5528

[5] Jauho A P 2000 *Time-dependent transport in interacting mesoscopic systems (pp. 250)* Progress in Nonequilibrium Green's Functions (World Scientific)

[6] Jauho A P 2003 *Nonequilibrium Green function modelling of transport in mesoscopic systems (pp. 181)* Progress in Nonequilibrium Green's Functions II (World Scientific)

[7] Geerligs L J, Anderegg V F, Holweg P A M, Mooij J E, Pothier H, Esteve D, Urbina C and Devoret M H 1990 *Phys. Rev. Lett.* **64** 2691

[8] Switkes M, Marcus C M, Campman K and Gossard A C 1999 *Science* **283** 1905

[9] Watson S K, Potok R M, Marcus C M and Umansky V 2003 *Phys. Rev. Lett.* **91** 258301

[10] Blumenthal M D, Kaestner B, Li L, Giblin S, Hanssen T J B M, Pepper M, Anderson D, Jones G, and Ritchie D A 2007 *Nat. Phys.* **3** 343

[11] Gabelli J, Fève G, J-M, Berroir, Placais B, Cavanna A, Etienne B, Jin Y and Glattli D C 2006 *Science* **313** 499

[12] Fève G, Mahé A, Berroir J M, Kontos T, Placais B, Cavanna A, Etienne B, Jin Y and Glattli D C 2007 *Science* **316** 1169

[13] Granger G, Eisenstein J P and Reno J L 2009 *Phys. Rev. Lett.* **102** 086803

[14] Arrachea L 2004 *Phys. Rev. B* **70** 155407

[15] Arrachea L, Yeyati A L and Martin-Rodero A 2008 *Phys. Rev. B* **77** 165326

[16] Arrachea L 2005 *Phys. Rev. B* **72** 125349

[17] Arrachea L and Moskalets M 2006 *Phys. Rev. B* **74** 245322

[18] Arrachea L, Moskalets M and Martin-Moreno L 2007 *Phys. Rev. B* **75** 245420
[19] Caso A, Arrachea L and Lozano G S 2010 *Phys. Rev. B* **81** 041301
[20] Wang J S, Wang J and Zeng N 2006 *Phys. Rev. B* **74** 033408
[21] Wang J S, Zeng N, Wang J and Gan C K 2007 *Phys. Rev. E* **75** 061128
[22] Wang J S, Wang J and Lü J T 2008 *EPJB* **62** 381
[23] Arrachea L, Mucciolo E R, Chamon C and Capaz R B 2012 *Phys. Rev. B* **86** 125424
[24] a recent review on cooling by active feedback see A Cho F 2010 *Science* **327** 516
[25] Zippilli S, Morigi G and Bachtold A 2009 *Phys. Rev. Lett.* **102** 096804
[26] Zippilli S, Bachtold A and Morigi G 2010 *Phys. Rev. B* **81** 205408
[27] McEniry E J, Todorov T N and Dundas D 2009 *J. Phys.: Condens. Matter* **21** 195304
[28] Galperin M, Saito K, Balatsky A V and Nitzan A 2009 *Phys. Rev. B* **80** 115427
[29] Pistolesi F 2009 *J. Low Temp. Phys.* **154** 199
[30] Santandrea F, Gorelik L Y, Shekhter R I and Jonson M 2011 *Phys. Rev. Lett.* **106** 186803
[31] Chamon C, Mucciolo E R, Arrachea L and Capaz R B 2011 *Phys. Rev. Lett.* **106** 135504
[32] Moskalets M and Büttiker M 2009 *Phys. Rev. B* **80** 081302
[33] Fermi E, Pasta J and Ulam S 1955 *Los Alamos technical report* LA–1940
[34] Lepri S, Livi R and Politi A 1997 *Phys. Rev. Lett.* **78** 1896
[35] Terraneo M, Peyrard M and Casati G 2002 *Phys. Rev. Lett.* **88** 094302
[36] Li N, Li B and Flach S 2010 *Phys. Rev. Lett.* **105** 054102
[37] Lepri S, Livi R and Politi A 2003 *Phys. Rep.* **377** 1
[38] Dhar A 2008 *Adv. Phys.* **57** 457
[39] Büttiker M 1988 *Phys. Rev. B* **38** 9375
[40] Büttiker M 1989 *Phys. Rev. B* **40** 3409
[41] Gramespacher T and Büttiker M 1997 *Phys. Rev. B* **56** 13026
[42] Segal D 2009 *J. Chem. Phys.* **130** 134510
[43] Agarwalla B K, Wang J S and Li B 2011 *Phys. Rev. E* **84** 041115
[44] Ren J, Hänggi P and Li B 2010 *Phys. Rev. Lett.* **104** 170601
[45] Kadanoff L and Baym G 1962 *Quantum Statistical Mechanics* (Benjamin/Cummings Publishing Group USA)
[46] Schwinger J 1961 *J. Math. Phys.* **2** 407
[47] Keldysh L V 1964 *Zh. Eksp. Teor. Fiz.* **47** 1515
[48] Arrachea L 2007 *Phys. Rev. B* **75** 035319
[49] Bonart J and Cugliandolo L F 2012 *Phys. Rev. A* **86** 023636

Progress in Nonequilibrium Green's Functions V (PNGF V) IOP Publishing
Journal of Physics: Conference Series **427** (2013) 012013 doi:10.1088/1742-6596/427/1/012013

Prediction of quantum interference in molecular junctions using a parabolic diagram: Understanding the origin of Fano and anti- resonances

Daijiro Nozaki[1]*, Stanislav M. Avdoshenko[1,2], Hâldun Sevinçli[1,3], Rafael Gutierrez[1], and Gianaurelio Cuniberti[1,4]

[1] Institute for Materials Science and Max Bergmann Center of Biomaterials, Dresden University of Technology, 01062 Dresden, Germany
[2] School of Materials Engineering, Purdue University, West Lafayette, IN 47907-2045, USA
[3] Department of Micro- and Nanotechnology, Technical University of Denmark, Kongens Lyngby, Denmark
[4] Division of IT Convergence Engineering and National Center for Nanomaterials Technology, POSTECH, Pohang 790-784, Republic of Korea

E-mail: `daijiro.nozaki@gmail.com`

Abstract. Recently the interest in quantum interference (QI) phenomena in molecular devices (molecular junctions) has been growing due to the unique features observed in the transmission spectra. In order to design single molecular devices exploiting QI effects as desired, it is necessary to provide simple rules for predicting the appearance of QI effects such as anti-resonances or Fano line shapes and for controlling them. In this study, we derive a transmission function of a generic molecular junction with a side group (T-shaped molecular junction) using a minimal toy model. We developed a simple method to predict the appearance of quantum interference, Fano resonances or anti- resonances, and its position in the conductance spectrum by introducing a simple graphical representation (parabolic model). Using it we can easily visualize the relation between the key electronic parameters and the positions of normal resonant peaks and anti-resonant peaks induced by quantum interference in the conductance spectrum. We also demonstrate Fano and anti-resonance in T-shaped molecular junctions using a simple tight-binding model. This parabolic model enables one to infer on-site energies of T-shaped molecules and the coupling between side group and main conduction channel from transmission spectra.

1. Introduction

Quantum interference (QI) effects, which are interference phenomena between waves observed in nanostructures, in *electron transport* systems have been broadly studied in the field of mesoscopic physics and quantum dots [1, 2, 3] and their appearance in molecular junctions had been proposed in the early 1990s from studies of electron transfer [4, 5, 6]. Recently QI effects in electron transport through molecular wires have increasingly gained interest due to the unique spectral features observed in conductance spectra [7, 8, 9, 10, 11, 12, 13, 14, 15]. This is not only of great interest from a fundamental point of view, but also from a practical one. Since QI effect can dramatically change the electron transmission at low energies (around the Fermi energy), it can be used for many applications such as molecular switches [14], sensors, and interferometers, and thermoelectric devices [16]. For instance, QI spectral features emerging around the Fermi energy may lead to dramatic changes in the Seebeck coefficient of a molecular junction and thus make it a relevant candidate for molecular scale thermoelectric applications [16]. Therefore, in order to tune QI related spectral features, it is crucial to be able to develop a basic understanding of the conditions under which QI effects appear as well as their dependence on the molecular electronic structure. Although this can be achieved via time-consuming first-principle calculations in each specific case, it is desirable (also for the experimentalists working in the field of nanoscale electronics) to provide simple yet accurate rules able to predict the emergence and type of QI based structures in molecular junctions.

In this report, we are able to give a rather simple answer to some fundamental questions: Why does quantum interference appear in certain molecules? Which particular features have T-shaped molecular junctions, which manifest in the existence of both Fano-type and antiresonance line shapes? What is the relationship between QI effects and the low energy electronic structure of the molecular systems [17]? To address these problems, we deal with a toy model of a T-shaped molecular junction and introduce a simple yet accurate graphical representation (which we called the parabolic diagram for the sake of simplicity) that allows for a clear visualization of the relationship between different key electronic structure parameters of the molecule and the existence of different QI related features in the conductance spectrum (Fano resonances and anti-resonances). Using this simple approach, we can predict which type of QI effect can be expected in T-shaped junctions according to the specific electronic structure of the molecular system.

This report is organized as follows. In the next section, we derive the transmission function of generic molecular junctions with a side functional group using a toy model, and introduce a parabolic model that we developed. In section 3, we test the parabolic model and demonstrate two types of QI-related line shapes (anti-resonance and Fano resonance) in a simple two site approximation. In section 4, we demonstrate Fano and anti-resonance in T-shaped molecular junctions using the Hückel model. We conclude this report in section 5.

2. Parabolic model

In this section, we derive an expression for the transmission functions of T-junctions made from molecular wires with side functional groups. The derivation is also given in [16, 8] in a similar way. Then we introduce a parabolic diagram to predict the position of QI-effect in the transmission spectra. This parabolic model enables one to visualize the relationship between electronic parameters and the positions of negative and positive peaks in the transmission spectra.

For simplicity, we consider a T-shaped molecule consisting of two sites coupled between two contacts as shown in Fig. 1. The molecule is attached to left and right contacts only through the central site, i.e. the side group do not directly couple to the contacts. Hamiltonian matrix and overlap matrix for the molecule are given by:

$$H_M = \begin{bmatrix} \varepsilon_0 & V \\ V & \varepsilon_1 \end{bmatrix}, S_M = \begin{bmatrix} 1 & S \\ S & 1 \end{bmatrix}. \tag{1}$$

The on-site energies for the central site and the side group are given by ε_0 and ε_1, respectively. Those are coupled by V. The retarded Green's function $G^r(E)$ for this system in orthogonal basis is given by

$$[G^r(E)] \begin{bmatrix} E + i\delta - \varepsilon_0 - \Sigma_{\rm L}(E) - \Sigma_{\rm R}(E) & -V \\ -V & E + i\delta - \varepsilon_1 \end{bmatrix} = \begin{bmatrix} 1 & 0 \\ 0 & 1 \end{bmatrix}, \tag{2}$$

where $\Sigma_{\rm L/R}(E) = \eta(E) + i\gamma(E) = {\rm Re}[\Sigma_{\rm L/R}(E)] + i{\rm Im}[\Sigma_{\rm L/R}(E)]$ are self-energy terms for left/right contacts and $i\delta$ is infinitesimal imaginary value. Equation (2) can be written in scalar value as:

$$G^r(E) = (E + i\delta) - \varepsilon_0 - \Sigma_{\mathrm{L}}(E) - \Sigma_{\mathrm{R}}(E) - \frac{V^2}{E - \varepsilon_1} \tag{3}$$

Using the Fisher-Lee relation [18], the transmission function $T(E)$ is given by

$$T(E) \equiv \mathrm{Tr}[\Gamma_{\mathrm{L}}(E)G^r(E)\Gamma_{\mathrm{R}}(E)G^a(E)] = \frac{4\gamma(E)^2}{(E - \varepsilon_0 - 2\eta(E) - \frac{V^2}{E-\varepsilon_1})^2 + 4\gamma(E)^2}, \tag{4}$$

where $\Gamma_{\mathrm{L/R}}(E)$ are the broadening functions defined as $\Gamma_{\mathrm{L/R}}(E) \equiv i[\Sigma_{\mathrm{L/R}} - \Sigma^{\dagger}_{\mathrm{L/R}}]$. If we apply the wide band limit (WBL), $\Sigma_{\mathrm{L/R}}(E) = -i\gamma$, with no energy dependence to $\gamma(E)$. The transmission function is given by

$$T(E) = \frac{4\gamma^2}{(E - \varepsilon_0 - \frac{V^2}{E-\varepsilon_1})^2 + 4\gamma^2}. \tag{5}$$

This transmission function $T(E)$ has Breit-Wigner-type resonances [19] showing the maxima ($T(E) = 1$) with broadening γ at the energies where the term in the denominator in eq. (5) satisfies $(E - \varepsilon_0 - \frac{V^2}{E-\varepsilon_1}) = 0$. These peaks correspond to eigen-states of the molecular system and two solutions of $(E - \varepsilon_0 - \frac{V^2}{E-\varepsilon_1}) = 0$ are given by

$$\varepsilon_{\mathrm{VB,CB}} = \frac{\varepsilon_0 + \varepsilon_1 \pm \sqrt{(\varepsilon_0 - \varepsilon_1)^2 + 4V^2}}{2}. \tag{6}$$

Interestingly, $T(E)$ shows a QI-induced negative peak (antiresonance) with a minimum ($T(E) = 0$) at $E = \varepsilon_1$ since the denominator in eq. (4) diverges at $E = \varepsilon_1$.

Here we introduce a parabolic diagram to visualize the relationship between electronic parameters (ε_0, ε_1, V) and the energetic positions of Breit-Wigner-type resonances ($E = \varepsilon_{\mathrm{VB}}, \varepsilon_{\mathrm{CB}}$) and the QI-induced anti-resonance ($E = \varepsilon_0$) in the transmission spectra [17]. The position of the two Breit-Wigner-type resonances can be easily estimated by depicting the parabolic curve, $y = (E - \varepsilon_1)(E - \varepsilon_0)$, as shown in Fig. 1(b). The two intersection points between the line $y = V^2$ and the parabolic curve are the energetic positions of Breit-Wigner-type resonances ($E = \varepsilon_{\mathrm{VB,CB}}$) since eq. (5) at $E = \varepsilon_{\mathrm{VB,CB}}$ gives $T(E) = 1$. As for the energetic position of the antiresonance in the transmission spectra, it is obvious that $E = \varepsilon_1$ gives a negative peak ($T(E) = 0$) from eq. (5) because of the divergence of denominator at $E = \varepsilon_1$.

This is how we can easily visualize the position of Breit-Wigner-type positive peaks and the QI-induced negative peak in transmission spectra from the parabolic model in Fig. 1(b). Figure 1(c) shows the transmission spectra of the T-shaped molecular junction in Fig. 1(a) from eq. (5) using the same parameters in Fig. 1(b); $\varepsilon_0 = 0.0$ eV, $\varepsilon_1 = 3.0$ eV, $V^2 = 2.0$ eV^2, and $\gamma = 0.01$. We can see that the position of the negative and positive peaks completely matches with the features of the parabolic diagram. The merit of the parabolic model is that if we observe the Breit-Wigner-type resonances and the anti-resonance in conductance spectra by measurement, it is possible to determine the energies of the conduction channels and the localized states at the side group and their coupling from the parabolic diagram. Even in the non-WBL approximation, this parabolic model is valid as long as the energy-dependence of $\eta(E)$ in (4) is small, since the $\eta(E)$ in (4) just shifts the position of ε_0 in the quadratic equation $y = (E - \varepsilon_0)(E - \varepsilon_1)$ to ε_0', where $\varepsilon_0' = \varepsilon_0 + \eta(E)$. Although the inclusion of the $\eta(E)$ shifts the one of Breit-Wigner resonances at $E = \varepsilon_{\mathrm{VB}}$, this does not affect the energetic position of QI-induced anti-resonance at $E = \varepsilon_1$.

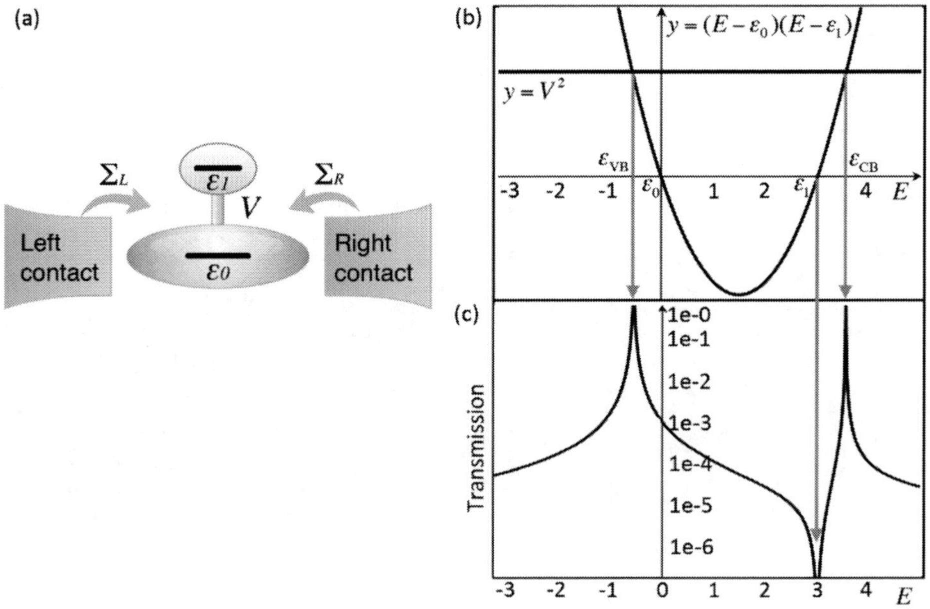

Figure 1: (a) Schematic image of a molecular junction with a side group considered in this work. A molecule consisting of two fragments is coupled between contacts. The energies of the two fragment molecular orbitals (MOs) are ε_0 and ε_1. The transfer integral between them is give by V. $\Sigma_{L/R}$ represents the self-energies due to coupling to the contacts. (b) Parabolic diagram describing the relation between electronic parameters in (a). (c) Transmission spectrum through the molecular device depicted schematically on panel (a). The quantum interference appears as a abrupt negative peak in transmission spectrum and its position can be predicted from the parabolic model in panel (b). Parameters in (b) and (c) are set to $\varepsilon_0 = 0.0$ eV, $\varepsilon_1 = 3.0$ eV, $V^2 = 2.0$ eV^2, and $\gamma = 0.01$ eV.

3. Demonstration of Anti- and Fano resonances from the parabolic model

Our parabolic model is general and thus can be applied to any systems having T-shaped topology as shown in eq. (1) or Fig. 1(a). The results obtained in the simple model were further validated using first-principle quantum transport methods in realistic T-shaped molecular junctions. The expansion to the realistic molecular junctions including more than 2 sites are shown elsewhere (see Ref. [17]). In this section we demonstrate two types of interference patterns, anti- and Fano resonances, in transmission spectra using the parabolic model. The condition necessary to realize Fano resonance is to have large $|\varepsilon_1 - \varepsilon_0|/|V|$, while the condition necessary to realize anti-resonance is to have small $|\varepsilon_1 - \varepsilon_0|/|V|$. Figure 2 presents the transmission functions of the T-shaped molecular junction shown in Fig. 1(a) with difference coupling strengths, V. When the coupling V is reduced, one of the Breit-Wigner resonances and the QI-induced anti-resonance come close to each other forming a Fano resonance. This behavior can be easily understood from the parabolic diagram in Fig. 2(a).

Next, we examine how the energy difference between two on-site energies, $\varepsilon_1 - \varepsilon_0$, affects the quantum interference in transmission spectra. Figure 3 presents the transmission functions of the T-shaped molecular junctions in Fig. 1(a) with different on-site energies, ε_1. It is notable that when the sign of the term, $\varepsilon_0 - \varepsilon_1$, is changed from negative to positive, the phase of the Fano resonances is flipped for π. This feature enables one to know whether the energy of the state from side group, ε_1, is higher than the conducting state, ε_0. These trends can be also interpreted from the parabolic model as shown in Fig. 3(a).

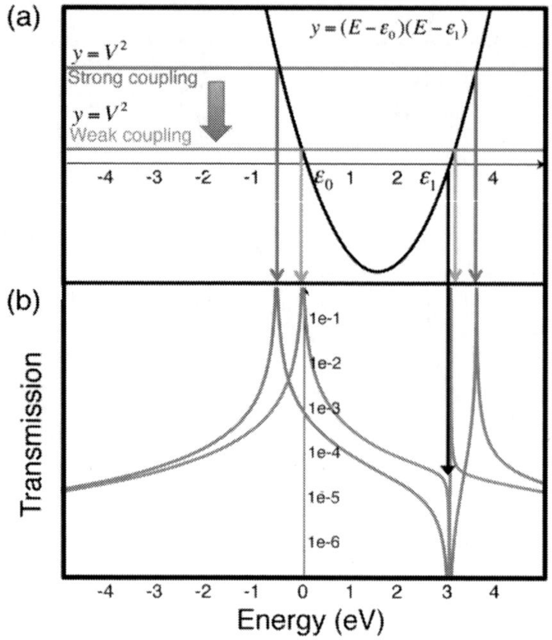

Figure 2: Coupling dependence of transmission functions in T-shaped molecular junctions in Fig. 1(a). Electronic parameters are set to $\varepsilon_0 = 0.0$ eV, $\varepsilon_1 = 3.0$ eV, $V^2 = 2.0$ eV^2 (strong coupling), $V^2 = 0.3$ eV^2 (weak coupling), and $\gamma = 0.01$ eV. (a) Parabolic diagram and (b) transmission spectrum through the molecular device depicted in Fig. 1(a). The quantum interference patterns gradually shift from anti-resonance to Fano resonance as the coupling V is reduced.

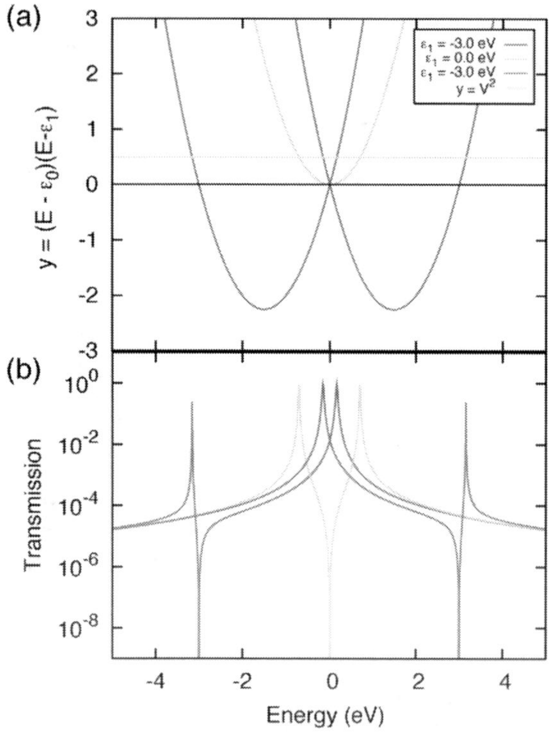

Figure 3: On-site energy dependence of transmission functions in T-shaped molecular junctions in Fig. 1(a). Electronic parameters are set to $\varepsilon_0 = 0.0$ eV, $V^2 = 0.5$ eV^2, and $\gamma = 0.01$ eV. The parameters for ε_1 are shown in the inset of (a). (a) Parabolic diagram and (b) transmission spectra through the molecular device depicted in Fig. 1(a) with different on-site energies for the side groups, ε_1. Depending on the sign of the term, $\varepsilon_0 - \varepsilon_1$, the phase of the Fano resonances is flipped for π.

Finally we investigate the density of states (DOS) of the T-shaped molecular devices. Fig. 4 shows the total DOS (TDOS) and projected DOS (PDOS) of the molecular system shown in Fig.1(a). The TDOS and two PDOS at $E = 0.0$ eV presents a normal Lorenzian-type resonance, while the PDOS plot of the conducting channel originating from the ε_0 state shows Fano resonance at $E = -3.0$ eV. We can see that the PDOS from the side group contributes to

Journal of Physics: Conference Series **427** (2013) 012013 doi:10.1088/1742-6596/427/1/012013

almost all of the TDOS of the resonant peak at $E = -3.0$ eV. Therefore the state ε_1, which is weakly coupled to main conduction channel ε_0, is localized at the side group and can disturb the electron transport through the main conducting state ε_0 yielding a Fano resonance in transmission spectra. This behavior shows a good agreement with the well-known interpretation of the Fano resonance where a localized state interferes with the continuous state.

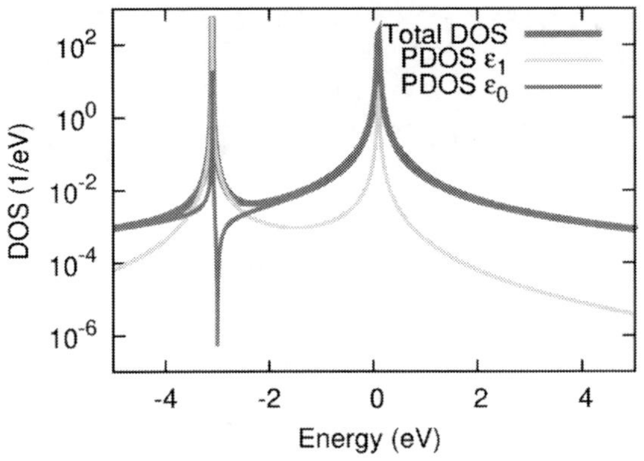

Figure 4: Density of states (DOS) in T-shaped molecular junctions in Fig. 1(a). Electronic parameters are set to $\varepsilon_0 = 0.0$ eV, $\varepsilon_1 = -3.0$ eV, $V^2 = 0.5$ eV^2, and $\gamma = 0.01$ eV. The peak at $E = -3.0$ eV in the TDOS plot originates from the localized state of the side group. Note that the projected DOS (PDOS) from the site ε_1 at $E = -3.0$ eV is much larger than the PDOS from the main conducting channel, ε_0. Because of the localized state from the side group, the conducting state in gray curve is subject to interference showing the Fano resonance at $E = -3.0$ eV.

4. Application of a parabolic model to molecular systems

Here we aim to demonstrate Fano and anti- resonances in the π-electron systems using a simple Hückel model. We modeled two T-shaped molecules named as model A and model B shown in Figs. 5 and Fig. 6 to realize Fano and anti- resonance, respectively. The condition to realize a Fano resonance in the transmission spectra is that the energy difference between the eigen energy of side fragment (ε_1) and that of main chain (ε_0) is large compared with the coupling V. This can be restated as the condition to realize a Fano resonance is that molecular junctions have localized local DOS (LDOS) at the side groups and those energies are separated enough from the energies of conducting states. On the other hand, if the energy difference $|\varepsilon_1 - \varepsilon_0|$) is small compared with the coupling V, molecular junctions will show an anti-resonance. Especially when an eigen energy of a side fragment is identical to that of the main chain, an anti-resonance will appear as we will see later.

In model A, pyrene is covalently attached to the main chain as a side group. The side group has HOMO and LUMO at $E = \pm 0.402\beta$, while the main chain has HOMO and LOMO at $E = \pm 0.215\beta$. The eigen energies of the side group are energetically separated from the ones of the main chain. The model A has large LDOS localized at the side group around $E = \pm 0.402\beta$ as shown in Fig. 5(d). Thus, model A satisfies the conditions for Fano resonances. Indeed, Fano resonances can be seen at $E = \pm 0.402\beta$ in Fig. 5(c). When the transfer integral V between the side group and the main chain is reduced, the Fano resonance gets sharper (see Fig. 7(a)).

In model B, the side group has almost same structure with the main chain. Thus, the eigen-energies of the side fragment are almost identical with that of the main chain. Thus, the resonant peaks of the main chain (pristine chain in 6(c)) in the transmission function split into two resonances after the attachment of the side group, showing an anti-resonance in the middle

Journal of Physics: Conference Series **427** (2013) 012013 doi:10.1088/1742-6596/427/1/012013

of them. In these levels, molecular orbitals delocalize because of the orbital hybridization. This delocalization can be seen in the LDOS plots in Fig. 6(d). When the coupling with the side group is enhanced the resonant levels split further (see Fig. 7(b)). These features show a good agreement with parabolic analysis discussed in the previous section (see Figs. 2 and 3).

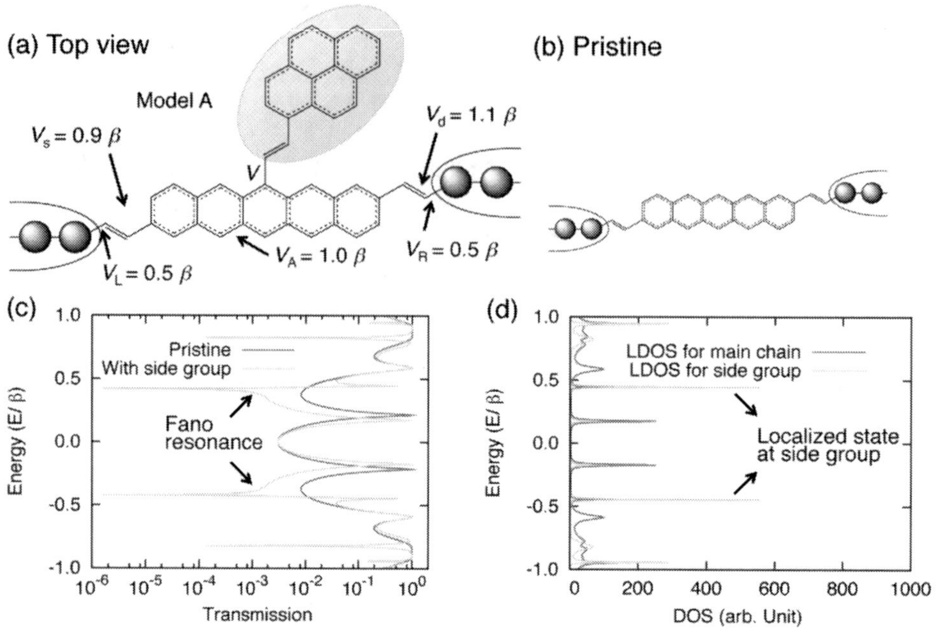

Figure 5: Demonstration of Fano resonance in a T-shaped molecular junction using the Hückel model: (a) Tight-binding scheme for a molecular junction (model A) with a side group using the Hückel model. Coupling between the molecule and left (right) 1D electrode is set to $V_{L/R} = 0.5\beta$. The transfer integrals for aromatic carbons, single bond, and double bonds are set to $V_A = 1.0\beta$, $V_s = 0.9\beta$, and $V_d = 1.1\beta$, respectively. The transfer integral V between the side chain and the main chain is set to $V = 0.9\beta$. All transfer integrals are normalized in β. Transfer integrals in the electrodes are set to β. All on-site energies are set to 0. Side group is highlighted in light blue. (b) Pristine molecular wire connected between two electrode for the comparison with the T-shaped molecular wires. (c) Transmission spectra through the molecular device depicted schematically on panel (a) and (b). Since the side fragment has eigenstates at $E = \pm 0.402\beta$ and these states localize at side group (see LDOS plot in (d)), Fano resonances appear at these energies. (d) local DOS plots for the side group and the main chain.

5. Conclusion

In order to understand the origin of quantum interference phenomena in electron transport systems with single side groups we have developed a simple but accurate toy model to predict the appearance and the positions of the quantum interference in transmission spectra. Introducing a parabolic diagram, the relationship between key electronic parameters and the line shape of the transmission spectra can be visualized. We have demonstrated two types of quantum interference phenomena namely, Fano and anti- resonances, in transmission spectra using the parabolic model. If observable peaks and dips of the conductance can be resolved experimentally, the conditions for quantum interference effects to appear can be visualized, thus the parabolic model potentially allows for tailoring the molecular system presenting Fano- or anti-resonance as desired.

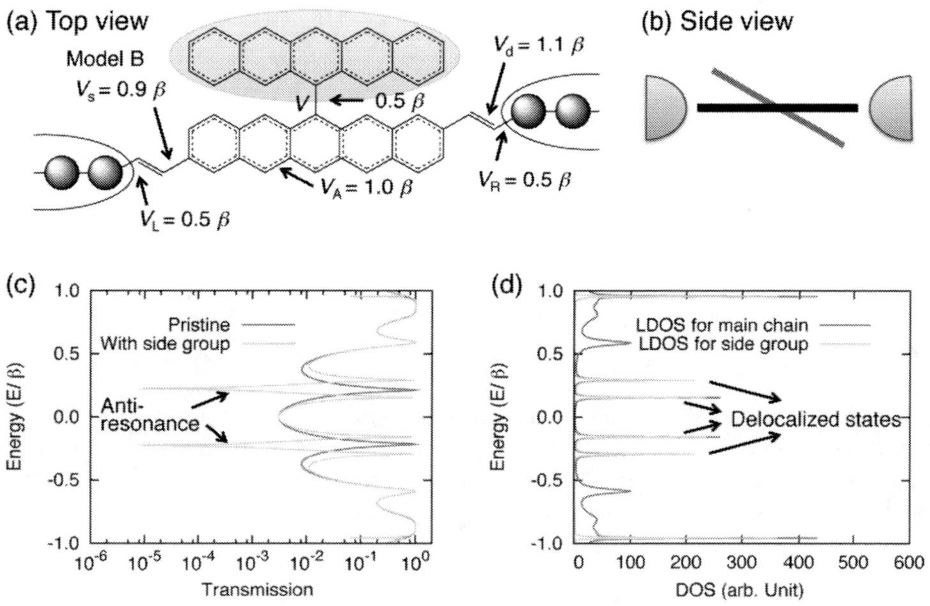

Figure 6: Demonstration of anti-resonance in a T-shaped molecular junction using the Hückel model: (a) Tight-binding scheme for a molecular junction (model B) with a side group using the Hückel model. Coupling between the molecule and left (right) 1D electrode is set to $V_{L/R} = 0.5\beta$. The transfer integrals for aromatic carbons, single bond, and double bonds are set to $V_A = 1.0\beta$, $V_s = 0.9\beta$, and $V_d = 1.1\beta$, respectively. The transfer integral V between the side chain and the main chain is reduced to 0.5β because of the torsion which could be caused by the steric repulsion as shown in panel (b). All transfer integrals are normalized in β. Transfer integrals in the electrodes are set to β. All on-site energies are set to 0. Side group is highlighted in light blue. (b) Side view of molecular junction. Because of the steric repulsion between the main chain and the side group, the side chain would be twisted. (c) Transmission spectra through the molecular device depicted schematically on panel (a) with pristine one. Since the eigen energies of the side fragment are almost identical with the ones of the main chain the resonant peaks of the main chain split into two resonance showing anti-resonances in the middle of resonant peaks. (d) local DOS plots for the side group and the main chain.

Acknowledgements

We gratefully acknowledge support from the German Excellence Initiative via the Cluster of Excellence EXC 1056 "Center for Advancing Electronics Dresden" (cfAED). We also gratefully acknowledge for funding by the European Union (ERDF) and the Free State of Saxony via TP A2 (MolFunc) of the cluster of excellence European Center for Emerging Materials and Processes Dresden (ECEMP), by the European Union (ERDF) and the Free State of Saxony via the ESF project 080942409 InnovaSens, and by World Class University program funded by the Ministry of Education, Science and Technology through the National Research Foundation of Korea (R31-10100).

References

[1] Miroshnichenko A E, Flach S and Kivshar Y S 2010 *Rev. Mod. Phys.* **82** 2257
[2] Kang K, Cho S Y, Kim J J and Shin S C 2001 *Phys. Rev. B* **63** 113304
[3] Ladrón de Guevara M L, Claro F and Orellana P A 2003 *Phys. Rev. B* **67** 195335
[4] Ratner M A 1990 *J. Phys. Chem.* **94** 4877
[5] D'Amato J L, Pastawski H M and Weisz J F 1989 *Phys. Rev. B* **39** 3544

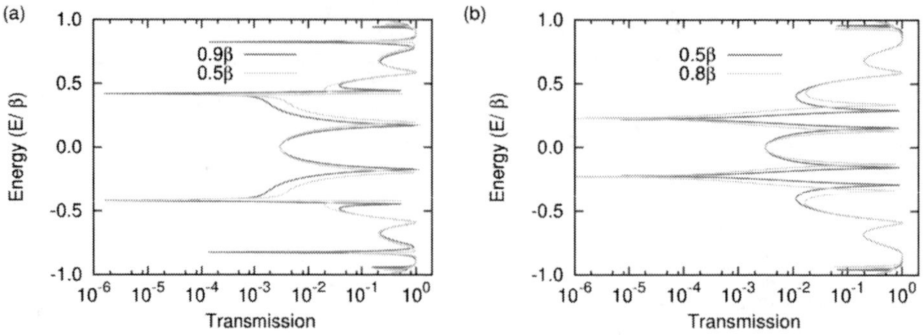

Figure 7: Dependence of line shape of transmission functions on the coupling strength between side groups and main chains in two model systems. (a) In model A, the reduction of the coupling V sharpens the Fano resonances. (b) In model B, the increase of the coupling strength V enhances the splitting of the two resonance peaks across the anti-resonance.

[6] Levstein P R, Pastawski H M and D'Amato J L 1990 *J. Phys.: Cond. Matt. 2* 1781
[7] Emberly E G and Kirczenow G 1999 *J. Phys.: Cond. Matt.* **11** 6911
[8] Papadopoulos T A, Grace I M and Lambert C J 2006 *Phys. Rev. B* **74** 193306
[9] Stafford C A, Cardamone D M and Mazumdar S 2007 *Nanotechnology* **18** 424014
[10] Solomon G C, Andrews D Q, Hansen T, Goldsmith R H, Wasielewski M R, Van Duyne R P and Ratner M A 2008 *J. Chem. Phys.* **129** 054701
[11] Solomon G C, Andrews D Q, Van Duyne R P and Ratner M A 2009 *ChemPhysChem* **10** 257
[12] Hansen T, Solomon G C, Andrews D Q and Ratner M A 2009 *J. Chem. Phys.* **131** 194704
[13] Guédon C M, Valkenier H, Markussen T, Thygesen K S, Hummelen J C and van der Molen S J 2012 *Nat. Nanotechnol.* **7** 305
[14] Tada T and Yoshizawa K 2002 *ChemPhysChem* **3** 1035
[15] Walczak K 2004 *Euro. J. Chem.* **2** 524
[16] Stadler R and Markussen T 2011 *J. Chem. Phys.* **135** 154109
[17] Nozaki D, Sevinçli H, Avdoshenko S M, Gutierrez R and Cuniberti G 2012 *Preprint* arXiv:1203.5269v3
[18] Fisher D S and Lee P A 1981 *Phys. Rev. B* **23** 6851
[19] Breit G and Wigner E 1936 *Phys. Rev.* **49** 519

Time-dependent Landauer–Büttiker formula for transient dynamics

Riku Tuovinen[1], **Robert van Leeuwen**[1,4], **Enrico Perfetto**[2] **and Gianluca Stefanucci**[2,3,4]

[1] Department of Physics, Nanoscience Center, FIN 40014, University of Jyväskylä, Jyväskylä, Finland

[2] Dipartimento di Fisica, Università di Roma Tor Vergata, Via della Ricerca Scientifica 1, I-00133 Rome, Italy

[3] Laboratori Nazionali di Frascati, Istituto Nazionale di Fisica Nucleare, Via E. Fermi 40, 00044 Frascati, Italy

[4] European Theoretical Spectroscopy Facility (ETSF)

E-mail: `riku.m.tuovinen@jyu.fi`

Abstract. We solve analytically the Kadanoff–Baym equations for a noninteracting junction connected to an arbitrary number of noninteracting wide-band terminals. The initial equilibrium state is properly described by the addition of an imaginary track to the time contour. From the solution we obtain the time-dependent electron densities and currents within the junction. The final results are analytic expressions as a function of time, and therefore no time propagation is needed — either in transient or in steady-state regimes. We further present and discuss some applications of the obtained formulae.

1. Introduction

The Landauer–Büttiker formula [1, 2] provides an intuitive physical picture of the steady-state current flowing in a multi-terminal junction and it is simple to implement. First one calculates the steady-state current $I_{\alpha\beta}$ in terminal β carried by the scattering states originating from terminal $\alpha \neq \beta$ and populated according to the electrochemical potential μ_α. Then one sums the difference $I_{\alpha\beta} - I_{\beta\alpha}$ between the currents flowing in and out terminal β over all terminals $\alpha \neq \beta$. This gives the steady-state current I_β in terminal β.

The first microscopic derivation (based on the time-dependent Schrödinger equation) of the Landauer–Büttiker formula was given by Caroli and co-workers [3, 4]. They considered the terminals initially uncontacted and in equilibrium at different chemical potentials. Then they switched on the contacts and derived the Landauer–Büttiker formula as the long-time limit of the expectation value at time t of the current operator. We will refer to this procedure as the *partitioned approach*.

An alternative approach, more akin to the the way the experiments are carried out, was proposed by Cini about a decade later [5]. He considered the system initially contacted and in equilibrium at a unique chemical potential and then drove the system out of equilibrium by applying a bias voltage between the terminals. We will refer to this procedure as the *partion-free approach*. In both approaches one recovers the Landauer–Büttiker formula due to the loss of memory of the initial preparation [6].

The microscopic derivation of the Landauer–Büttiker formula requires the evaluation of the expectation value $I_\beta(t) = \langle\Psi(t)|\hat{I}_\beta|\Psi(t)\rangle$ where $|\Psi(t)\rangle$ is the many-body state of the system at time t and $\hat{I}_\beta$ is the current operator. Since the electrons are noninteracting this expectation value can be rewritten as the sum over all occupied one-particle states $|\psi_k(t)\rangle = \mathrm{e}^{-i\hat{h}t}|\psi_k\rangle$ of $\langle\psi_k(t)|\hat{I}_\beta|\psi_k(t)\rangle$. Here $\hat{h}$ is the Hamiltonian of the contacted and biased system whereas $|\psi_k\rangle$ are the eigenstates of the Hamiltonian $\hat{h}_0$ which describes either the non-biased uncontacted system (in the partitioned approach) or the non-biased contacted system (in the partition-free approach). For the evaluation of $\langle\psi_k(t)|\hat{I}_\beta|\psi_k(t)\rangle$ one could naively insert a complete set of eigenstates $|\phi_q\rangle$ of $\hat{h}$ and evaluate the overlaps $\langle\psi_k|\phi_q\rangle$. This procedure is, however, numerically lengthy and unstable due to the singular δ-like contribution to the overlaps. The calculation of $I_\beta(t)$ is most easily carried out using nonequilibrium Green's functions [7, 8]. This mathematical tool when applied to quantum transport in multi-terminal junctions provides a natural framework to calculate the current at all times and not only at the steady state.

In fact, there have been several attempts to generalize the Landauer–Büttiker formula to the time domain. Here we mention the work of Pastawski who derived a formula for $I_\beta(t)$ using the partitioned approach in the linear response and adiabatic regime [9]. An important step forward in the calculation of $I_\beta(t)$ was done by Jauho et al. [8]. These authors used the partitioned approach to write $I_\beta(t)$ as a double integral (over time and energy) of the trace over the junction degrees of freedom of a calculable combination of Green's functions in the same region. In the special case of terminals with a wide band and of junctions with one single level it is possible to perform the time-integral and obtain a time-dependent version of the Landauer–Büttiker formula. This formula was then derived in Ref. [6] using the partition-free approach, thus confirming the loss of memory of the initial preparation.

The derivation of a time-dependent Landauer–Büttiker formula for arbitrary junctions would be extremely useful to interpret the oscillations and damping times typically observed in the transient current after the sudden switch on of a bias. A progress in this direction was done in Ref. [10] where the authors derived a time-dependent Landauer–Büttiker formula for the spin current of a single-level junction.

In this work we generalize the results of Ref. [10] to junctions of any shape and dimensions using the wide-band limit approximation (WBLA) for noninteracting electrons (Secs. 2 and 3). Furthermore we also derive a general formula for the time-dependent one-particle density matrix which can be used to calculate the local density and current density. We will work in the partition-free approach which is conceptually easier since it does not involve the subtle issue of different chemical potentials in equilibrium. The final formulae for the current and the one-particle density matrix have the merit of elucidating the relative importance of the electronic transitions at a certain time. As an illustration we will use these formulae to calculate the transient response of a ring-shaped junction (Sec. 4).

2. Assumptions and set-up

We investigate the following quantum transport setup: An arbitrary number of metallic leads (α) acting as charge-carrier reservoirs are connected to a lattice network acting as a molecular device (C). We assume that the electron transport is ballistic and therefore neglect the electron–electron interactions. We will also assume that the energy eigenvalues of the Hamiltonian of the molecular device are well inside the continuous energy spectrum of the leads and use the WBLA.

The described set-up is characterized by the following Hamiltonian:

$$\hat{H} = \sum_{k\alpha,\sigma} \epsilon_{k\alpha}\hat{n}_{k\alpha,\sigma} + \sum_{mn,\sigma} T_{mn}\hat{d}^\dagger_{m,\sigma}\hat{d}_{n,\sigma} + \sum_{mk\alpha,\sigma}\left[T_{mk\alpha}\hat{d}^\dagger_{m,\sigma}\hat{d}_{k\alpha,\sigma} + T_{k\alpha m}\hat{d}^\dagger_{k\alpha,\sigma}\hat{d}_{m,\sigma}\right]. \quad (1)$$

The first term accounts for the leads with $k\alpha$ indexing the k:th basis function of the $\alpha = 1, 2, 3, \ldots$

Journal of Physics: Conference Series **427** (2013) 012014 doi:10.1088/1742-6596/427/1/012014

lead. The single-particle spectrum of the leads is $\epsilon_{k\alpha}$ and the number operator in the leads is expressed in terms of the creation and annihilation operators as $\hat{n}_{k\alpha,\sigma} = \hat{d}^\dagger_{k\alpha,\sigma}\hat{d}_{k\alpha,\sigma}$, with σ the spin index. The second term is for the molecular device, or central region, (indices m and n) with creation and annihilation operators $\hat{d}^\dagger_{m,\sigma}$ and $\hat{d}_{m,\sigma}$ and hoppings T_{mn} between sites m and n. The last term is for the coupling between the central region and the leads with hoppings $T_{mk\alpha}$.

At times $t < t_0$ the system is in thermal equilibrium at inverse temperature β and chemical potential μ, the density matrix having the form $\hat{\rho} = \frac{1}{\mathcal{Z}}\mathrm{e}^{-\beta(\hat{H}-\mu\hat{N})}$ where $\mathcal{Z}$ is the grand-canonical partition function. At $t = t_0$ the lead energy levels are suddenly shifted by some constant value, $\epsilon_{k\alpha} \to \epsilon_{k\alpha} + V_\alpha$, to model the sudden switch-on of an external bias voltage in the α:th lead. This means that the system is driven out of equilibrium and charge carriers start to flow through the central region. To calculate the time-dependent current we use the equations of motion for the one-particle Green's function on the Keldysh contour γ_{K}. This quantity is defined as the ensemble average of the contour-ordered product of particle creation and annihilation operators in the Heisenberg picture

$$G_{rs}(z,z') = -\mathrm{i}\langle \mathcal{T}_{\gamma_{\mathrm{K}}}[\hat{d}_{r,\mathrm{H}}(z)\hat{d}^\dagger_{s,\mathrm{H}}(z')]\rangle \tag{2}$$

where the indices r, s can be either indices in the leads or in the central region and the variables z, z' run on the contour[1]. The matrix $\boldsymbol{G}$ with matrix elements G_{rs} satisfies the equations of motion

$$\left[\mathrm{i}\frac{\mathrm{d}}{\mathrm{d}z} - \boldsymbol{h}(z)\right]\boldsymbol{G}(z,z') = \delta(z,z')\mathbf{1}\ , \tag{3}$$

$$\boldsymbol{G}(z,z')\left[-\mathrm{i}\frac{\overleftarrow{\mathrm{d}}}{\mathrm{d}z'} - \boldsymbol{h}(z')\right] = \delta(z,z')\mathbf{1}\ , \tag{4}$$

with Kubo–Martin–Schwinger (KMS) boundary conditions. Here $\boldsymbol{h}(z)$ is the single-particle Hamiltonian. In the basis $k\alpha$ and m the matrix $\boldsymbol{h}$ has the following block structure

$$\boldsymbol{h} = \begin{pmatrix} h_{11} & 0 & 0 & \cdots & h_{1C} \\ 0 & h_{22} & 0 & \cdots & h_{2C} \\ 0 & 0 & h_{33} & \cdots & h_{3C} \\ \vdots & \vdots & \vdots & \ddots & \vdots \\ h_{C1} & h_{C2} & h_{C3} & \cdots & h_{CC} \end{pmatrix}, \tag{5}$$

where $(h_{\alpha\alpha'})_{kk'} = \delta_{\alpha\alpha'}\delta_{kk'}\epsilon_{k\alpha}$ corresponds to the leads, $(h_{\alpha C})_{km} = T_{k\alpha m}$ is the coupling part, and $(h_{CC})_{mn} = T_{mn}$ accounts for the central region. As the system is initially in thermal equilibrium we have that for z on the vertical track of the contour $\epsilon_{k\alpha}(z) = \epsilon_{k\alpha} - \mu$, $T_{k\alpha m}(z) = T_{k\alpha m}$ and $T_{mn}(z) = T_{mn} - \mu\delta_{mn}$. On the other hand for z on the horizontal branches we have $\epsilon_{k\alpha}(z) = \epsilon_{k\alpha} + V_\alpha$, $T_{k\alpha m}(z) = T_{k\alpha m}$ and $T_{mn}(z) = T_{mn}$. Due to the coupling between the central region and the leads the matrix $\boldsymbol{G}$ has nonvanishing entries everywhere

$$\boldsymbol{G} = \begin{pmatrix} G_{11} & \cdots & G_{1C} \\ \vdots & \ddots & \vdots \\ G_{C1} & \cdots & G_{CC} \end{pmatrix}. \tag{6}$$

In the next Section we solve the equations of motion (3) and (4) for the Green's function G_{CC} projected in the central region.

[1] The contour has a forward and a backward branch on the real-time axis, $[t_0, \infty[$, and also a vertical branch on the imaginary axis, $[t_0, t_0 - \mathrm{i}\beta]$ with inverse temperature β, see e.g. [11].

3. Derivation of the time-dependent density and current

3.1. Projecting the equation of motion

We project the equation of motion (3) onto regions CC and αC. The equation for $G_{\alpha C}$ can be integrated using the Green's function $g_{\alpha\alpha}(z,z')$ of the isolated α:th reservoir. This Green's function solves the equation of motion $\left[\mathrm{i}\frac{\mathrm{d}}{\mathrm{d}z} - h_{\alpha\alpha}\right] g_{\alpha\alpha}(z,z') = \delta(z,z')$ as well as the adjoint equation with KMS boundary conditions. Introducing the embedding self-energy (with indices in region C)

$$\Sigma_{\mathrm{em}}(z,z') = \sum_{\alpha} \Sigma_{\alpha}(z,z') \quad ; \quad \Sigma_{\alpha}(z,z') = h_{C\alpha} g_{\alpha\alpha}(z,z') h_{\alpha C} \tag{7}$$

we obtain the equation of motion for the Green's function of the central region (the subscripts CC are omitted from now on)

$$\left[\mathrm{i}\frac{\mathrm{d}}{\mathrm{d}z} - h\right] G(z,z') = \delta(z,z') + \int_{\gamma_{\mathrm{K}}} \mathrm{d}\bar{z}\, \Sigma_{\mathrm{em}}(z,\bar{z}) G(\bar{z},z') \tag{8}$$

The adjoint equation of motion can be derived similarly and read [11]

$$G(z,z') \left[-\mathrm{i}\frac{\overleftarrow{\mathrm{d}}}{\mathrm{d}z'} - h\right] = \delta(z,z') + \int_{\gamma_{\mathrm{K}}} \mathrm{d}\bar{z}\, G(z,\bar{z}) \Sigma_{\mathrm{em}}(\bar{z},z') \ . \tag{9}$$

The embedded equations of motion for G have the same structure as the Kadanoff–Baym equations (KBE), the difference being that the many-body self-energy is replaced by the embedding self-energy. In the case of interacting electrons with an interaction only in the central region Eqs. (8) and (9) are modified by the addition of the many-body self-energy Σ to the embedding self-energy Σ_{em}, i.e., $\Sigma_{\mathrm{em}} \to \Sigma_{\mathrm{em}} + \Sigma$. Since $\Sigma = \Sigma[G]$ is a functional of the Green's function in region C the embedded equations of motion in the interacting case constitute a closed set of integro-differential, nonlinear equations for G [11]. The simplification brought by the absence of interactions is that the KBE (8) and (9) are linear in G since the embedding self-energy is completely specified by the parameters of the Hamiltonian.

The density and current density can be extracted from the lesser component of the Green's function at equal time. We denote by $z = t_-$ the contour point on the forward branch, $z = t_+$ the contour point on the backward branch and $z = t_0 - \mathrm{i}\tau$ the contour point on the vertical track. The Keldysh components lesser ($<$), greater ($>$), retarded (R), advanced (A), left ($\lceil$), right ($\rceil$) and Matsubara (M) of a function $k(z,z')$ on the contour are defined according to [12]

$$\begin{aligned}
k^{<}(t,t') &= k(t_-,t'_+) && (10)\\
k^{>}(t,t') &= k(t_+,t'_-) && (11)\\
k^{\mathrm{R}}(t,t') &= +\theta(t-t')\left[k^{>}(t,t') - k^{<}(t,t')\right] && (12)\\
k^{\mathrm{A}}(t,t') &= -\theta(t'-t)\left[k^{>}(t,t') - k^{<}(t,t')\right] && (13)\\
k^{\lceil}(\tau,t') &= k(t_0 - \mathrm{i}\tau, t') && (14)\\
k^{\rceil}(t,\tau) &= k(t, t_0 - \mathrm{i}\tau) && (15)\\
k^{\mathrm{M}}(\tau,\tau') &= k(t_0 - \mathrm{i}\tau, t_0 - \mathrm{i}\tau') && (16)
\end{aligned}$$

To generate an equation for $G^{<}$ we subtract Eq. (9) from Eq. (8) and set $z = t_-$, $z' = t'_+$. Taking into account that $h(z) = h$ is independent of z for z on the horizontal branches we get at equal time

$$\begin{aligned}
\mathrm{i}\frac{\mathrm{d}}{\mathrm{d}t} G^{<}(t,t) - \left[h, G^{<}(t,t)\right] &= \left[\Sigma^{\mathrm{R}}_{\mathrm{em}} \cdot G^{<} - G^{\mathrm{R}} \cdot \Sigma^{<}_{\mathrm{em}} + \Sigma^{<}_{\mathrm{em}} \cdot G^{\mathrm{A}} - G^{<} \cdot \Sigma^{\mathrm{A}}_{\mathrm{em}}\right](t,t)\\
&+ \left[\Sigma^{\rceil}_{\mathrm{em}} \star G^{\lceil} - G^{\rceil} \star \Sigma^{\lceil}_{\mathrm{em}}\right](t,t) \ , && (17)
\end{aligned}$$

Journal of Physics: Conference Series **427** (2013) 012014 doi:10.1088/1742-6596/427/1/012014

where we defined $[f \cdot g](t,t') = \int_{t_0}^{\infty} d\bar{t} f(t,\bar{t}) g(\bar{t},t')$ and $[f \star g](t,t') = -i\int_0^{\beta} d\tau f(t,\tau) g(\tau,t')$. Equation (17) can also be written as

$$i\frac{d}{dt}G^<(t,t) - [h, G^<(t,t)] = -\left[G^R \cdot \Sigma_{em}^< + G^< \cdot \Sigma_{em}^A + G^\rceil \star \Sigma_{em}^\lceil\right](t,t) + \text{h.c.} \tag{18}$$

where we used the properties of G and Σ_{em} under complex conjugation [12].

Let us comment Eq. (18) briefly. Setting the right-hand side to zero we see that Eq. (18) reduces to the Liouville equation for the one-particle density matrix $\rho = -iG$ of the isolated central region. Thus the embedding self-energy accounts for the openness of region C. The first term inside the square brackets is a convolution between the propagator in region C, G^R, and $\Sigma_{em}^<$. Since $\Sigma_{em}^<$ is proportional to the probability of finding an electron in the leads this term can be interpreted as a source term, i.e., a term that describes the injection of electrons into region C. The second term has the opposite structure: a propagator in the leads, Σ_{em}^A, is convoluted with $G^<$ which is propotional to the probability of finding an electron in region C. Thus this term can be interpreted as a drain term and is responsible for damping and equilibration effects. The last term inside the square brackets accounts for the initial preparation of the system. In the partioned approach this term would be zero since the hopping integrals $T_{k\alpha m} = 0$ in equilibrium. However, in the partition-free approach this term is nonzero and accounts for the initial coupling of the central region to the leads.

More generally convolutions along the vertical track carry information on the initial preparation of the system. For instance for a system of interacting electrons we can either start with a noninteracting system and then switch on the interaction in real time or we can start with a system already interacting. In the latter case the many-body self-energy is nonvanishing on the vertical track and the convolution $G \star \Sigma$ accounts for the effects of initial correlations.

3.2. Self-energy and Green's function calculations

The solution of Eq. (18) requires first to calculate the Matsubara component G^M, and then from G^M the right and left component $G^\rceil$ and $G^\lceil$. The Matsubara component G^M can be determined from the retarded/advanced components by analytic continuation, see below. Since the equations for G^M, $G^\rceil$ and $G^\lceil$ contain the embedding self-energy the preliminary step is to obtain an expression for Σ_{em}.

Having a time-independent Hamiltonian (on the horizontal branches of the contour) the retarded/advanced components of the self-energy depend only on the time difference

$$\Sigma_{\alpha,mn}^A(t,t') = \int \frac{d\omega}{2\pi} e^{-i\omega(t-t')} \sum_k T_{mk\alpha}\, g_{k\alpha}^A(\omega)\, T_{k\alpha n} \tag{19}$$

where $g_{k\alpha}$ is the diagonal element of the Green's function $g_{\alpha\alpha}$ of the isolated α:th lead, see Eq. (7). The retarded component of the self-energy is found by conjugating $\Sigma_\alpha^R(t,t') = [\Sigma_\alpha^A(t',t)]^\dagger$. According to the WBLA the Fourier transform of Σ_α^A is frequency independent

$$\Sigma_{\alpha,mn}^A(\omega) = \sum_k T_{mk\alpha}\, g_{k\alpha}^A(\omega)\, T_{k\alpha n} = \sum_k T_{mk\alpha} \frac{1}{\omega - \epsilon_{k\alpha} - V_\alpha - i\eta} T_{k\alpha n} = \frac{i}{2}\Gamma_{\alpha,mn} \tag{20}$$

which implies that Σ_α^A is also independent of the external bias voltage V_α. The time-dependent self-energy of Eq. (19) is therefore

$$\Sigma_{\alpha,mn}^A(t,t') = \int \frac{d\omega}{2\pi} e^{-i\omega(t-t')} \Sigma_{\alpha,mn}^A(\omega) = \frac{i}{2}\Gamma_{\alpha,mn}\delta(t-t') . \tag{21}$$

Journal of Physics: Conference Series **427** (2013) 012014 doi:10.1088/1742-6596/427/1/012014

Within the WBLA we can easily calculate the two other self-energy components in Eq. (18) (see Appendix A)

$$\Sigma^{\lceil}_{\alpha,mn}(\tau,t) = \Gamma_{\alpha,mn}\frac{1}{-\mathrm{i}\beta}\sum_q \mathrm{e}^{-\omega_q\tau}\int\frac{\mathrm{d}\omega}{2\pi}\frac{\mathrm{e}^{\mathrm{i}(\omega+V_\alpha)t}}{\omega_q-\omega+\mu}, \tag{22}$$

$$\Sigma^{<}_{\alpha,mn}(t,t') = \mathrm{i}\Gamma_{\alpha,mn}\int\frac{\mathrm{d}\omega}{2\pi}f(\omega-\mu)\mathrm{e}^{-\mathrm{i}(\omega+V_\alpha)(t-t')}, \tag{23}$$

where the sum over q is a sum over the Matsubara frequencies $\omega_q = \frac{(2q+1)\pi}{-\mathrm{i}\beta}$, and the function f is the Fermi function, $f(\omega) = 1/[\mathrm{e}^{\beta\omega}+1]$.

Having the explicit form of the self-energy components we can derive expressions for the Green's function. For the following calculations it is convenient to define the *effective* Hamiltonian $h_{\mathrm{eff}} = h - \frac{\mathrm{i}}{2}\Gamma \Rightarrow h^{\dagger}_{\mathrm{eff}} = h + \frac{\mathrm{i}}{2}\Gamma$, where $\Gamma = \sum_\alpha \Gamma_\alpha$. This effective Hamiltonian is therefore non-hermitean. The two Green's function components in the square brackets of Eq. (18) read (see Appendix B)

$$G^{\rceil}(t,\tau) = \mathrm{e}^{-\mathrm{i}h_{\mathrm{eff}}t}\left[G^{\mathrm{M}}(0,\tau) - \int_0^t \mathrm{d}t'\mathrm{e}^{\mathrm{i}h_{\mathrm{eff}}t'}\int_0^\beta \mathrm{d}\bar{\tau}\,\Sigma^{\rceil}_{\mathrm{em}}(t',\bar{\tau})G^{\mathrm{M}}(\bar{\tau},\tau)\right], \tag{24}$$

$$G^{\mathrm{R}}(t,t') = -\mathrm{i}\theta(t-t')\mathrm{e}^{-\mathrm{i}h_{\mathrm{eff}}(t-t')}, \tag{25}$$

with G^{M} the Matsubara Green's function. G^{M} can be obtained from G^{R} and G^{A} by analytic continuation since $G^{\mathrm{M}}(\omega_q) = G^{\mathrm{R}}(\omega_q+\mu)$ if $\mathrm{Im}[\omega_q] > 0$ and $G^{\mathrm{M}}(\omega_q) = G^{\mathrm{A}}(\omega_q+\mu)$ if $\mathrm{Im}[\omega_q] < 0$, see Appendix B.

Now we have all ingredients to calculate the convolutions in Eq. (18). We report here the final results and refer to Appendix C for details. The three terms read

$$\left[G^{\mathrm{R}}\cdot\Sigma^{<}_{\mathrm{em}}\right](t,t) = \mathrm{i}\sum_\alpha\int\frac{\mathrm{d}\omega}{2\pi}f(\omega-\mu)\left[1-\mathrm{e}^{\mathrm{i}(\omega+V_\alpha-h_{\mathrm{eff}})}\right]G^{\mathrm{R}}(\omega+V_\alpha)\Gamma_\alpha, \tag{26}$$

$$\left[G^{<}\cdot\Sigma^{\mathrm{A}}_{\mathrm{em}}\right](t,t) = \frac{\mathrm{i}}{2}G^{<}(t,t)\Gamma, \tag{27}$$

$$\left[G^{\rceil}\star\Sigma^{\lceil}_{\mathrm{em}}\right](t,t) = \mathrm{i}\int\frac{\mathrm{d}\omega}{2\pi}f(\omega-\mu)\sum_\alpha \mathrm{e}^{\mathrm{i}(\omega+V_\alpha-h_{\mathrm{eff}})t}G^{\mathrm{R}}(\omega)\Gamma_\alpha. \tag{28}$$

3.3. Solving Eq. (18) *for* $G^<(t,t)$

We insert Eqs. (26), (27) and (28) into Eq. (18) and get

$$\begin{aligned}
&\mathrm{i}\frac{\mathrm{d}}{\mathrm{d}t}G^<(t,t) - \left[h, G^<(t,t)\right] \\
&= -\left\{\mathrm{i}\sum_\alpha\int\frac{\mathrm{d}\omega}{2\pi}f(\omega-\mu)\left[1-\mathrm{e}^{\mathrm{i}(\omega+V_\alpha-h_{\mathrm{eff}})t}\right]G^{\mathrm{R}}(\omega+V_\alpha)\Gamma_\alpha + \frac{\mathrm{i}}{2}\Gamma G^<(t,t)\right. \\
&\quad\left. + \mathrm{i}\int\frac{\mathrm{d}\omega}{2\pi}f(\omega-\mu)\sum_\alpha \mathrm{e}^{\mathrm{i}(\omega+V_\alpha-h_{\mathrm{eff}})t}G^{\mathrm{R}}(\omega)\Gamma_\alpha\right\} + \mathrm{h.c.}
\end{aligned} \tag{29}$$

This is a nonhomogeneous, linear, first-order differential equation for $G^<(t,t)$ and, therefore, can be solved explicitly. The solution is worked out in Appendix D and reads

$$\begin{aligned} -\mathrm{i}G^<(t,t) &= \int\frac{\mathrm{d}\omega}{2\pi}f(\omega-\mu)\sum_\alpha\Big\{ A_\alpha(\omega+V_\alpha) \\ &+ V_\alpha\left[\mathrm{e}^{\mathrm{i}(\omega+V_\alpha-h_{\mathrm{eff}})t}G^{\mathrm{R}}(\omega)A_\alpha(\omega+V_\alpha)+\mathrm{h.c.}\right] \\ &+ V_\alpha^2\mathrm{e}^{-\mathrm{i}h_{\mathrm{eff}}t}G^{\mathrm{R}}(\omega)A_\alpha(\omega+V_\alpha)G^{\mathrm{A}}(\omega)\mathrm{e}^{\mathrm{i}h_{\mathrm{eff}}^\dagger t}\Big\}, \end{aligned} \tag{30}$$

where we introduced the partial spectral function as

$$A_\alpha(\omega)=G^{\mathrm{R}}(\omega)\Gamma_\alpha G^{\mathrm{A}}(\omega)\,. \tag{31}$$

The full nonequilibrium spectral function is $A(\omega)=\sum_\alpha A_\alpha(\omega)$.

Given the original complexity of the problem the final result is surprisingly compact. Equation (30) is an explicit closed formula for the equal-time $G^<$ or, equivalently, for the reduced one-particle density matrix. All the quantities inside the integral can be calculated separately, and no time-propagation nor self-consistency algorithms are needed. Also, we may extract several physical properties:

(i) With no external bias, $V_\alpha=0$, only the first row contributes. This term correctly gives the equilibrium value of the equal-time $G^<$ since at zero bias $\sum_\alpha A_\alpha(\omega)$ is the equilibrium spectral function.

(ii) Both the second and the third row vanish exponentially in the long-time limit, and the equal-time $G^<$ approaches a unique steady-state value.

(iii) The transient dynamics is given by the second and the third row. By inserting a complete set of eigenstates of the effective Hamiltonian h_{eff} we notice that:
   - (a) The second row gives rise to oscillations with frequency $\omega_j=|\mu+V_\alpha-\epsilon_j^{\mathrm{eff}}|$ where $\epsilon_j^{\mathrm{eff}}$ is the real part of the j:th complex eigenvalue of h_{eff}. These oscillations correspond to transitions between the biased Fermi level of the leads and the resonant levels of the central molecule.
   - (b) The third term accounts for intramolecular transitions and leads to oscillations with frequency $\omega_{jk}=|\epsilon_j^{\mathrm{eff}}-\epsilon_k^{\mathrm{eff}}|$. These oscillations are visible only if the effective Hamiltonian h_{eff} does *not* commute with Γ_α. In the case that $[h_{\mathrm{eff}},\Gamma_\alpha]=0$ the time dependence of the third term is of the form $\mathrm{e}^{-\mathrm{i}h_{\mathrm{eff}}t+\mathrm{i}h_{\mathrm{eff}}^\dagger t}=\mathrm{e}^{-\Gamma t}$.

3.4. Current calculation

The time-dependent current through the interface between the central region and the α:th reservoir is calculated from the following equation [11]

$$I_\alpha(t)=4q\,\mathrm{Re}\left\{\mathrm{Tr}\left[\Sigma_\alpha^<\cdot G^{\mathrm{A}}+\Sigma_\alpha^{\mathrm{R}}\cdot G^<+\Sigma_\alpha^{\rceil}\star G^{\lceil}\right](t,t)\right\}\,. \tag{32}$$

For the terms inside Eq. (32) we proceed in the same manner as we did previously to obtain the results in Eqs. (26), (27) and (28):

$$\left[\Sigma_\alpha^<\cdot G^{\mathrm{A}}\right](t,t) = \mathrm{i}\int\frac{\mathrm{d}\omega}{2\pi}f(\omega-\mu)\Gamma_\alpha G^{\mathrm{A}}(\omega+V_\alpha)\left[1-\mathrm{e}^{-\mathrm{i}(\omega+V_\alpha-h_{\mathrm{eff}}^\dagger)t}\right], \tag{33}$$

$$\left[\Sigma_\alpha^{\mathrm{R}}\cdot G^<\right](t,t) = -\frac{\mathrm{i}}{2}\Gamma_\alpha G^<(t,t)\,, \tag{34}$$

$$\left[\Sigma_\alpha^{\rceil}\star G^{\lceil}\right](t,t) = \mathrm{i}\int\frac{\mathrm{d}\omega}{2\pi}f(\omega-\mu)\Gamma_\alpha G^{\mathrm{A}}(\omega)\mathrm{e}^{-\mathrm{i}(\omega+V_\alpha-h_{\mathrm{eff}}^\dagger)t}\,. \tag{35}$$

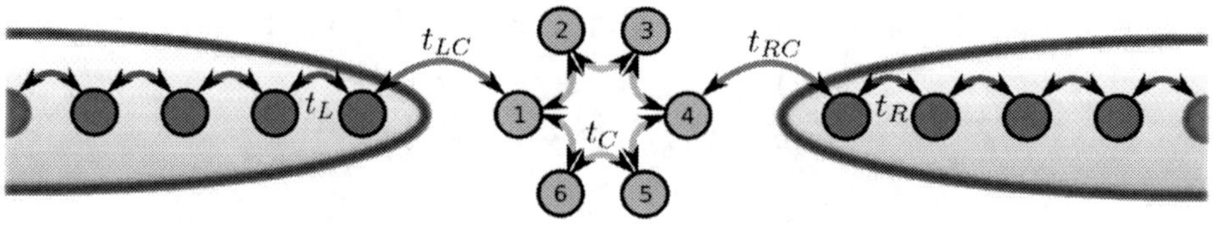

Figure 1: Six-site ring coupled symmetrically to one-dimensional TB leads.

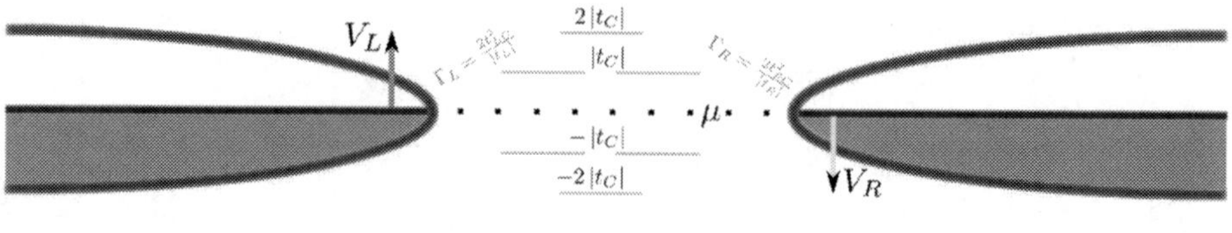

Figure 2: Schematic of the transport setup through the energy levels of the molecule.

Inserting these results into Eq. (32) and taking into account the explicit expression for $G^<(t,t)$ in Eq. (30) we get

$$\begin{aligned} I_\alpha(t) &= -2\int\frac{d\omega}{2\pi}f(\omega-\mu)\sum_\beta \mathrm{Tr}\Big\{ \\ & \Gamma_\alpha G^{\mathrm{R}}(\omega+V_\beta)\Gamma_\beta G^{\mathrm{A}}(\omega+V_\beta) - \Gamma_\alpha G^{\mathrm{R}}(\omega+V_\alpha)\Gamma_\beta G^{\mathrm{A}}(\omega+V_\alpha) \\ &+ V_\beta\left[\Gamma_\alpha e^{\mathrm{i}(\omega+V_\beta-h_{\mathrm{eff}})t}G^{\mathrm{R}}(\omega)\left(-\mathrm{i}\delta_{\alpha\beta}G^{\mathrm{R}}(\omega+V_\beta)+A_\beta(\omega+V_\beta)\right)+\mathrm{h.c.}\right] \\ &+ V_\beta^2\Gamma_\alpha e^{-\mathrm{i}h_{\mathrm{eff}}t}G^{\mathrm{R}}(\omega)A_\beta(\omega+V_\beta)G^{\mathrm{A}}(\omega)e^{\mathrm{i}h^\dagger_{\mathrm{eff}}t}\Big\}. \end{aligned} \tag{36}$$

The physical interpretation of the terms in Eq. (36) is similar to the one after Eq. (30). We have a steady-state part given by the first row, and a time-dependent part given by the second and the third rows. The time-dependent part vanishes exponentially in the long-time limit and the oscillations in the current have the same structure as in the reduced one-particle density matrix.

4. Results

Let us consider a six-site tight-binding ring connected to two tight-binding, semi-infinite, one-dimensional leads as shown schematically in Figs. 1 and 2. The parameters according to the notation in the figures are $t_C = -2.0$ (hopping in the molecule), chemical potential $\mu = 0$ and zero temperature ($\beta \to \infty$). We choose the hopping $t_L = t_R$ in the left/right lead to be much larger than any other energy scale. Then $\Sigma^{\mathrm{A}}_\alpha(\omega\sim\mu) = \mathrm{i}t^2_{\alpha C}/t_\alpha + O(1/t^2_\alpha)$ where $t_{\alpha C}$ is the hopping between the molecule and the leads. For this situation the WBLA with $\Gamma_\alpha = 2t^2_{\alpha C}/t_\alpha$ is a very good approximation. We study the weak coupling case $\Gamma_{L,R} = 0.1$ and drive the system out of equilibrium by the sudden switch-on of a bias $V_L = V = -V_R$. We analyze the contribution of different terms in the charge current corresponding to different physical features as discussed below Eq. (30). In Eq. (36) the first row is 'steady state', the second row consists of '1st term, a' and '1st term, b' and the third row is '2nd term'. The second row is divided into two parts [$\sim G^{\mathrm{R}}(\omega)G^{\mathrm{R}}(\omega+V_\beta)$ and $\sim G^{\mathrm{R}}(\omega)A_\beta(\omega+V_\beta)$] since they give rise to different features.

In Figs. 3 and 4 we plot the current through the right interface and see that weakly biased leads, $V = 0.5$, give a negligible steady-state current. Transitions between the biased leads and the molecule are captured by the '1st term, a'. This is confirmed by the peak in the Fourier spectrum at $\omega_j = \epsilon^{\mathrm{eff}}_j \pm V$. Transitions between the molecular levels are accounted for by the

Journal of Physics: Conference Series **427** (2013) 012014 doi:10.1088/1742-6596/427/1/012014

'2nd term', as it can be seen in the Fourier transform with a peak at $\omega = 6$. In addition to our previous observations: (1) the '1st term, b' also gives rise to intramolecular transitions and (2) there seems to be no intramolecular transitions at $\omega = 2$ or $\omega = 4$. By expanding Eq. (36) in the eigenbasis of the effective Hamiltonian and manipulating the terms further one can show that the the '1st term, b' contains a term of the form $\mathrm{e}^{-\mathrm{i}(\epsilon_j^{\mathrm{eff}}-\epsilon_k^{\mathrm{eff}})t}$ which explains the first finding. The second finding suggests that there is some underlying selection rule for some of the energy levels and hence that some levels do not participate to the transport process.

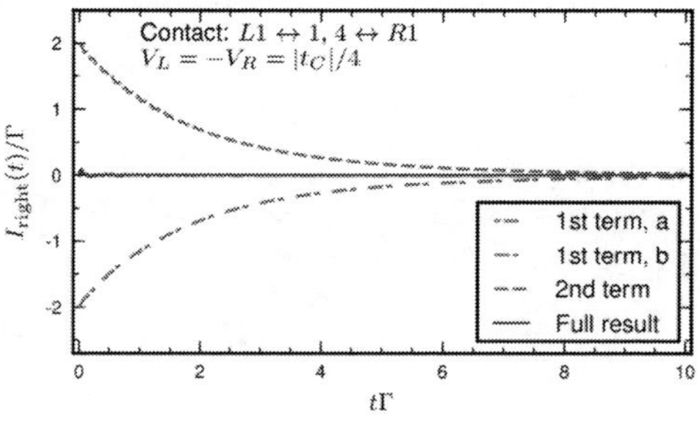

Figure 3: Different terms of the time-dependent current in units of Γ through the right interface with symmetric coupling and weak bias.

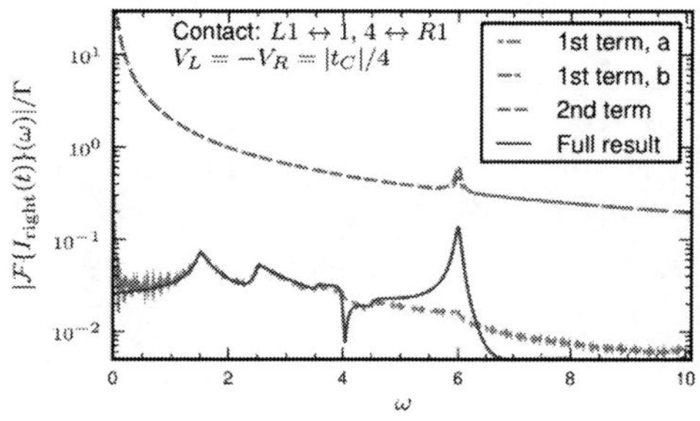

Figure 4: Absolute value of the Fourier transform of the terms of the current in units of Γ.

If we increase the bias window to cover the first molecular levels, $V = 2.5$, then we see in Fig. 5 that the current has a non-zero steady-state value. Similar findings, as with weaker bias, for the possible transitions are seen in Fig. 6. We also see that there is a small bump at $\omega = 4$ in '1st term, b' and '2nd term'. Given that the setup is completely identical to the previous case, this fact is due to a second (or higher) order response since the same symmetry arguments apply.

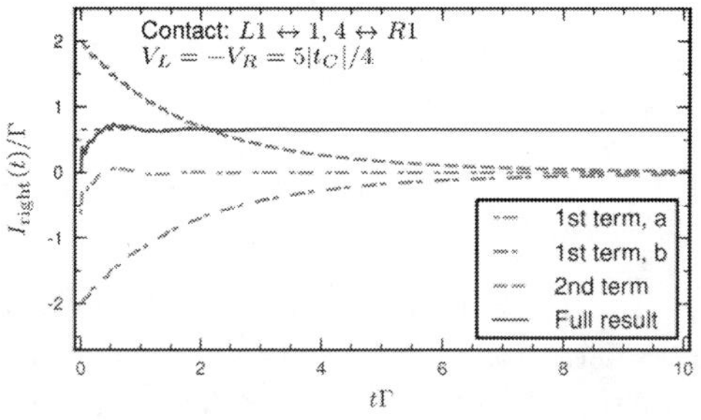

Figure 5: Different terms of the time-dependent current in units of Γ through the right interface with symmetric coupling and strong bias.

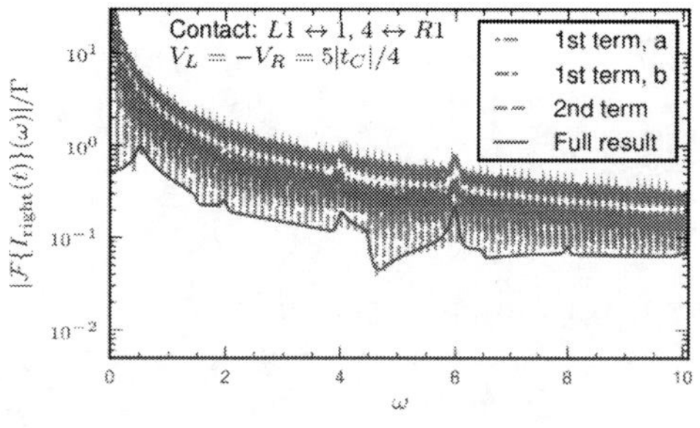

Figure 6: Absolute value of the Fourier transform of the terms of the current in units of Γ.

If we, however, distort the symmetry of the junction then also the intramolecular transitions with lower energies become visible. This is clearly seen in Figs. 7 and 8 where we connect the molecule asymmetrically to the leads (1st site to the left and 3rd site to the right, see Fig. 1). We can also break the symmetry by deforming the molecule with, for instance, one hopping (between

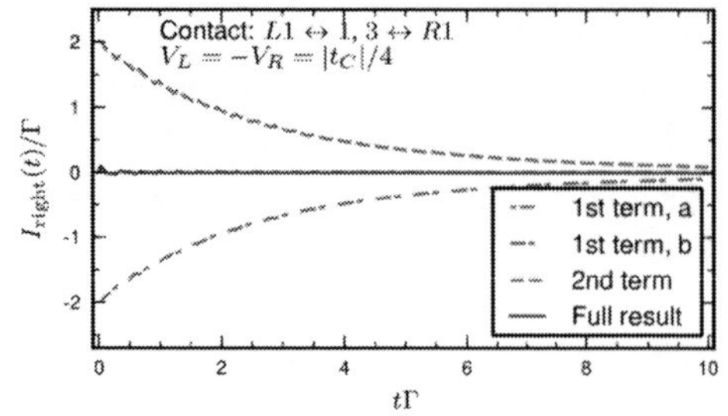

Figure 7: Different terms of the time-dependent current in units of Γ through the right interface with asymmetric coupling and weak bias.

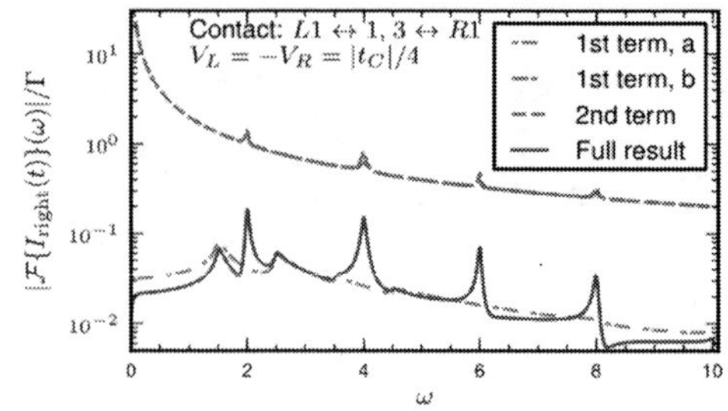

Figure 8: Absolute value of the Fourier transform of the terms of the current in units of Γ.

sites 1 and 6) being $2t_C$. This splits the degenerate levels in Fig. 2 and also the corresponding intramolecular transitions can be seen in Figs. 9 and 10.

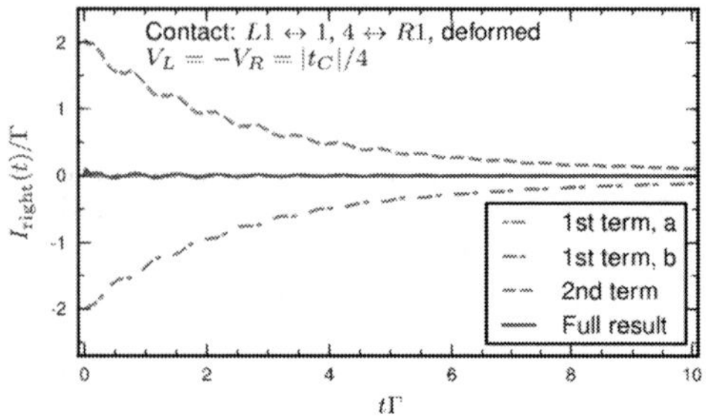

Figure 9: Different components of the time-dependent current in units of Γ through the right interface of a deformed molecule with symmetric coupling and weak bias.

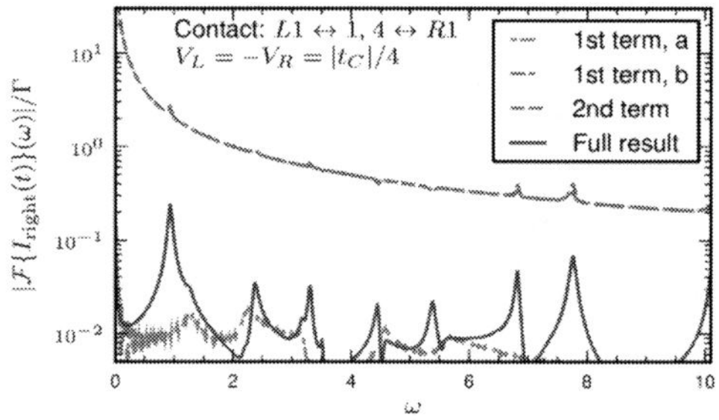

Figure 10: Absolute value of the Fourier transform of the terms of the current in units of Γ.

As the contributions from different terms sum up to the total current we can plot the full results for, e.g., the right current of the symmetrically coupled molecule against, e.g., the bias or the coupling strength. In Figs. 11, 12, 13 and 14 we display the full results. The transient dynamics is visualized better but distinguishing between the different contributions is more complicated. In Fig. 13 and 14 the axes are not scaled due to varying Γ. It is clear that by increasing the bias window more levels open up for transport and therefore the steady-state current grows. The oscillation frequencies corresponding to transitions between molecular levels remain unchanged while the oscillation frequencies corresponding to transitions between the molecule and the leads vary (peak shift). By increasing Γ, and hence by widening the resonances, electrons can flow even with intermediate bias voltages. Correspondingly, the steady-state value of the current increases, the relaxation time decreases whereas the oscillation frequencies remain invariant.

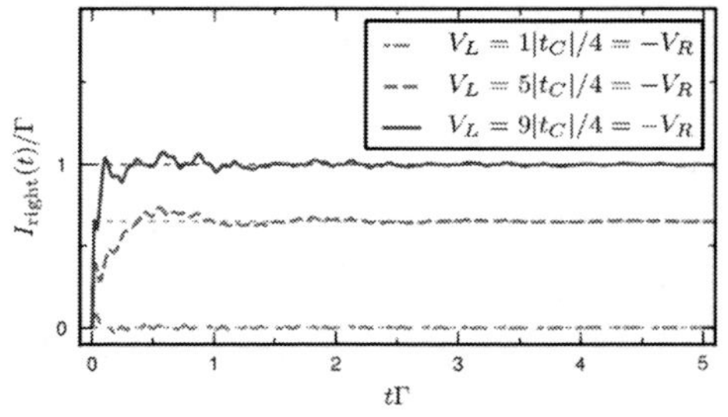

Figure 11: Time-dependent current in units of Γ through the right interface with symmetric coupling ($\Gamma = 0.1$) and varying bias. (Dotted lines refer to steady-state values.)

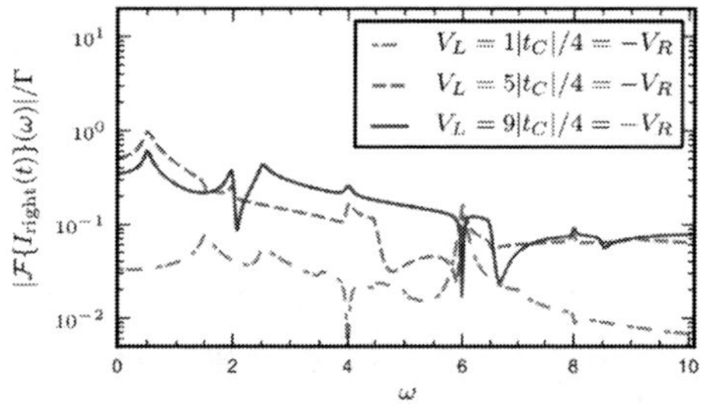

Figure 12: Absolute value of the Fourier transformed right current in units of Γ.

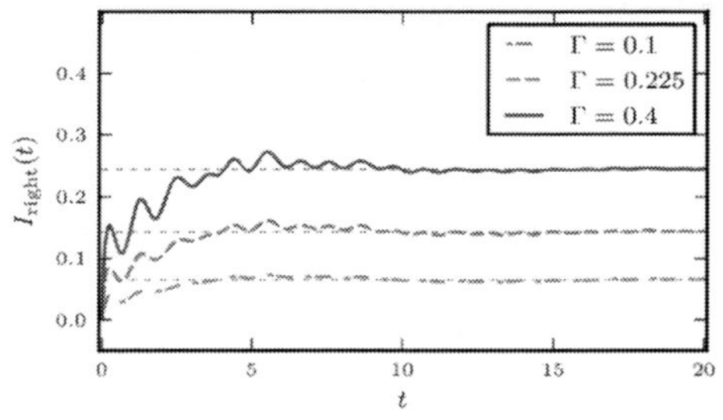

Figure 13: Time-dependent current through the right interface with symmetric coupling (varying strength) and bias $V_L = -V_R = 2.5$. (Dotted lines refer to steady-state values.)

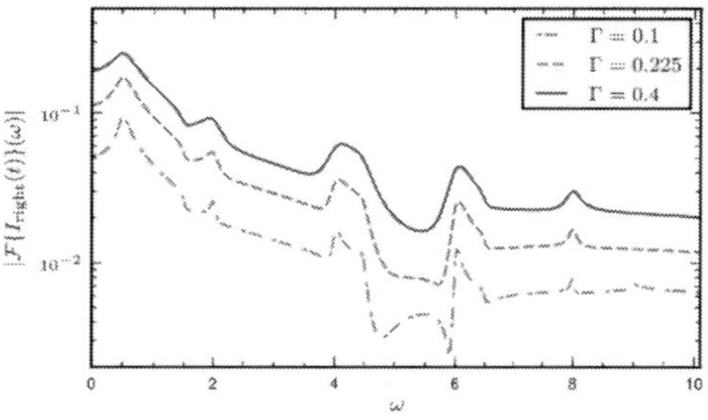

Figure 14: Absolute value of the Fourier transformed right current.

5. Conclusions

In conclusion we solved the Kadanoff–Baym equations for the Green's function of an open noninteracting system by properly taking into account the initial contacts between the system and the reservoirs. We used the analytic solution for the time-dependent density matrix to derive a time-dependent version of the Landauer–Büttiker formula. As an application we considered a tight-binding benzene-shaped junction and calculated the time-dependent current flowing through it. The advantages of having an explicit solution are that the numerical effort is drastically reduced and that the transient behavior can easily be interpreted in terms of virtual transitions and decay rates. Our time-dependent Landauer–Büttiker formula holds promise for studying the transient behavior of large junctions like, e.g., wide nanoribbons or large-diameter nanotubes, as well as disordered junctions where a large number of simulations is required to perform the average over different configurations.

Acknowledgments

R.T. wishes to thank Ellen and Artturi Nyyssönen's foundation for financial support and CSC — the Finnish IT Center for Science — for providing computing resources. We also acknowledge

Journal of Physics: Conference Series **427** (2013) 012014 doi:10.1088/1742-6596/427/1/012014

Petri Myöhänen, Anna-Maija Uimonen, Niko Säkkinen and Markku Hyrkäs for productive discussions.

Appendix A. Self-energy calculations

The Matsubara self-energy is an antiperiodic function (we are studying fermions) with period given by the inverse temperature β. For the calculation of the Fourier coefficients we can use the relation: $\Sigma^{\rm M}_\alpha(\omega \pm {\rm i}\eta) = \Sigma^{\rm R/A}_\alpha(\omega + \mu)$. Therefore the Matsubara self-energy is simply given by

$$\Sigma^{\rm M}_{\alpha,mn}(\tau,\tau') = \frac{1}{-{\rm i}\beta}\sum_q {\rm e}^{-\omega_q(\tau-\tau')}\Sigma^{\rm M}_{\alpha,mn}(\omega_q) \quad (A.1)$$

where

$$\Sigma^{\rm M}_{\alpha,mn}(\omega_q) = \begin{cases} -\frac{\rm i}{2}\Gamma_\alpha & {\rm Im}[\omega_q] > 0 \\ +\frac{\rm i}{2}\Gamma_\alpha & {\rm Im}[\omega_q] < 0 \end{cases} \quad (A.2)$$

and $\omega_q = (2q+1)\pi/(-{\rm i}\beta)$ are the Matsubara frequencies. For the isolated Green's function of the biased α:th reservoir we have

$$g^{\rm R}_{k\alpha}(t,t_0) = -{\rm i}\theta(t-t_0){\rm e}^{-{\rm i}(\epsilon_{k\alpha}+V_\alpha)(t-t_0)}\ , \quad (A.3)$$

$$g^{\rm M}_{k\alpha}(\tau,\tau') = \frac{1}{-{\rm i}\beta}\sum_q \frac{{\rm e}^{-\omega_q(\tau-\tau')}}{\omega_q - \epsilon_{k\alpha} + \mu}\ . \quad (A.4)$$

Without loss of generality we take the time t_0 at which the bias is switched on to be zero. Then we can write

$$g^{\rceil}_{k\alpha}(t,\tau) = {\rm i}g^{\rm R}_{k\alpha}(t,0)g^{\rm M}_{k\alpha}(0,\tau) = {\rm e}^{-{\rm i}(\epsilon_{k\alpha}+V_\alpha)t}g^{\rm M}_{k\alpha}(0,\tau)\ , \quad (A.5)$$

$$g^{\lceil}_{k\alpha}(\tau,t) = -{\rm i}g^{\rm M}_{k\alpha}(\tau,0)g^{\rm A}_{k\alpha}(0,t) = {\rm e}^{{\rm i}(\epsilon_{k\alpha}+V_\alpha)t}g^{\rm M}_{k\alpha}(\tau,0)\ . \quad (A.6)$$

By using Eqs. (A.5) and (A.6) we can calculate the right and left embedding self-energies

$$\begin{aligned}\Sigma^{\rceil}_{\alpha,mn}(t,\tau) &= \frac{1}{-{\rm i}\beta}\sum_q {\rm e}^{\omega_q\tau}\sum_k T_{m,k\alpha}\frac{{\rm e}^{-{\rm i}(\epsilon_{k\alpha}+V_\alpha)t}}{\omega_q-\epsilon_{k\alpha}+\mu}T_{k\alpha,n} \\ &= \Gamma_{\alpha,mn}\frac{1}{-{\rm i}\beta}\sum_q {\rm e}^{\omega_q\tau}\int\frac{{\rm d}\omega}{2\pi}\frac{{\rm e}^{-{\rm i}(\omega+V_\alpha)t}}{\omega_q-\omega+\mu}\ , \end{aligned} \quad (A.7)$$

$$\begin{aligned}\Sigma^{\lceil}_{\alpha,mn}(\tau,t) &= \frac{1}{-{\rm i}\beta}\sum_q {\rm e}^{-\omega_q\tau}\sum_k T_{m,k\alpha}\frac{{\rm e}^{{\rm i}(\epsilon_{k\alpha}+V_\alpha)t}}{\omega_q-\epsilon_{k\alpha}+\mu}T_{k\alpha,n} \\ &= \Gamma_{\alpha,mn}\frac{1}{-{\rm i}\beta}\sum_q {\rm e}^{-\omega_q\tau}\int\frac{{\rm d}\omega}{2\pi}\frac{{\rm e}^{{\rm i}(\omega+V_\alpha)t}}{\omega_q-\omega+\mu}\ , \end{aligned} \quad (A.8)$$

where we used $\Gamma_{\alpha,mn} = 2\pi\sum_k T_{mk\alpha}\delta(\omega-\epsilon_{k\alpha}-V_\alpha)T_{k\alpha n}$. It only remains to calculate the *lesser* component. We have

$$g^<_{k\alpha}(t,t') = {\rm i}f(\epsilon_{k\alpha}-\mu){\rm e}^{-{\rm i}(\epsilon_{k\alpha}+V_\alpha)(t-t')} \quad (A.9)$$

and therefore

$$\begin{aligned}\Sigma^<_{\alpha,mn}(t,t') &= \sum_k T_{m,k\alpha}{\rm i}f(\epsilon_{k\alpha}-\mu){\rm e}^{-{\rm i}(\epsilon_{k\alpha}+V_\alpha)(t-t')}T_{k\alpha,n} \\ &= {\rm i}\Gamma_{\alpha,mn}\int\frac{{\rm d}\omega}{2\pi}f(\omega-\mu){\rm e}^{-{\rm i}(\omega+V_\alpha)(t-t')}\ . \end{aligned} \quad (A.10)$$

Appendix B. Green's function calculations

The Fourier coefficient of the Matsubara Green's function reads

$$G^{\mathrm{M}}(\omega_q) = \frac{1}{\omega_q - h - \Sigma^{\mathrm{M}}_{\mathrm{em}}(\omega_q) + \mu} = \begin{cases} \frac{1}{\omega_q - h + \frac{\mathrm{i}}{2}\Gamma + \mu} & \mathrm{Im}[\omega_q] > 0 \\ \frac{1}{\omega_q - h - \frac{\mathrm{i}}{2}\Gamma + \mu} & \mathrm{Im}[\omega_q] < 0 \end{cases} = \begin{cases} \frac{1}{\omega_q - h_{\mathrm{eff}} + \mu} & \mathrm{Im}[\omega_q] > 0 \\ \frac{1}{\omega_q - h^{\dagger}_{\mathrm{eff}} + \mu} & \mathrm{Im}[\omega_q] < 0 \,, \end{cases} \tag{B.1}$$

where we defined $h_{\mathrm{eff}} = h - \frac{\mathrm{i}}{2}\Gamma$. The *right* component of the Green's function can be derived from the equation of motion

$$\left[\mathrm{i}\frac{\mathrm{d}}{\mathrm{d}t} - h\right] G^{\rceil}(t,\tau) = \int_0^{\infty} \mathrm{d}\bar{t}\,\Sigma^{\mathrm{R}}_{\mathrm{em}}(t,\bar{t})G^{\rceil}(\bar{t},\tau) - \mathrm{i}\int_0^{\beta} \mathrm{d}\bar{\tau}\,\Sigma^{\rceil}_{\mathrm{em}}(t,\bar{\tau})G^{\mathrm{M}}(\bar{\tau},\tau)\,. \tag{B.2}$$

The insertion of the retarded self-energy from Eq. (20) leads to

$$G^{\rceil}(t,\tau) = \mathrm{e}^{-\mathrm{i}h_{\mathrm{eff}}t}\left[G^{\mathrm{M}}(0,\tau) - \int_0^t \mathrm{d}t'\mathrm{e}^{\mathrm{i}h_{\mathrm{eff}}t'} \int_0^{\beta} \mathrm{d}\bar{\tau}\,\Sigma^{\rceil}_{\mathrm{em}}(t',\bar{\tau})G^{\mathrm{M}}(\bar{\tau},\tau)\right]\,, \tag{B.3}$$

where we noticed that $G^{\mathrm{M}}(0,\tau) = G^{\rceil}(0,\tau)$.

Finally the retarded Green's function in Fourier space reads

$$G^{\mathrm{R}}(\omega) = \frac{1}{\omega - h - \Sigma^{\mathrm{R}}_{\mathrm{em}}(\omega)} = \frac{1}{\omega - h + \frac{\mathrm{i}}{2}\Gamma} \tag{B.4}$$

and Fourier transforming back in the time domain we recover Eq. (25).

Appendix C. The three terms in Eq. (18)

For the first term we use Eqs. (25) and (A.10) to obtain

$$\begin{aligned}\left[G^{\mathrm{R}} \cdot \Sigma^{<}_{\mathrm{em}}\right](t,t) &= \int_0^{\infty} \mathrm{d}\bar{t}\,G^{\mathrm{R}}(t,\bar{t})\Sigma^{<}_{\mathrm{em}}(\bar{t},t) \\ &= \mathrm{i}\sum_{\alpha}\int \frac{\mathrm{d}\omega}{2\pi} f(\omega-\mu)\left[1 - \mathrm{e}^{\mathrm{i}(\omega+V_\alpha-h_{\mathrm{eff}})}\right] G^{\mathrm{R}}(\omega+V_\alpha)\Gamma_\alpha\,. \end{aligned} \tag{C.1}$$

Also the second term is readily calculated by using Eq. (21)

$$\left[G^{<} \cdot \Sigma^{\mathrm{A}}_{\mathrm{em}}\right](t,t) = \int_0^{\infty} \mathrm{d}\bar{t}\,G^{<}(t,\bar{t})\Sigma^{\mathrm{A}}_{\mathrm{em}}(\bar{t},t) = \frac{\mathrm{i}}{2}G^{<}(t,t)\Gamma\,. \tag{C.2}$$

The third term involves somewhat more trickery because of the rather complicated form of the right Green's function. Inserting the expressions from Eqs. (B.3) and (A.8) we get

$$\begin{aligned}\left[G^{\rceil} \star \Sigma^{\lceil}_{\mathrm{em}}\right](t,t) &= -\mathrm{i}\int_0^{\beta} \mathrm{d}\tau\,G^{\rceil}(t,\tau)\Sigma^{\lceil}_{\mathrm{em}}(\tau,t) \\ &= \mathrm{e}^{-\mathrm{i}h_{\mathrm{eff}}t}\left\{\left[G^{\mathrm{M}} \star \Sigma^{\lceil}_{\mathrm{em}}\right](0,t) - \mathrm{i}\int_0^t \mathrm{d}t'\mathrm{e}^{\mathrm{i}h_{\mathrm{eff}}t'}\left[\Sigma^{\rceil}_{\mathrm{em}} \star G^{\mathrm{M}} \star \Sigma^{\lceil}_{\mathrm{em}}\right](t',t)\right\}. \end{aligned} \tag{C.3}$$

By using $\int_0^\beta \mathrm{d}\tau \mathrm{e}^{(\omega_q-\omega_{q'})\tau} = \beta\delta_{qq'}$ for the Matsubara frequencies we may manipulate Eq. (C.3) further. Inserting Eqs. (A.7), (A.8) and (B.1) we obtain for the double convolution

$$\begin{aligned}\left[\Sigma^{\rceil}_{\mathrm{em}} \star G^{\mathrm{M}} \star \Sigma^{\lceil}_{\mathrm{em}}\right](t',t) &= -\mathrm{i}\int_0^\beta \mathrm{d}\tau(-\mathrm{i})\int_0^\beta \mathrm{d}\bar{\tau}\Sigma^{\rceil}_{\mathrm{em}}(t',\bar{\tau})G^{\mathrm{M}}(\bar{\tau},\tau)\Sigma^{\lceil}_{\mathrm{em}}(\tau,t) \\ &= \int\frac{\mathrm{d}\omega}{2\pi}\int\frac{\mathrm{d}\omega'}{2\pi}\sum_{\alpha,\alpha'}\Gamma_\alpha\frac{1}{-\mathrm{i}\beta}\sum_q \frac{\mathrm{e}^{-\mathrm{i}(\omega+V_\alpha)t'}}{\omega_q-\omega+\mu}G^{\mathrm{M}}(\omega_q)\frac{\mathrm{e}^{\mathrm{i}(\omega'+V_{\alpha'})t}}{\omega_q-\omega'+\mu}\Gamma_{\alpha'}\ .\end{aligned} \tag{C.4}$$

The integration with respect to ω can be done by closing the contour in lower-half plane (LHP) because of the exponential convergence factor, whereas the integration with respect to ω' can be done by closing the contour in the upper-half plane (UHP). However, the corresponding poles are located on different half planes, and this makes the double integral to vanish for every ω_q. Hence

$$\left[\Sigma^{\rceil}_{\mathrm{em}} \star G^{\mathrm{M}} \star \Sigma^{\lceil}_{\mathrm{em}}\right](t',t) = 0\ , \tag{C.5}$$

and in Eq. (C.3) we are left with

$$\begin{aligned}\left[G^{\mathrm{M}} \star \Sigma^{\lceil}_{\mathrm{em}}\right](0,t) &= -\mathrm{i}\int_0^\beta \mathrm{d}\tau G^{\mathrm{M}}(0,\tau)\Sigma^{\lceil}_{\mathrm{em}}(\tau,t) \\ &= \int\frac{\mathrm{d}\omega}{2\pi}\frac{1}{-\mathrm{i}\beta}\sum_q\frac{G^{\mathrm{M}}(\omega_q)\mathrm{e}^{\eta\omega_q}}{\omega_q-\omega+\mu}\sum_\alpha \Gamma_\alpha \mathrm{e}^{\mathrm{i}(\omega+V_\alpha)t}\ ,\end{aligned} \tag{C.6}$$

where on the last line a convergence factor $\mathrm{e}^{\eta\omega_q}$ was added to account for correct limiting behaviour when $t \to 0$. The sum over Matsubara frequencies can be performed using the Luttinger–Ward trick [13] and yields

$$\begin{aligned}\frac{1}{-\mathrm{i}\beta}\sum_q\frac{G^{\mathrm{M}}(\omega_q)\mathrm{e}^{\eta\omega_q}}{\omega_q-\omega+\mu} &= \int_\gamma \frac{\mathrm{d}\omega'}{2\pi}f(\omega')\mathrm{e}^{\eta\omega'}\frac{G^{\mathrm{M}}(\omega')}{\omega'-\omega+\mu} \\ &= \int_{-\infty}^{\infty}\frac{\mathrm{d}\omega'}{2\pi}f(\omega')\left[\frac{G^{\mathrm{M}}(\omega'-\mathrm{i}\delta)}{\omega'-\omega+\mu-\mathrm{i}\delta} - \frac{G^{\mathrm{M}}(\omega'+\mathrm{i}\delta)}{\omega'-\omega+\mu+\mathrm{i}\delta}\right]\ .\end{aligned} \tag{C.7}$$

By inserting Eq. (C.7) into Eq. (C.6) we get

$$\left[G^{\mathrm{M}} \star \Sigma^{\lceil}_{\mathrm{em}}\right](0,t) = \int\frac{\mathrm{d}\omega}{2\pi}\int\frac{\mathrm{d}\omega'}{2\pi}f(\omega')\left[\frac{G^{\mathrm{M}}(\omega'-\mathrm{i}\delta)}{\omega'-\omega+\mu-\mathrm{i}\delta} - \frac{G^{\mathrm{M}}(\omega'+\mathrm{i}\delta)}{\omega'-\omega+\mu+\mathrm{i}\delta}\right]\sum_\alpha \Gamma_\alpha \mathrm{e}^{\mathrm{i}(\omega+V_\alpha)t}\ . \tag{C.8}$$

Now the integral over ω can be done by closing the contour in the UHP. The first term in square brackets integrates to zero because of the pole in the LHP. The pole of the second term occurs at $\omega = \omega' + \mu + \mathrm{i}\delta$, and therefore

$$\begin{aligned}\left[G^{\mathrm{M}} \star \Sigma^{\lceil}_{\mathrm{em}}\right](0,t) &= \mathrm{i}\int\frac{\mathrm{d}\omega'}{2\pi}f(\omega')G^{\mathrm{M}}(\omega'+\mathrm{i}\delta)\sum_\alpha \mathrm{e}^{\mathrm{i}(\omega'+\mu+\mathrm{i}\delta+V_\alpha)t} \\ \left.\begin{matrix}\delta\to 0^+\\ \omega'=\omega-\mu\end{matrix}\right\}\to\quad &= \mathrm{i}\int\frac{\mathrm{d}\omega}{2\pi}f(\omega-\mu)\underbrace{G^{\mathrm{M}}(\omega-\mu)}_{=G^{\mathrm{R}}(\omega)}\sum_\alpha \Gamma_\alpha \mathrm{e}^{\mathrm{i}(\omega-\mu+\mu+V_\alpha)t} \\ &= \mathrm{i}\int\frac{\mathrm{d}\omega}{2\pi}f(\omega-\mu)G^{\mathrm{R}}(\omega)\sum_\alpha \Gamma_\alpha \mathrm{e}^{\mathrm{i}(\omega+V_\alpha)}\ .\end{aligned} \tag{C.9}$$

Journal of Physics: Conference Series **427** (2013) 012014 doi:10.1088/1742-6596/427/1/012014

This can be inserted into Eq. (C.3) to obtain

$$\left[G^{\rceil} \star \Sigma^{\lceil}_{\text{em}}\right](t,t) = \mathrm{i}\int \frac{\mathrm{d}\omega}{2\pi} f(\omega-\mu) \sum_{\alpha} \mathrm{e}^{\mathrm{i}(\omega+V_\alpha-h_{\text{eff}})t} G^{\mathrm{R}}(\omega)\Gamma_\alpha \,. \tag{C.10}$$

Appendix D. Derivation of Eq. (30)

We first state some useful identities for the retarded/advanced Green's function to be used later:

$$G^{\mathrm{R/A}}(\omega+V_\alpha) = G^{\mathrm{R/A}}(\omega) - V_\alpha G^{\mathrm{R/A}}(\omega) G^{\mathrm{R/A}}(\omega+V_\alpha)\,, \tag{D.1}$$

which can be checked directly by using Eq. (B.4) and its adjoint. From Eq. (D.1) it follows that

$$V_\alpha^2 G^{\mathrm{R}}(\omega)G^{\mathrm{A}}(\omega)G^{\mathrm{R}}(\omega+V_\alpha)G^{\mathrm{A}}(\omega+V_\alpha) = \left[G^{\mathrm{R}}(\omega) - G^{\mathrm{R}}(\omega+V_\alpha)\right]\left[G^{\mathrm{A}}(\omega) - G^{\mathrm{A}}(\omega+V_\alpha)\right]. \tag{D.2}$$

Taking into account Eq. (D.1) in Eq. (29) we get

$$\begin{aligned}
&\mathrm{i}\frac{\mathrm{d}}{\mathrm{d}t}G^<(t,t) - h_{\text{eff}}G^<(t,t) + G^<(t,t)h^\dagger_{\text{eff}} \\
=\;& -\mathrm{i}\int\frac{\mathrm{d}\omega}{2\pi}f(\omega-\mu)\sum_\alpha\left\{\mathrm{e}^{\mathrm{i}(\omega+V_\alpha-h_{\text{eff}})t}V_\alpha G^{\mathrm{R}}(\omega)G^{\mathrm{R}}(\omega+V_\alpha) + G^{\mathrm{R}}(\omega+V_\alpha)\right\}\Gamma_\alpha \\
+\;& \mathrm{i}\int\frac{\mathrm{d}\omega}{2\pi}f(\omega-\mu)\sum_\alpha\Gamma_\alpha\left\{G^{\mathrm{A}}(\omega+V_\alpha)G^{\mathrm{A}}(\omega)V_\alpha\mathrm{e}^{-\mathrm{i}(\omega+V_\alpha-h^\dagger_{\text{eff}})t} + G^{\mathrm{A}}(\omega+V_\alpha)\right\}.
\end{aligned} \tag{D.3}$$

It is convenient to rewrite the Green's function as $G^<(t,t) = \mathrm{e}^{-\mathrm{i}h_{\text{eff}}t}\tilde{G}^<(t,t)\mathrm{e}^{\mathrm{i}h^\dagger_{\text{eff}}t}$. In this way the left-hand side of Eq. (D.3) becomes

$$\begin{aligned}
&\mathrm{i}\frac{\mathrm{d}}{\mathrm{d}t}\left[\mathrm{e}^{-\mathrm{i}h_{\text{eff}}t}\tilde{G}^<(t,t)\mathrm{e}^{\mathrm{i}h^\dagger_{\text{eff}}t}\right] - h_{\text{eff}}\mathrm{e}^{-\mathrm{i}h_{\text{eff}}t}\tilde{G}^<(t,t)\mathrm{e}^{\mathrm{i}h^\dagger_{\text{eff}}t} + \mathrm{e}^{-\mathrm{i}h_{\text{eff}}t}\tilde{G}^<(t,t)\mathrm{e}^{\mathrm{i}h^\dagger_{\text{eff}}t}h^\dagger_{\text{eff}} \\
=\;& \mathrm{e}^{-\mathrm{i}h_{\text{eff}}t}\mathrm{i}\frac{\mathrm{d}}{\mathrm{d}t}\tilde{G}^<(t,t)\mathrm{e}^{\mathrm{i}h^\dagger_{\text{eff}}t}\,.
\end{aligned} \tag{D.4}$$

Then the right-hand side of Eq. (D.3) can be multiplied from left by $\mathrm{e}^{\mathrm{i}h_{\text{eff}}t}$ and from right by $\mathrm{e}^{-\mathrm{i}h^\dagger_{\text{eff}}t}$ to give

$$\begin{aligned}
\mathrm{i}\frac{\mathrm{d}}{\mathrm{d}t}\tilde{G}^<(t,t) \;=\;& -\mathrm{i}\int\frac{\mathrm{d}\omega}{2\pi}f(\omega-\mu)\sum_\alpha\mathrm{e}^{\mathrm{i}h_{\text{eff}}t}\left[G^{\mathrm{R}}(\omega+V_\alpha)\Gamma_\alpha - \Gamma_\alpha G^{\mathrm{A}}(\omega+V_\alpha)\right]\mathrm{e}^{\mathrm{i}h^\dagger_{\text{eff}}t} \\
& -\mathrm{i}\int\frac{\mathrm{d}\omega}{2\pi}f(\omega-\mu)\sum_\alpha V_\alpha\left[G^{\mathrm{R}}(\omega)G^{\mathrm{R}}(\omega+V_\alpha)\Gamma_\alpha\mathrm{e}^{\mathrm{i}(\omega+V_\alpha-h^\dagger_{\text{eff}})t}\right. \\
& \left. -\mathrm{e}^{-\mathrm{i}(\omega+V_\alpha-h_{\text{eff}})t}\Gamma_\alpha G^{\mathrm{A}}(\omega+V_\alpha)G^{\mathrm{A}}(\omega)\right].
\end{aligned} \tag{D.5}$$

Now we are ready to integrate both sides over t to obtain

$$\begin{aligned}
\tilde{G}^<(t,t) - \underbrace{\tilde{G}^<(0,0^+)}_{=G^<(0,0^+)=G^{\mathrm{M}}(0,0^+)} \;=\;& -\int\frac{\mathrm{d}\omega}{2\pi}f(\omega-\mu)\sum_\alpha\int_0^t \mathrm{d}t' \\
& \times\left\{\mathrm{e}^{\mathrm{i}h_{\text{eff}}t'}\left[G^{\mathrm{R}}(\omega+V_\alpha)\Gamma_\alpha - \Gamma_\alpha G^{\mathrm{A}}(\omega+V_\alpha)\right]\mathrm{e}^{-\mathrm{i}h^\dagger_{\text{eff}}t'}\right. \\
& -V_\alpha\left[G^{\mathrm{R}}(\omega)G^{\mathrm{R}}(\omega+V_\alpha)\Gamma_\alpha\mathrm{e}^{\mathrm{i}(\omega+V_\alpha-h^\dagger_{\text{eff}})t'}\right. \\
& \left.\left. -\mathrm{e}^{-\mathrm{i}(\omega+V_\alpha-h_{\text{eff}})t'}\Gamma_\alpha G^{\mathrm{A}}(\omega+V_\alpha)G^{\mathrm{A}}(\omega)\right]\right\}.
\end{aligned} \tag{D.6}$$

The integration over t' for the second term in Eq (D.6) can easily be done. For the first term we need the following result: Given two arbitrary matrices A and B

$$\int_0^t \mathrm{d}t' \mathrm{e}^{\mathrm{i}At'} \left[\frac{1}{x-A}B - B\frac{1}{x-A^\dagger}\right] \mathrm{e}^{-\mathrm{i}A^\dagger t'} = -\mathrm{i}\mathrm{e}^{\mathrm{i}At'}\frac{1}{x-A}B\frac{1}{x-A^\dagger}\mathrm{e}^{-\mathrm{i}A^\dagger t'} , \tag{D.7}$$

which can directly be verified by differentiating the right-hand side with respect to t'. Applying this result to Eq. (D.6) we obtain

$$\begin{aligned} \tilde{G}^<(t,t) = \mathrm{i}\int\frac{\mathrm{d}\omega}{2\pi}f(\omega-\mu)\sum_\alpha \Big\{ & G^{\mathrm{R}}(\omega)\Gamma_\alpha G^{\mathrm{A}}(\omega) + \mathrm{e}^{ih_{\mathrm{eff}}t}G^{\mathrm{R}}(\omega+V_\alpha)\Gamma_\alpha G^{\mathrm{A}}(\omega+V_\alpha)\mathrm{e}^{-ih^\dagger_{\mathrm{eff}}t} \\ & -G^{\mathrm{R}}(\omega+V_\alpha)\Gamma_\alpha G^{\mathrm{A}}(\omega+V_\alpha) \\ & +V_\alpha G^{\mathrm{R}}(\omega)G^{\mathrm{R}}(\omega+V_\alpha)\Gamma_\alpha G^{\mathrm{A}}(\omega)G^{\mathrm{A}}(\omega+V_\alpha)\mathrm{e}^{\mathrm{i}(\omega+V_\alpha-h^\dagger_{\mathrm{eff}})t} \\ & -V_\alpha G^{\mathrm{R}}(\omega)G^{\mathrm{R}}(\omega+V_\alpha)\Gamma_\alpha G^{\mathrm{A}}(\omega)G^{\mathrm{A}}(\omega+V_\alpha) \\ & +V_\alpha G^{\mathrm{R}}(\omega+V_\alpha)\mathrm{e}^{-\mathrm{i}(\omega+V_\alpha-h_{\mathrm{eff}})t}\Gamma_\alpha G^{\mathrm{A}}(\omega+V_\alpha)G^{\mathrm{A}}(\omega) \\ & -V_\alpha G^{\mathrm{R}}(\omega+V_\alpha)\Gamma_\alpha G^{\mathrm{A}}(\omega+V_\alpha)G^{\mathrm{A}}(\omega)\Big\} . \end{aligned} \tag{D.8}$$

Then the definition for $\tilde{G}$ can be inserted into the left-hand side, and multiply accordingly with $\mathrm{e}^{-ih_{\mathrm{eff}}t}$ from left and with $\mathrm{e}^{ih^\dagger_{\mathrm{eff}}t}$ from right. Combining terms according to Eqs. (D.1) and (D.2) we find Eq. (30).

References

[1] Landauer R 1957 *IBM J. Res. Dev.* **1** 233
[2] Büttiker M 1986 *Phys. Rev. Lett.* **57** 1761
[3] Caroli C, Combescot R, Nozières P and Saint-James D 1971 *J. Phys. C* **4** 916
[4] Caroli C, Combescot R, Lederer D, Nozières P and Saint-James D 1971 *J. Phys. C* **4** 2598
[5] Cini M 1980 *Phys. Rev. B* **22** 5887
[6] Stefanucci G and Almbladh C O 2004 *Phys. Rev. B* **69** 195318
[7] Mier Y and Wingreen N S 1992 *Phys. Rev. Lett.* **68** 2512
[8] Jauho A P, Wingreen N S and Mier Y 1994 *Phys. Rev. B.* **50** 5528
[9] Pastawski H M 1992 *Phys. Rev. B.* **46** 4053
[10] Perfetto E, Stefanucci G and Cini M 2008 *Phys. Rev. B* **78** 155301
[11] Myöhänen P, Stan A, Stefanucci G and van Leeuwen R 2009 *Phys. Rev. B* **80** 115107
[12] Stefanucci G and van Leeuwen R 2013 *Nonequilibrium Many-Body Theory of Quantum systems: A Modern Introduction* (Cambridge University Press)
[13] Luttinger J M and Ward J C 1960 *Phys. Rev.* **118** 1417

AUTHOR INDEX

Institute of Physics
Dirac House, Temple Back
Bristol BS1 6BE UK

CURRAN ASSOCIATES INC
proceedings.com